CW00539485

DO

Advances in Irrigation Agronomy

Plantation Crops

Irrigation has been used for thousands of years to maximise the performance, efficiency and profitability of crops, and it is a science that is constantly evolving. This potential for improved crop yields has never been more important as population levels and demand for food continue to grow.

Recognising the need for a coherent and accessible review of international irrigation research, this book examines the factors influencing water productivity in individual crops. It focuses on nine key plantation/industrial crops on which millions of people in the tropics and subtropics depend for their livelihoods (banana, cocoa, coconut, coffee, oil palm, rubber, sisal, sugar cane and tea). Linking crop physiology, agronomy and irrigation practices, this is a valuable resource for planners, irrigation engineers, agronomists and producers concerned with the international need to improve water productivity in agriculture in the face of increased pressure on water resources.

MIKE CARR is Emeritus Professor of agricultural water management at Cranfield University, UK. He has over 45 years of experience in the management and delivery of international research, education, training and consultancy in agriculture and natural resource management. He is former editor of the Cambridge University Press journal *Experimental Agriculture*.

Advances in Irrigation Agronomy

Plantation Crops

M. K. V. CARR
Cranfield University, UK

With contributions from
ROB LOCKWOOD
Independent consultant, Kent, UK

JERRY KNOX
Cranfield University, UK

CAMBRIDGE UNIVERSITY PRESS
Cambridge, New York, Melbourne, Madrid, Cape Town,
Singapore, São Paulo, Delhi, Mexico City

Cambridge University Press
The Edinburgh Building, Cambridge CB2 8RU, UK

Published in the United States of America by
Cambridge University Press, New York

www.cambridge.org
Information on this title: www.cambridge.org/9781107012479

First published 2012

Printed in the United Kingdom at the University Press, Cambridge

A catalogue record for this publication is available from the British Library

Library of Congress Cataloging-in-Publication Data

Carr, M. K. V.
  Advances in irrigation agronomy: plantation crops / M.K.V. Carr; with contributions from
  Rob Lockwood and Jerry Knox.
    p. cm.
  Includes index.
  ISBN 978 1 107 01247 9 (Hardback)
  1. Tropical crops. 2. Crops and water. 3. Irrigation farming. I. Lockwood, Rob.
  II. Knox, J. W. (Jerry W.) III. Title. IV. Title: Plantation crops.
  SB111.C37 2012
  633–dc23 2011042624

ISBN 978 1 107 01247 9 Hardback

This book is dedicated to:
My mother and father
My wife
My children
My grandchildren and
Plantation workers and smallholders everywhere

Contents

The colour plates are between pages 144 and 145.

Foreword

Water scarcity is one of the most pressing issues facing humanity. Globally, 70% of all water withdrawn from rivers and groundwater is consumed by agriculture and yet, surprisingly, water for producing food hardly gets onto the international development agenda except when there is drought and famine such as in the Horn of Africa in 2011. Despite the media attention there still seems to be little appreciation of water's critical role in producing food and fibre at a time when about 20% of the world's rivers run dry before reaching the sea, and more than 1.4 billion people live in water-stressed river basins – a situation that is set to worsen. By 2050 the world needs to produce as much as 70% more food on less land, using less water, energy, fertiliser and pesticides while at the same time bringing down sharply the level of greenhouse gas emissions. Climate change is yet another dark cloud on the horizon.

This is a daunting challenge but history tells us that we should be optimistic. Irrigated farming was, for example, one of the agricultural and engineering success stories of the twentieth century, witness the Asian 'green revolution' in the 1960s and 70s. Food production has more than doubled over the past 50 years in response to a doubling of the world's population and agricultural productivity has risen steadily over the past 40 years. Our understanding of the importance of water ecosystem services has also grown and so too has our appreciation of the need for sustainability and for a balance between the ever-conflicting water demands of people, industry, food and the environment.

Good science has underpinned this success. Research has served us well in the past and it will continue to play a crucial role in the future. Understanding the important relationships between crops and water will be a vital part of this process if we are to make the best use of our limited water resources.

In this context this book is most timely. It focuses on plantation crops which make significant contributions to both food security and the economic life of many of the less developed countries. Annual world sugar production, for example, is estimated to be over 160 million tonnes with a value in excess of US$20 billion.

Plantations also have a long tradition of good quality research on which their profitability depends. Mike Carr has tapped into this rich seam. He has searched globally and brought together widely scattered published and grey literature on the links between water and yield response for a range of plantation crops. I have known and worked with Mike for over 30 years following his early career in East Africa researching the water requirements of tea – his favourite plantation crop. This book is typical of the thorough and uncompromising way he approaches his work. He offers us considered synthesis of (often) conflicting research results from a range of sources of varying reliability. He provides us with good practical advice for crop and water management professionals and offers an excellent foundation on which future researchers can continue to build sound knowledge of crop water relationships without having to 'reinvent the wheel'. What more can you ask?

In the current jargon – we need to become more 'water smart' in growing food crops. This book will help us to do just that.

Melvyn Kay[1]

---

[1] Melvyn Kay is a Chartered Civil Engineer with over 40 years' international experience in irrigation engineering and agricultural water management. Following a career in teaching, research and consultancy with Cranfield University, he now works as an independent consultant with UK farmers and various international organisations such as FAO and IFAD on water, agriculture and food production.

Preface

My first job was with the Tea Research Institute of East Africa. This institute served the tea industries in three neighbouring countries, Kenya, Tanzania and Uganda, each with diverse ecological conditions. The results of the research, which was largely funded by the industry, had to be interpreted and applied to areas distant from the location of the field experiments, and then communicated to smallholders as well as large estates in appropriate ways. There was little opportunity to undertake basic research as the industry wanted answers to immediate practical questions, but if the results were to have generic value answers to fundamental questions were also needed.

My second job was in the UK with Wye College (University of London) where the task was to support the introduction of maize, a 'new crop', into British agriculture. This might seem very different from the job in East Africa, but the challenge was the same. To undertake and report research of immediate value to enterprising farmers concerned about profitability, yet supported by good science. It also meant working with a multidisciplinary team of agronomists, engineers and economists.

My third job was with Cranfield University at Silsoe (in the UK) where I was responsible for teaching postgraduate students from all over the world soil/plant/water relations and irrigation agronomy. This was done alongside engineers teaching irrigation engineering and related topics. The students came with first degrees in agriculture, geography or engineering and the challenge was to bridge the disciplinary gap, as well as to ensure that the MSc courses were relevant to any part of the world where irrigation is practised. Whilst at Cranfield, I was able to continue working internationally with the plantation sector as a researcher, trainer and consultant, particularly with the tea industry.

The common themes in all three jobs were to convert science into practice and to facilitate effective communication of the outcomes from research to the stakeholders whether they were students, smallholders, estate managers, engineers, policy makers, or other researchers or consultants. Traditionally scientists are most comfortable when working within their own, often narrow, discipline, but the world does not work like that. Bridges need to be built!

That therefore is my background from which this book has evolved. Each of the nine core chapters covers one so-called plantation crop, although most of these crops are now primarily grown by smallholders, produced in the tropics or subtropics. The topics reviewed extend from the centre of origin of the crop and its development stages, through fundamental water relations, water requirements and water productivity, to irrigation systems and scheduling (where appropriate), discussed in the context of the current farming system(s). Each chapter follows a common format, including interim summaries, and is designed to contribute towards putting science into practice by bringing together information from a diverse range of sources (over 800 references have been accessed and cited). Each chapter concludes with a summary of its content. In addition, there is an introductory chapter at the beginning of the book, and a synthesis at the end.

With one exception (sisal, for which there is limited research), each chapter is based on a paper that has already been published in *Experimental Agriculture*, an established refereed journal published by Cambridge University Press. The coffee chapter (originally published as a paper in 2001) has been updated. The chapters on cocoa and sugar cane were both co-authored. The other crops covered are banana, coconut, oil palm, rubber and tea. My hope is that people from a wide range of backgrounds will find the book useful and of interest.

Acknowledgements

This book would not have been possible without the help of many people. I begin by acknowledging the role played by (the late) Professor J.P. Hudson who inspired me as an undergraduate student at the University of Nottingham (UK), and subsequently to (the late) Ernest Hainsworth, Director of the Tea Research Institute of East Africa, who had the confidence to employ me as a 22-year-old to set up and run a small irrigation research unit in Tanzania. Without their guidance my career would have been very different, and this book would not have happened. Later, Cranfield University at Silsoe (UK) allowed me to continue my involvement with the international plantation sector, an involvement that continues to this day.

Continuing support and encouragement to write this book has come from Dr Rob Lockwood and Dr Hereward Corley. Many people provided helpful feedback on drafts of the papers on which the chapters in this book are based, as did anonymous referees for *Experimental Agriculture*. Specific help was provided on banana (by Dr D.W. Turner), cocoa (Dr G. Lockwood, co-author; Dr F. Amoah and Dr A.J. Daymond), coconut (Dr X. Bonneau, Dr R.H.V. Corley and Dr. G. Lockwood), coffee (colleagues at Cranfield), oil palm (Dr R.H.V. Corley and Dr I.E. Henson), rubber (Dr R.H.V. Corley and Dr F. Do), sugar cane (Dr J.W. Knox, co-author, and Dr D.J. Nixon), and tea (Dr P.J. Burgess and Dr D.J. Nixon).

Many people contributed to the research at the Ngwazi Tea Research Unit/ Station summarised in Chapter 10, at all levels. Most of them are cited in the references listed. Their commitment and support is gratefully recognised. I name three, all sadly no longer with us: the late Galus Myinga, Julio Lugusi and Badan Sanga. The tea industry in Tanzania and the UK Department for International Development funded the work at NTRU.

Several libraries kindly facilitated access to their journals and textbooks, specifically Cranfield University, Warwick University, the Royal Agricultural College, Stratford upon Avon Library and Information Centre, and Google. Much 'grey' literature was provided by colleagues listed above.

Photographs have come from a number of different sources, including Luiz Minisola (LM – banana and coconut); Andrew Daymond (AJD – cocoa); Xavier Bonneau, CIRAD, PT Multi-Agro Corporation (XB – coconut); Hereward Corley (RHVC – oil palm, cocoa); Anisio Henrique Leite Santana

(AHLS – sisal). Specific images are acknowledged by these initials in the text. I thank them all.

I thank too my long-time colleague and friend, Melvyn Kay, for kindly writing the Foreword to this book.

I am grateful to the staff of Cambridge University Press for the supportive way in which they facilitated the production process.

Finally, my wife, Dr Susan Carr, kindly read and edited the whole script (several times!) with great patience and skill, but any remaining errors are all mine!

1 Introduction

There are few easily identifiable or accessible sources where the results of international irrigation research have been brought together and interpreted in coherent and useful ways for individual crops. This is in part due to the diversity of sources, and also to the difficulty of reconciling the results of research conducted in contrasting situations, often with insufficient supporting information, to allow the results to be extrapolated to new situations with confidence (Carr, 2000a).

A scientific understanding of the role that water plays in the growth and development of crops is essential, but this knowledge needs to be interpreted and presented as practical advice in a language that can assist planners, irrigation engineers, irrigation agronomists and producers to allocate and use water, whether rainfall or irrigation, effectively and profitably. Communication between the professions attempting to improve irrigation water management for the benefit of the commercial producer and the wider community can always be improved. Field experiments must be designed and managed to quantify with precision the (marketable) yield responses of crops to water. Adequate supporting measurements need to be taken to enable the results to be interpreted and applied with confidence to other locations, or at other times, where the climate, weather and/or soils may be different. Site specific, single discipline, empirical studies should normally be avoided. *But, to minimise duplication of effort, existing information on the water relations and irrigation need of individual crops first needs to be collated and interpreted in practically useful ways.* This is especially true for plantation crops having international commercial importance (Carr, 2000a).

What is plantation agriculture?

Plantation agriculture can be broadly distinguished from other farming systems in the following ways (Tiffen and Mortimore, 1990; Stephens *et al.*, 1998):

- Plantations involve the cultivation of one crop or perhaps two crops from a restricted range.
- In the tropics and subtropics, these include tree crops such as cocoa, coconut, coffee (Figure 1.1.), oil palm, rubber and tea and perennial field crops such as bananas, sugar cane and sisal.

Figure 1.1 Irrigated coffee estate – Kilimanjaro Region, northern Tanzania (MKVC). (See also colour plate.)

- These are mainly crops that require prompt initial processing (added value) and for which there is an export market (source of foreign exchange).
- This involves a high fixed capital investment cost: for example, for planting material, and processing facilities and infrastructure, including roads and housing.
- Work on plantations often continues throughout the year, requiring a large permanent labour force and associated social facilities.
- The size of a plantation can vary from <100 ha to several thousand hectares.
- Plantations require specialised management teams to run an industrial-scale business, and to be responsible for quality control at all stages in the production process.
- Plantation agriculture is very vulnerable to fluctuations in global commodity prices: there is always a risk of overproduction and there are usually limited opportunities for a rapid change in the product or processes.
- Plantations are dependent on access to the results of good-quality research if they are to remain profitable. Specialised plantation crop research institutes, often funded by the industry, exist in many countries.

Increasingly, so-called plantation crops are grown by smallholders on a much smaller scale, either for local consumption or for export when they may be linked, for example, to a central estate and/or a processing facility through an

Figure 1.2 A mosaic of smallholder crops (tea, bananas) –Tukuyu, southern Tanzania (MKVC).

outgrower-scheme or a cooperative (Figure 1.2). In these cases an individual cropped area may be less than 0.3 ha.

Well-managed estates offer some advantages as a production system due to economies of scale when combined with adequate capitalisation (Figure 1.3). If management is poor the estate loses much of its economic advantage over smallholder production. Organisational structures range from autocratic bureaucracy on some large estates through to loose democratic alliances of smallholders. Although the early plantations in America and elsewhere were often oppressive, a modern plantation is a progressive commercial enterprise competing for labour and capital with other industries in an open market (Stephens *et al.*, 1998).

The business objectives of a plantation are to maximise profit in the short and long term, to optimise the use of resources (e.g. land/climate, people, capital) and to ensure long-term sustainability (wealth creation, employment, environment).

Research

Until the 1960s, agricultural research in the tropics was focused on plantation crops grown for export, at the expense of food crops research. Specialist research institutes were established focusing on individual crops (e.g. tea research institutes, coffee research foundations, rubber research institutes). Many of these still exist. They were often funded, at least in part, by a cess (levy) paid by the producer. This meant that the scientists were closely integrated with the industry they were serving, which developed as a result from a strong science base. Research focused

Figure 1.3 A well-managed irrigated tea estate surrounded by indigenous forest – Mufindi, Tanzania (MKVC). (See also colour plate.)

on plantation crops today is largely outside the international research networks (CGIAR[1]) and continues to be funded by producers, national governments and willing donors (those who appreciate that successful plantation crops can make a major contribution to poverty alleviation). Many of these specialist crop research institutes remain small, and physically (and often intellectually) isolated. The challenge to the researcher with limited resources is how best to reconcile the competing, sometimes conflicting, demands of the progressive estate and the resource-poor smallholder (Figure 1.4). This applies particularly to research on crop water relations and irrigation where much could be gained by an international coordinated approach on topics of generic interest. The same need applies to public concerns about the environmental impact of plantation agriculture in the tropics. With a focus on soil erosion, soil fertility decline, pollution, biodiversity and carbon sequestration, research on this subject has been reviewed by Hartemink (2005).

Purpose

Average yields of all plantation crops, even the best commercial yields, are still far below the potential yields. Water is just one of many limiting factors, but in some locations it is the major one. One purpose of this book is to collate all the published information on the water relations of the important plantation crops

Figure 1.4 A tea irrigation experiment – Ngwazi Tea Research Station, Mufindi, Tanzania (MKVC).

in order to quantify where possible the yield losses due to water stress or, where appropriate, the likely benefits from irrigation or other approaches to drought mitigation as an aid to planning. Another purpose is to provide an entry point for researchers on these topics wishing to build on what is already known rather than duplication of effort. A third purpose is to compare and contrast different plantation crops since, because of the structure of the plantation industry and its research support, there is often little cross-fertilisation of knowledge between crops. A fourth purpose is to make a contribution to the need, frequently stated, to use water efficiently in the face of increasing competition for a scarce resource. The uncertainties associated with climate change make this even more of an imperative. Finally the book is intended to be a source of reference for students wishing to know more about tropical agriculture and its continuing challenges.

Content

Each chapter is devoted to a single crop and follows a similar format. Following an introduction, the centre of origin of the crop is described, together with the current centres of production and production trends. There then follows a description of the key crop development stages, including root growth, with an emphasis on how water availability influences each stage. A detailed review of research on fundamental plant water relations, crop water requirements and water productivity then follows. Where appropriate, irrigation systems suitable for the crop are

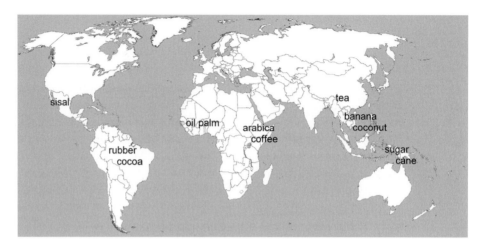

Figure 1.5 World map showing centres of origin of plantation crops.

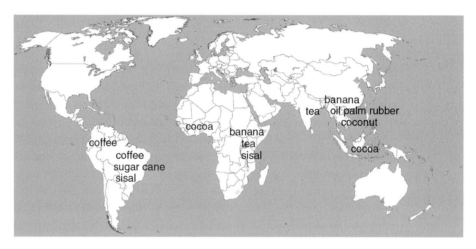

Figure 1.6 World map showing principal regions of plantation crop production.

then considered, together with irrigation scheduling methods. Not all crops have been researched to the same level of detail. As background, the key generic issues and terminology are now summarised here using the same headings.

Centres of origin and production

Understanding the conditions under which a species originated and evolved is central to understanding the climatic and soil conditions under which it is likely to perform well and/or the likely limiting factors in areas where it is being cultivated, which may be thousands of kilometres away from its region of origin (Figures 1.5 and 1.6). Several of the crops considered here evolved in rainforests under shade, for example cocoa, coffee and tea. Does this mean that they need shade when grown commercially? If a crop originated where there is a regular dry season, does it mean it is drought tolerant? What attributes contribute to its capacity to withstand dry conditions?

The transfer of crops across the world is a continuing process driven by economic and social forces. For example, rubber is now being grown in drier regions of South-east Asia as competition from oil palm for land, and urbanisation, displace it from its traditional areas of production. Tea is now being grown as an irrigated crop in dry areas of eastern Africa as public pressure makes it mandatory to protect the rainforest, the previous site of first choice for tea plantations. The threat of climate change will increase the need for managers to be as well informed as possible on plant water requirements when making decisions about where best to plant long-term perennial crops.

Crop development stages

Among the plantation crops being considered in this book, there is tremendous variety in the so-called useful product, which ranges from young shoots (tea), through fruits (banana), seeds (coffee, cocoa) to sucrose (sugar cane), oil (coconut, oil palm), fibre (sisal) and latex (rubber). Some crops have a diverse range of products such as the coconut, the so-called tree of life, in which nearly the whole tree contributes something to livelihoods although the primary product is copra (for oil).

The yield of any crop (Y) can be considered in terms of the efficiency of successive stages in the conversion of solar energy (S) to the economic or useful product (Monteith, 1972). Thus:

$$Y = S \times f \times e \times HI$$

where S is the total solar energy received at the surface of the crop, f is the fraction of the energy intercepted by the leaf canopy, e is the conversion ratio (or efficiency) of solar radiation to dry matter, and HI is the ratio of energy in the economic product to the total energy fixed by the crop (or in non-oil-bearing crops the dry matter ratio is often used instead of energy). Typical annual incident solar radiation totals range from around 55 TJ ha^{-1} in the high-rainfall tea-growing areas in Bangladesh and Assam to 63 TJ ha^{-1} in oil-palm-growing areas of Malaysia to 70 TJ ha^{-1} in parts of East Africa which have clear skies during long dry seasons (Stephens *et al.*, 1998).

The leaf area required to intercept a given proportion of solar radiation depends largely on the canopy geometry. Crops with erect leaves held in clumps (e.g. palms) require a larger leaf area index (L) to intercept a given proportion of radiation than those with horizontal, uniformly spaced leaves (e.g. tea). The aim with most crops is to seek to achieve full crop cover as soon as possible after planting in order to intercept as much radiation as possible during the lifetime of the crop. The duration of the immature phase varies between crops from, for example, three to four years for tea to seven to eight years for coconuts (talls). Growing trees under shade obviously reduces the proportion of light intercepted by the crop canopy. The conversion ratio is expressed in units of g (dry matter) MJ^{-1} (intercepted radiation) and values range from 0.2 g MJ^{-1} for tea to 0.8 g MJ^{-1} for oil palm. For comparison the corresponding figure for a temperate cereal crop is 1.4 g MJ^{-1}. Excessively high leaf temperatures and/or dry air (low humidity or large saturation deficit), soil water stress and nutrient stress can all reduce the photosynthetic efficiency. Losses of dry matter as a result of respiration by a large standing biomass (e.g. the trunks of a palm) in a warm climate are another reason for apparent low conversion efficiencies.

The aim of plant breeders and others is to maximise the amount of dry matter (or energy) in the plant that is allocated to the useful product. This is known as the harvest index and varies considerably between species. For example, Corley (1983) listed the harvest indices (above-ground dry matter) recorded for a selection of a well-managed plantation crops as 0.42 for oil palm to 0.20 for cocoa.

Using this analytical approach it is possible to calculate the potential yield of a crop and, by comparing this with the actual yield, seek to identify possible reasons for a yield deficit. Corley (1983, 1985) did such an analysis for a selection of plantation crops. Water stress can influence each of the growth processes described above, including crop establishment, leaf expansion (light interception), photosynthesis (conversion efficiency), flower formation, pollination, fruit development, and the harvest index, while root growth, depth and distribution affect the amount of water easily available to the crop.

Plant water relations

Water deficits in plants develop as a consequence of water loss from the leaves as the *stomata* open to allow the ingress of carbon dioxide from the atmosphere for *photosynthesis* and the egress of water vapour (*transpiration*). This is referred to as a *gaseous exchange* process (water vapour for carbon dioxide). Stomata are found on either the adaxial (upper) surface of the leaf or the abaxial (lower) surface or both. The water lost by *transpiration* from the leaf mesophyl cells is replaced by water drawn from the soil into the roots, and then up the stems and through the leaves along the *xylem vessels*. Water moves along a gradient of *water potential* from a relatively wet soil (high potential) to a relatively dry air (low potential). The energy driving this process comes primarily from solar radiation, which is providing the latent heat needed to evaporate water (*transpiration*). The energy status of the water is described in terms of its *water potential*, which in the plant has two principal components: the *osmotic potential* (due to the presence of salts in solution) and the *pressure potential* (or *turgor pressure*). In the soil the principal component is the *matric potential* (a result of the *capillary forces* in the soil pores, and the *attraction* of water molecules to soil particles) and, if there are salts in solution, the *osmotic potential*.

A *pressure bomb* is commonly used to measure the *leaf water status* (*leaf water potential* and its components), while a *tensiometer* measures the *matric potential* in the soil. A *porometer* (there are several types) measures the *stomatal conductance* (a measure of the degree of stomatal opening). *Infrared gas analysers* are used to measure *photosynthesis* and *instantaneous transpiration* rates (Monteith *et al.*, 1981; Squire *et al.*, 1981).

Crop water requirements

Actual crop water use (*ET*) can be measured (by means of water balance, sap flow, micrometeorology) or estimated (by calculation) in a number of ways. The *water balance approach* involves measuring the change in water content (volumetric) of the soil profile (ΔW) over a period of time after allowing for rainfall (*P*), run-off (*R*) and deep drainage (*D*), and finding *ET* by difference:

$$ET = P - R - D \pm \Delta W$$

This can be done at different scales – from a whole catchment, when comparing changes in land use from, for example, rainforest to tea or oil palm, or from an individual tree grown in a large container (known as a lysimeter). Changes in soil water content can be measured *gravimetrically* or with a *neutron probe* or a *capacitance probe*.

Evapotranspiration (*ET*) has two components: *transpiration* (*T*) and *evaporation* (*E*) from the soil (and crop) surface. Both processes occur simultaneously, and there is no easy way of distinguishing between the two. When the crop is small, water is predominantly lost by evaporation from the soil surface (while it remains wet), but once the crop canopy covers the ground *T* becomes the main process (Allen *et al.*, 1998).

The *sap flow method* (of which there are several variations) involves measuring the rate of flow of water up the stem using a heat pulse. It is a direct measure of *T*. It is well suited to tree crops and has been tried, for example, on tea, coconut and rubber.

Micrometeorological methods, namely the Bowen Ratio and eddy-flux methods, involve measuring the flux of water vapour above a crop (*ET*) using an array of sensors. These methods have been used with, for example, oil palm and sugar cane.

In most practical situations, *potential crop evapotranspiration* (ET_c) is estimated using a formula such as the Penman equation or the Penman–Monteith equation, both of which require standard weather data, or a well-sited evaporation pan such as the USWB Class A pan (E_{pan}). These give estimates of evaporation from a standard crop surface, usually taken to be short grass or alfalfa, well supplied with water, now known as *reference crop evapotranspiration* (ET_0) (Allen *et al.*, 1998).

To convert this to potential water use by a specific crop (ET_c) a crop factor (K_c) is needed. This varies with the stage of development of the crop.

$$ET_c = K_c \times ET_0$$

A pan factor (K_p), its value depending on the siting of the pan, is needed to convert E_{pan} to ET_0, thus:

$$ET_0 = K_p \times E_{pan}$$

Unfortunately few researchers define precisely the methods they have used to calculate crop water use (there are several versions of the Penman equation). This can sometimes lead to confusion. The guidelines provided by Allen et al. (1998) are intended to help to standardise the approaches used internationally.

Water productivity

There are several ways in which water productivity can be defined and again it is necessary to be very precise in order to compare like with like. The term *transpiration efficiency* is used to describe dry matter production per unit of transpiration. Alternatively, *water-use efficiency* describes dry matter production per unit of water lost by *evaporation* (from the soil and crop surface) and by *transpiration*. For commercial purposes, it is often easier to compare the water-use efficiency on the basis of the *commercial yield* per unit of *evapotranspiration* (evaporation plus transpiration) or per unit of rainfall and/or irrigation. It is important to be able to differentiate between these descriptors when making comparisons; they are rarely defined precisely. *Water productivity* is a generic term covering all these terms (Turner, 1986; Carr and Stephens, 1992).

As an example, for a tea crop yielding 5000 kg ha^{-1} y^{-1} in an area where the annual evapotranspiration (ET) is 1250 mm, of which transpiration is 1050 mm, the *water-use efficiency – ET* (for yield) – is 4 kg ha^{-1} mm^{-1} (5000/1250), and the *transpiration efficiency* is 4.8 kg ha^{-1} mm^{-1} (5000/1050). If the total annual rainfall is 1700 mm, the *water-use efficiency for rain* is 2.9 kg ha^{-1} mm^{-1} (5000/1700). If 300 mm of supplementary irrigation increases yields by 1500 kg ha^{-1}, the incremental *yield response to irrigation* is 5 kg ha^{-1} mm^{-1} (1500/300). Water productivity values like these are a valuable way of evaluating the effectiveness of various agronomic or drought-mitigation practices, or for assessing in crop yield and financial terms the worthwhileness of irrigation. They can also act as a benchmark against which to judge good practice (Carr and Stephens, 1992).

Another way of specifying the yield response to water is that proposed by Doorenbos and Kassam (1979), using the following relationship:

$$(1 - Y_a/Y_m) = K_y(1 - ET_a/ET_m)$$

where Y_a is the actual harvested yield, Y_m is the maximum harvested yield, ET_a is the actual evapotranspiration, and ET_m is the maximum evapotranspiration. K_y is the slope of the linear relationship (assumed) between the *relative yield decrease* and the *relative evapotranspiration deficit*, known as the '*yield response factor*'. The higher the value of K_y the more sensitive the crop is to water stress.

   Based on an analysis of the published results of experiments, Doorenbos and Kassam (1979) developed *yield response functions* for the total growing period for a selection of crops (including banana and sugar cane), and for individual development stages of these crops. The K_y values so obtained were intended to help optimise the planning, design and operation of an irrigation project, taking into account the effect of different water regimes on crop production. It is not known how widely used or successful this approach has been. The K_y values for banana and sugar cane were, for example, 1.2–1.35 and 1.2 respectively, implying 'high sensitivity' to water stress in both cases. By contrast pineapple was classified as having 'low sensitivity'. The target *water use efficiencies* (irrigation) for banana were presented as 3.5–6.0 kg (fruit, 70% water, ratoon crop[2]) m^{-3} and for sugar cane 5–8 kg (cane, 80% water) m^{-3}, and 0.6–1.0 kg (sucrose, dry) m^{-3}.

   These relationships all depend on the fundamental link between *dry matter production* (W) and *transpiration* (T) through the gaseous exchange process:

$$W \alpha \ e_w (\Sigma T)$$

where e_w is the *transpiration efficiency*, but e_w does not have a constant value. It varies with the inverse of the *saturation deficit* of the air (D):

$$e_w \alpha 1/D.$$

What this means in practice is that the water productivity is always less in situations where the air is dry compared with humid conditions. Irrigation can never completely compensate for rain!

Irrigation systems

Successful irrigation depends on being able to apply the right quantity of water at the right time as uniformly as possible across what may be a large area. As the scale of production of plantation crops varies considerably – as well as the topography, soils and the financial resources and skills available – so the methods of irrigation adopted vary. They can broadly be classified into three categories: flood irrigation, sprinkler irrigation and trickle (or drip) irrigation.

- *Flood irrigation*. This category includes furrow and basin irrigation. *Furrow irrigation* is commonly used in row crops, including sugar cane (Figure 1.7). Careful grading of the soil surface is necessary to obtain relatively uniform distribution of water across a field, with the water discharge rate matched to the slope and length of the furrow and the infiltration rate of the soil. Deep seepage (unseen) and excess run-off can lead to water wastage with the risk of waterlogging and salinity. Furrow irrigation is still practised on bananas (although the ratoon crops develop in different positions from

Figure 1.7 Evaluating furrow irrigation of sugar cane – Swaziland (MKVC).

the plant crop). *Basin irrigation* (small bunded, flat areas surrounding individual trees) is suitable for coconuts.

- *Sprinkler irrigation*. Low-pressure *microsprinklers* (under-tree) are suitable for bananas, coconut and oil palm; *conventional sprinklers* on high risers are used in sugar cane, as are high-pressure *rainguns*; sprinklers on *draglines* are used in sugar cane and tea; *centre pivots* have been used on coffee and tea.
- *Drip irrigation*. This involves the precise application of water to the soil surface through a network of plastic pipes and emitters. It has been successfully used on coffee, tea, bananas, oil palm and sugar cane, but it requires very good management to be fully effective.

No one system is necessarily better than another. All can be made to work well given the right situation and good management.

Irrigation scheduling

Irrigation scheduling is the process of deciding when to irrigate and how much water to apply. The objective is to maintain optimum soil water conditions for crop growth in order to meet crop yield and quality targets with minimum water wastage. Although many approaches have been promoted over the years for all crops, it remains the case that only a minority of farmers use an objective (scientific) method of scheduling irrigation, and most still rely solely on their judgement or intuition. The concept of a *soil water deficit* is an important component of irrigation scheduling, virtually independent of the method of scheduling used. It is a measure of how much water is needed to bring the soil profile back to *field capacity* (the maximum depth of water that soil can retain against gravity). A *limiting (or allowable) deficit* is the critical deficit beyond which a crop will begin to suffer water stress. *Deficit irrigation* is when only a proportion of the water needed to rewet the soil to field capacity is applied at each irrigation event. This can sometimes be more economic than full irrigation.

Rising energy, labour and water costs, the need to increase water productivity, less water available for abstraction due to expansion of cropped areas, intensification of existing plantations, increasing competition for limited resources, climate change risks and demands for greater environmental protection are now the driving forces influencing technology choice in irrigated crop production. In this context a good understanding of soil–plant–water relations is important, and more accurate scheduling may prove to be a useful adaptation strategy.

Endnotes

1 Consultative Group on International Agricultural Research.
2 Second (or subsequent) crop that develops from buds at the base of the first or plant crop.

2   Banana

Introduction

The centre of origin of the wild banana *Musa* species, a giant perennial herb, is believed to be in South-east Asia, where it opportunistically exploits breaks in the rainforest such as river margins (Simmonds, 1962). Wild bananas are jungle weeds, pioneers in the succession to rainforest, and intolerant of shade (Price, 1995). From here, the banana is believed to have spread outwards into the Pacific and then westwards, reaching sub-Saharan Africa about two thousand years ago.[1] The banana became a staple crop in upland East/Central Africa, which is still the greatest centre of cultivation. In the sixteenth century, early European travellers may have taken the banana from West Africa across the Atlantic Ocean to the Americas where it was quickly adopted (Simmonds, 1995).

Bananas can now be found throughout the tropics and subtropics. They are mostly grown between latitudes 30° N and 30° S of the equator, at altitudes up to about 1500 m, except in central and eastern Africa where some clones are grown up to 2000 m (Stover and Simmonds, 1987). In the tropics, they are grown for subsistence and as a cash crop, and play a very important role in diets and in local domestic economies (Figure 2.1). The fruit can be eaten either fresh or after cooking, used for brewing, sold in local markets or exported. The leaves and leaf sheaths are variously used as animal feed, to wrap food for steaming, for thatching, for making mats and ropes, and in handicrafts. In these diverse ways bananas support the livelihoods of millions of people (Fleuret and Fleuret, 1985; van Asten *et al.*, 2011).

Commercial cultivation of the crop for export of the fruit to Europe and North America began in the nineteenth century, mostly in tropical America but also to a lesser extent in West Africa (Simmonds, 1998). Dessert crops grown for export are now found principally in Central and South America, the Caribbean, West Africa (Côte d'Ivoire and Cameroon) and the Philippines. India, China, Brazil and the Philippines are the world's largest producers while Ecuador is the world's largest producer for export. Dessert bananas are also grown commercially in the subtropics and in Mediterranean climates, including Israel and other eastern Asia countries, for internal consumption or local export. The total cropped area in 2009 was estimated to be 4.8 million ha, producing 95 million t of fruit (FAO, 2011a). According to Turner (1995),

Figure 2.1 Banana. In the tropics, bananas are grown for subsistence and as a cash crop by smallholders, here alongside maize and tea – southern Tanzania (MKVC). (See also colour plate.)

water is probably the most limiting non-biological factor affecting banana production. Although up-to-date data are difficult to obtain, Stover and Simmonds (1987) believed that more than two thirds of the bananas grown in the world for export were irrigated. Robinson (1995, 1996) summarised those regions where irrigation is widely practised, including (1) areas of the tropics with pronounced dry seasons, (2) the semi-arid subtropics, and (3) Mediterranean environments. In the smallholder-owned rain-fed banana systems found in the highlands of East Africa, drought is a major yield-limiting factor (van Asten *et al.*, 2011).

A comprehensive review of the eco-physiology of bananas has been published by Turner *et al.* (2007) to which the reader is referred for further detail. This followed earlier reviews of the environmental physiology of the banana also by Turner (1994, 1998).

Genotypes

Many cultivars of *Musa* exist, although the genetic base for international banana production is very narrow. They have been selected for diverse agro-ecological areas, and for different culinary purposes. Hybridisation is thought to have occurred between various subspecies of *M. acuminata* and this has led to a range of diploid cultivars. These in turn are believed to have hybridised with *M. balbisiana,* another diploid species found in the Indian subcontinent (Simmonds and Shepherd, 1955). The fruits of both these wild ancestors are inedible because they are full of seeds. Triploid cultivars of *M. acuminata*, and of the hybrids, also occur and these make up the vast majority of cultivated types, including the well-known cultivars now grown for export, such as Grand Nain, Williams and Dwarf Cavendish. Parthenocarpy and the absence of seeds in the fruits are presumed to have been the major attributes selected by cultivators. This in turn led to the requirement for vegetative propagation by 'suckers' (lateral shoots) and to the development of clones (Simmonds, 1995).

The relative expression of *M. acuminata* (A) and *M. balbisiana* (B) characteristics and the ploidy level are together used to classify genotypes as, for example, AAA, AAB or ABB (Simmonds and Shepherd, 1955; Purseglove, 1972; Stover and Simmonds, 1987; Simmonds, 1995; Daniells *et al.*, 2001). Although a distinction is frequently made between bananas, a sweet fruit eaten raw, and plantains, a fruit low in sugar and eaten cooked, this distinction is not justified botanically. The term 'banana' is now preferred by some as a descriptor for the fruits of all *Musa* cultivars (Fleuret and Fleuret, 1985; Gowen, 1988). The proportion of starch to sugar is reduced as the fruits ripen. This process is slower in the cooking cultivars than in those eaten raw. Some of the cultivar groups with the *balbisiana* (B) genome are considered to be relatively tolerant of seasonal drought (Purse-glove, 1972; Price, 1995; Simmonds, 1995). Although there was apparently no published experimental evidence to support this perception (Turner, 1995), the results of subsequent laboratory-based leaf gas exchange studies (see below) are consistent with the view that the presence of the B genome contributes to drought tolerance (Thomas *et al.*, 1998).

In recent papers, Ploetz *et al.* (2007) have highlighted the complexity of, and diversity of cultivars within, the *Musa* genus in the Pacific region alone, whilst Heslop-Harrison and Schwarzacher (2007) have reviewed the domestication and genomics of this fascinating crop.

Crop development

Pseudostem and leaf canopy

The apparent aerial shoot of the banana, known as the pseudostem, is actually composed of the rolled and enclosing bases of a large number of leaves. This aerial shoot is borne on a subterranean stem (rhizome), which grows slowly and horizontally to a length of about 0.3 m. Aerial shoots (suckers or followers, banana is a ratoon crop) arise from lateral buds on the rhizome and are deter-minate (Norman *et al.*, 1984). Mechanically the aerial stem (a sympodium) is completely dependent on the leaf sheaths for support, reaching a height of 2–8 m in cultivated varieties (Figure 2.2). The leaf primordia are initiated at the apex of the rhizome and develop into leaves that rapidly encircle the apex. The growth in these lateral organs overshadows that in the vegetative apex to create an immense photosynthetic structure. Successive internodes are difficult to discern since they are close together. This condition changes on flowering

Figure 2.2 Banana. A clump or 'mat' showing the mother plant with suckers or followers that have arisen from lateral buds on the rhizome – Swaziland (MKVC).

when a massive true erect stem forms (Barker and Steward, 1962; Simmonds, 1966; Karamura and Karamura, 1995).

The area of individual leaves of dessert cultivars can reach 2–3 m^2 and the total leaf area of a plant 17–25 m^2 (Stover and Simmonds, 1987). In the following ratoon crops, the leaf area index (L) varies from two to more than six depending on the cultivar, season and plant density. A value for L of 4.5 is associated in bananas with the maximum use of photosynthetically active radiation (Turner, 1990, quoting others), and corresponds to about 90% of the incoming solar radiation being intercepted by the leaf canopy (Turner *et al.*, 2007). Splitting of the leaf surface, or lamina, along the veins is common, but vascular connections between the midrib and the leaf margin remain intact. This tearing of the leaves (into strips <100 mm wide) has been shown to increase the cooling effects of wind, under hot dry conditions, and to increase the efficiency of leaf water use (Taylor and Sexton, 1972). More recently, Eckstein *et al.* (1996) have shown how, in subtropical South Africa, severe leaf tearing (strips 12.5 and 25 mm wide) can result in smaller plants and leaf areas, and a reduction in rates of photosynthesis, together leading to a 22% smaller fruit mass compared with plants with less severe leaf damage.

In Western Australia (25° S), Hoffman and Turner (1993) found that the elongation rate of leaves emerging from the pseudostem (up to 80 mm d^{-1}) of container-grown plants (cv. Williams, AAA Cavendish subgroup) was a sensitive measure of the response of the plant to a drying soil (range of soil water potential -10 to -80 kPa).

Figure 2.3 Banana. The floral phase is considered to extend from floral initiation to inflorescence (bunch) emergence from the throat of the pseudostem (shown here), the start of the fruiting phase – cv. Prata Anã (AAB Prata subgroup) – Espirito Santo state, Brazil (LM). (See also colour plate.)

Inflorescence

In lowland tropical areas, floral initiation occurs after 30–40 leaves have been produced. Like leaf initiation, this occurs at the apex of the rhizome, which is located within 0.3 m of the soil surface, following which the aerial stem begins to elongate, pushing through the centre of the pseudostem until it is visible at the top (a process known as shooting). The floral phase is considered to extend from floral initiation to inflorescence (bunch) emergence from the throat of the pseudostem (the start of the fruiting phase), although the fruits actually exist before they appear above the leaves of the plant (Simmonds, 1966; Karamura and Karamura, 1995). The flower induction stimulus is not yet known, and flowering can occur at any time of the year (Figure 2.3).

The inflorescence is a spike with stout peduncles on which flowers are arranged in nodal clusters in two rows on transverse cushions (crown). The first nodes of flowers are female, while the last nodes are male. The fruits grow from the female flowers without pollination (Purseglove, 1972). In cultivated varieties, the inflorescence on emergence behaves geotropically, which results in each banana bunch (or stem) hanging in a pendulous position. Mechanically the aerial stem is entirely dependent on the surrounding leaf sheaths for its support.

Holder and Gumbs (1982) reported a study of the effects of the timing of irrigation during floral initiation on female flower production (cv. Robusta, AAA Cavendish subgroup) on a clay loam soil in St Lucia (14° N). A continuous non-limiting supply of water during the 'dry' season from December to July increased the number of flowers produced on the plant crop (crop planted 1 December, all treatments pre-irrigated). Relieving water deficits after 120 d, for a period of 60 d, during the period of estimated floral initiation and early differentiation had a similar effect, but had the additional benefit of increasing the number of female flowers per inflorescence compared with the continuously watered treatment. There was evidence of enhanced growth rates (pseudostem girth), for a period of 50 d, following the relief of the water deficit, which made up for the earlier effects of drought. Positive linear correlations were observed between the final (240 d after planting) pseudostem diameter (range 50–58 mm at 0.60 m above the ground) and the number of female flowers per inflorescence (range 100–160; $r^2 = 81\%$, $n = 9$). Similarly, female flower production was correlated with the rate of increase in the diameter of the pseudostem (1–4 mm d^{-1}) during the period 150 to 181 d after planting ($r^2 = 85\%$; $n = 9$).

Fruit

The interval between inflorescence (bunch) emergence and harvest is seasonally variable, ranging in subtropical South Africa, for example, from 108 d after flowering in early December (summer) to 200 d in early May (autumn), and leading to corresponding uneven fruit production during the year. By tagging inflorescence emergence every two weeks, Robinson and Nel (1984) found that it was possible to predict harvest dates on the basis of historical records. There were also differences in bunch mass depending on the time of flowering, linked to the minimum temperatures prevailing at the time of flower initiation. For example, Robinson (1982), again in South Africa, reported how inflorescences that emerged from plants in November produced bunches that were small and malformed. This was explained as the effect of low temperatures at the time the flowers were being initiated in June.

Components of yield

In the subtropics, an individual mean bunch mass of 50 kg is possible from ratoon crops (cv. Williams, AAA Cavendish subgroup,). By comparison the corresponding weight in the tropics is 30–35 kg (Robinson and Nel, 1989). A bunch contains a cluster of fruits at a node (known as a hand), which in turn is composed of individual fruits, or fingers (Figure 2.4). The effects of drought at this time are illustrated by the results of an experiment in the Canary Islands (28° 37′ N). When water was withheld for 63 d from flower emergence before rewatering (cv. Grand Nain, AAA Cavendish subgroup) finger number remained constant (not surprisingly) compared with the control, but finger length and diameter were both

Figure 2.4 Banana. A fruit bunch contains a cluster of fruits at a node (known as a hand), which in turn is composed of individual fruits (or fingers) – cv. Prata Anã (AAB) – Espirito Santo state, Brazil (LM). (See also colour plate.)

reduced (by 9%). Total bunch fresh weights were reduced by 41%, from 30 to 17 kg plant^{-1}. Drought also delayed fruit maturity (Mahouachi, 2007).

Plant density and arrangement

The optimum plant density is difficult to define for specific cultivars and conditions, and there does not seem to have been any research reported which attempts to identify different densities depending on the hydrological regime under which the banana is grown. A universal observation with bananas is the extension of the crop cycle at high densities. Since floral initiation can occur at any time of the year, the harvesting period can, as a result, become similarly extended. Increased competition also leads to greater plant-to-plant variability and can extend the duration of the harvest. This may be a result in part of the reduced pseudostem temperature in plants grown at high density (Robinson and Nel, 1988, 1989). Attempts have been made to specify the target leaf area index (L) that will maximise solar radiation interception, production and profitability. Differences in the duration of crop cycles between places and between seasons complicate this approach, and for true comparisons *annual yields* need to be specified. Using this method, Robinson and Nel (1989) identified the optimum density for fruit production for an irrigated crop (cv. Williams, AAA Cavendish subgroup) in South Africa to be in the range 1800–2000 ha^{-1} ($L = 6$) or 1666 ha^{-1} ($L = 5$) if other management factors, including the timing of harvests, and cost are taken into account. A further complication is that the optimum density for a plant crop differs from that for a ratoon crop.

In tropical banana areas, nearly all new plantations are established on a hexagonal system. In the subtropics (South Africa) rectangular, or hedgerow, systems have been preferred because it is easier to identify the rows, but now double rows or 'tramline systems' have become popular, particularly in Australia, because these permit easy access by machinery and for irrigation (Robinson et al., 1989; Robinson, 1995).

Roots[2]

The subterranean parts of bananas follow the same overall arrangement as those of many other monocotyledons. Arising from the upper parts of the rhizome are the root axes (nodal roots, also known as cord roots) that produce primary (first order) laterals from which may develop secondary (second order) laterals. An axis with its laterals is considered to be a root. A single shoot can produce several hundred roots (Price, 1995). Root hairs up to 2 mm in length occur behind the root apex. Cord roots are relatively straight and cylindrical, reaching lengths of 3–5 m, and diameters of 4–10 mm (Draye et al., 2005). In South Africa, Robinson (1987) found that primary laterals remained functional (based on appearance) for about eight weeks, secondary laterals for five weeks and root hairs for three weeks. Roots are produced continuously until flowering.

In the Jordan Valley of Israel (c. 32° N; alt. −210 m), early studies of the roots of banana (cv. Dwarf Cavendish, AAA Cavendish subgroup) were reported by Shmueli (1953). Using trenches and a water jet to wash away the soil, and direct core sampling, Shmueli recorded a dense mass of, mainly horizontal, roots from a depth of 0.02 m down to 0.4–0.45 m. There was then a sharp decrease in the number of roots, with a few vertical roots below 0.6 m, but with some extending to 0.7–0.8 m by the time fruit was forming. Roots of suckers with 13–14 visible leaves reached depths of about 0.3 m. Roots extended laterally up to 3.3 m from the mother plant and sucker. Secondary roots of two types, depending on diameter and length, formed only on primary roots of at least 0.2 m in length. The soil was high in calcium carbonate (35–55%) with a combined clay plus silt fraction of 55–75%. The crop was irrigated and the water table was below 3 m depth at all times. The soil was mulched and not cultivated.

In South India (11° 00′ N; alt. 420 m), Krishnan and Shanmugavelu (1980) studied the effects of three irrigation regimes on root distribution (cv. Robusta; AAA Cavendish subgroup). The maximum depth of rooting, observed at harvest, was similar in all three treatments (c. 0.7 m), while the corresponding lateral spread was about 3 m. Infrequent irrigation resulted in slightly deeper and wider root systems than were found with more frequent applications of water. There were no effects of irrigation regime on the dry mass of roots, or on the total root number or length. Root depths at the time of shooting were similar to those observed at harvest, but lateral spread was 0.5–0.6 m less.

In New South Wales, Australia (28° 50′ S), Trochoulias and Murison (1981) found that 70% of the roots (live cord roots with diameters greater than 1 mm counted in 0.2 × 0.2 m squares on an exposed soil profile) were concentrated in the top 0.4 m of a clay loam soil, regardless of the supplementary (drip) irrigation treatment. In addition, 40% of the root system of 10–12-month-old plants (cultivar. not specified) was found in a 0.6 × 0.6 m square close to the base of the plant.

In the northern coastal plain of Israel (33° 16′ N), Lahav and Kalmar (1981) compared the effects of several drip irrigation treatments on root growth by counting the number of root cords exposed on a brown grummisol soil. In all cases, root distribution (cv. Williams, AAA Cavendish subgroup) was uniform at distances up to 2.5 m from the planting station (or 'mat'). Roots extended to maximum depths of nearly 1.0 m, with 75% in the top 0.6 m.

Summarising the results of root studies in South Africa, Robinson (1996) stated that 88% of the primary roots were at 0–0.3 m depth (cv. Williams, AAA Cavendish subgroup), 9% from 0.3–0.4 m and only 3% below 0.4 m. Water was extracted mainly (87%) from the top 0.3 m of soil when irrigated (Robinson and Alberts, 1989). The 'effective' rooting depth was considered to be only 0.3 to 0.4 m. By contrast, Daniells (1986) in northern Queensland (17° 38′ S) found that although roots (cv. Williams, AAA, and cv. Corne Plantain, AAB) extended to depths of 0.6 m in an alluvial clay loam soil, water was extracted to a depth of 0.9 m (the limit of measurement, with tensiometers) and laterally from within 1.5 m of the stem.

In a detailed study designed to establish guidelines for irrigation water management of bananas (cv. Pacovan, AAB Prata subgroup) in north-east Brazil (9° 9′ S; alt. 365 m), root distribution was evaluated over a 33-month period using the soil profile method supplemented by digital image analysis (Bassoi et al., 2004a). The 'effective' rooting depth (embracing 90% of the total length of roots greater than c. 1.0 mm diameter), in a medium-textured latosol soil, increased from 0.40 m at three months after planting to about 0.60 m after one year, 80 days after the end of flowering, when the rooting depth was about 1.0 m (the maximum depth of observation). Roots reached the midway between plants (spaced 3.0 m apart) six months after planting. Total root length increased throughout the period of observation, but at a declining rate. The recommended soil depth for monitoring and scheduling irrigation at this site was 0.40 m for the first nine months from planting and afterwards 0.60 m, with the instrumentation installed 0.60 m away from the stem.

In Costa Rica (10° 12′ N; alt. 268 m), Araya et al. (1998) and Araya (2005) compared the effects of several factors on root distribution of two cultivars ('Valery' and 'Grand Nain', both AAA Cavendish subgroup) grown in commercial banana plantations. Root fresh weight decreased sharply with soil depth independent of distance from the pseudostem with more than 65% being found in the 0–0.30 m depth increment. The roots of 'vigorous' plants reached depths of 1.2 m and those of 'poor vigour' 0.75 m.

In south-eastern Nigeria (4° 42′ N; alt. 10 m), Blomme *et al*. (2005) reported that 'most roots', as observed by core sampling of four cultivars, were to be found within a 0.60-m radius of the plant and to depths of about 0.70 m. Previously, Blomme and Ortiz (2000) had compared the root systems of 11 genotypes from six diverse *Musa* groups at the same high rainfall site. There was considerable variability in root development both between and within the groups. For most genotypes the majority of roots were in the 0–0.15-m layer, and no roots were found below 0.8 m.

In a controlled environment experiment in the UK, Kalorizou *et al*. (2007) found that resistance to nematodes (a major constraint to banana reduction in some places as a result of the damage they cause to primary roots) was associated with two clones (AAA and AABB) with greater root mass and more and larger primary roots than the other five cultivars tested. In Uganda (5° 13′ N; alt. 1200 m), Kashaija *et al*. (2004) found that legume intercrops had no effect on the vertical and horizontal distribution of roots of banana (AAA), or on nematode incidence. Approximately 90% of the root biomass was in the upper 0.30 m while there was none below 0.70 m. Within 0.50 m of the banana 'mat', the majority of the root mass was in the surface 0.15 m but outside this radius more were found at depths from 0.15–0.30 m. In a review, Delvaux (1995) confirmed the sensitivity of banana roots to physical constraints as they rarely explore compacted soils. For example, on a range of soils in Martinique (*c*. 14° 47′ N), root density, defined as the number of roots per unit of soil surface, increased by a factor of about six as the dry bulk density of the soil declined from about 1.2 to 0.7 g cm^{-3}.

Robinson and Bower (1988) recorded seasonal root growth (cv. Williams, AAA Cavendish subgroup) over a 12-month period in an underground root laboratory in Nelspruit, South Africa (25° 28′ S; alt. 700 m). In this subtropical region, maximum extension rates of primary roots were observed in late summer (mean 160 mm week^{-1}, maximum 275 mm week^{-1}) before declining during the autumn. In the tropics (Ivory Coast; 5° 37′ N; alt. 78 m), Lassoudiere (1978) observed similar peak rates of root extension. Extension ceased completely in the winter when soil temperatures, at 0.2 m depth, were 10–13 °C (at 0800 h). Leaf development was continuing at this time at a rate of about one leaf month^{-1}, compared with five leaves month^{-1} in the summer. Root hairs developed on primary roots after about one week and continued growing for a further three weeks, after which they died back. In a subsequent paper, Robinson and Alberts (1989) illustrated a highly significant relationship (r^2 = 84%; n = 43) between root extension rates and soil temperature (over the range 11.5 to 25 °C) recorded at a depth of 0.2 m (at 0800 h). Extension rates were erratic at temperatures above about 22°C. These differences in seasonal root activity were thought to explain, at least in part, corresponding differences in transpiration rates and irrigation need, although the links are tenuous.

As Draye *et al*. (2005) observed, 'the volume of soil where the uptake of water and nutrients occurs is constantly changing ... Unfortunately, these time-related

Figure 2.5 Banana. The apparent aerial shoot, known as the pseudostem, is composed of the rolled and enclosing bases of a large number of leaves – cv. Prata Anã, AAB – Espirito Santo state, Brazil (LM).

components of root distribution remain poorly addressed.' Detailed quantitative studies of the root system of the banana have, however, been undertaken and reported by Lecompte and his co-workers in Guadaloupe (15°09′ N; alt. 485 m) in order to facilitate the development of models of root system architecture (e.g. Lecompte *et al.*, 2002; Lecompte and Pagès, 2007).

Summary: crop development

1. The banana is a giant perennial herb. The apparent aerial shoot of the banana, known as the pseudostem, is actually composed of the rolled and enclosing bases of a large number of leaves (Figure 2.5).
2. This aerial shoot is born on a subterranean stem (rhizome), which grows slowly and horizontally to a length of about 0.3 m.
3. Aerial shoots arise from lateral buds on the rhizome.
4. In lowland tropical areas, floral initiation occurs after 30–40 leaves have been produced. This occurs at the apex of the rhizome, which is located within 0.3 m of the soil surface.
5. The aerial stem then begins to elongate, pushing through the centre of the pseudostem until it is visible at the top.

6. The interval between inflorescence (bunch) emergence and harvest is season-ally variable, ranging in subtropical South Africa, for example, from 108 to 200 d after flowering.
7. Roots spread laterally 2–3 m from the pseudostem, and sometimes up to 5 m. However, most roots are usually found within 0.6 m of the base of the plant.
8. Roots extend to depths of 1.0–1.5 m, providing that there are no physical restrictions. Relatively few are found below 0.6–0.8 m. Root extension virtu-ally ceases at flowering.
9. The 'effective' depth of rooting (however defined) is often taken to be 0–0.40 m, sometimes extending to 0.60 m.
10. In the subtropics, root growth is seasonal, and temperature dependent.
11. It appears that irrigation has only a small effect, if any, on the size and distribution of the root system.
12. There is some (limited) evidence that small differences between cultivars in root distribution may exist. These need to be explored further 'particularly for non-commercial cultivars grown in subsistence agriculture' (Price, 1995).

Plant water relations

Robinson and Bower (1987) commented that, despite the sensitivity of bananas to water deficits, there was very little published work on the physiological indicators of these deficits. This was due, in part, to difficulties in measurement, the presence of latex, and the very efficient stomatal regulation of transpiration, first reported by Shmueli (1953) in Israel. Subsequently, Turner and Thomas (1998) showed how measurements of the volumetric (relative leaf water content) or thermo-dynamic tissue water status (leaf water, osmotic and pressure potentials) of a laticiferous plant like the banana, using a range of traditional methods, do not help to explain gas exchange processes. In a short-term pot experiment in Western Australia (32° S), they found a better association between both stomatal conduct-ance and net photosynthesis and soil water potential, over the range −1 to −60 kPa, than with leaf water status. The Scholander pressure chamber technique for measuring leaf water potential is not suitable for use with the banana because of the difficulty of distinguishing the exudation of xylem sap from that of the latex. The method described by Milburn et al. (1990), which is based on measurements of the refractive index of exuded latex, was preferred and its reliability subse-quently confirmed by Thomas and Turner (2001). Some changes in the osmotic components (30%) of leaf water potential were observed during the day but differences caused by drought, in the leaf and the root, over the five days were small. The water potential of well-watered plants was found to cycle diurnally within the remarkably narrow range of 0 to −0.35 MPa. In fact, the rate of extension of the youngest leaf may be the most sensitive indicator of plant water status (Kallarackal et al., 1990), providing it is not too hot (Thomas and

Turner, 1998). Under hot, arid conditions, leaf folding is not considered to be a reliable plant-based indicator of when to irrigate (Thomas and Turner, 1998).

Stomata

Although stomata are present on both the adaxial (upper) and abaxial (lower) surfaces of the leaf, their density is up to four times greater on the lower surface (150–170 mm^{-2}) than the upper surface (35–50 mm^{-2}) (Brun, 1961; Chen, 1971). In subtropical South Africa (25° 7′ S), Eckstein and Robinson (1995a) found that the density of the stomata on either surface (c. 130/54 mm^{-2}) was not affected by the season (cv. Grand Nain, AAA Cavendish subgroup).

In pioneering studies in the Jordan Valley of Israel (c. 32°N; alt. −210 m), Shmueli (1953) estimated relative stomatal opening using the infiltration liquid method (based on kerosene) to try and identify the optimum irrigation interval. Measurements were made on the third fully expanded leaf, counting from the top, exposed to sunlight on plants (cv. Dwarf Cavendish, AAA Cavendish subgroup) irrigated at intervals of either five or about 10 days. In addition, one plot was allowed to dry to permanent wilting point. The relative infiltration index was first checked against direct microscopic measurement of stomatal apertures.

During the first four days after irrigation the stomata on both leaf surfaces remained wide open throughout most of the day. Partial closure began a few hours before sunset. Stomata on the upper surface were usually wider open than those on the lower surface. When about one third of the available water was depleted there was a gradual reduction in the degree of stomatal opening from midday or earlier, with stomata on the lower surface closing first. When the available soil water was depleted by about two thirds the pattern of opening of the stomata changed again. There was then one low peak early in the morning followed by a slow decline in the degree of opening during the rest of the day. Stomata on both surfaces of the leaf then behaved in similar ways.

Measurements were also made on one day in a similar crop watered by flood irrigation. The pattern of stomatal opening was observed to differ from that of a sprinkler-irrigated crop. In particular, stomata on the lower leaf surface were wider open than those on the upper surface, at all three levels of available soil water tested. Measurements made between 1200 and 1300 h showed clearly how the stomata began to close, in both sprinkler and flood irrigated crops, when about 20% of the available water in the 0.9 m deep soil profile had been depleted.

Stomata on leaves shaded from the sun were always wider open on the lower surface than the upper one, the opposite response from that observed on sunlit leaves. The diurnal pattern of opening was also different, and these stomata were generally less sensitive to changes in soil water status than those on leaves exposed to the sun.

In Taiwan (c. 24°N), Chen (1971) found that stomata were always closed at night but began to open at dawn, reaching maximum apertures in the mid-morning before starting to close during the afternoon with complete closure by

1800 h. On cloudy days the same pattern was followed but the maximum apertures reached were less than half those recorded on sunny days. Ke (1979) reported the results of a similar study, also in Taiwan, with similar findings. The stomata opened early in the morning and continued to open as the light intensity increased until about noon when they began to close rapidly, regardless of light intensity, with full closure by 1800 h throughout the year. Stomata on the abaxial leaf surface tended to open more quickly in the morning and to remain open longer during the day than those on the adaxial surface. The stomata were sensitive to changes in soil water availability.

In subtropical South Africa, Robinson and Bower (1988) reported that stomatal conductances of well-watered plants (cv. Williams, AAA Cavendish subgroup), at midday and in full sunshine, were about four times greater on the lower surface than the upper one (2.26 ± 0.07 and 0.57 ± 0.04 cm s^{-1} respectively). There was a similar difference in the relative partitioning of transpiration from the upper and lower surfaces of the leaf (see below).

Photosynthesis

In South Africa, Eckstein and Robinson (1995a) showed that maximum physiological activity, including photosynthesis, occurred in fully expanded leaves, numbers two to five (cv. Grand Nain, AAA Cavendish subgroup), counted from the top of 1.1-m-tall plants (the maximum leaf number at the time of measurement was nine). No differences in activity (photosynthesis, transpiration or stomatal conductance) could be discerned between measurements made in the proximal, medial and distal third of leaves two to five. It was concluded that any measurement point on these leaves could therefore be considered to be representative of the entire leaf. The photosynthetic potential of these leaves was, however, suppressed during winter months at this subtropical site (25° 7′ S; alt. 720 m).

The decline in rates of photosynthesis in leaves six to nine, by up to 38% relative to the maximum, indicated the relatively short lifespan of the photosynthetic apparatus in banana, a C3 species. The rate of photosynthesis of the youngest leaf nearest the bunch was also low. Although green leaf pruning is not recommended as a management technique, a standard practice in the subtropics is to remove the lower leaves (those hanging below the horizontal) in order to increase light penetration, pseudostem temperature and leaf emergence rate, and to reduce the ratoon cycle time (Robinson et al., 1992).

Eckstein and Robinson (1995a) also found that photosynthesis rates were reduced by unfavourable climatic conditions in the autumn and spring, and more on the adaxial surface than on the abaxial surface of the leaf. In the summer, photosynthetic rates were at their highest, since leaves three to five were replaced every month. The cessation of leaf production after flowering and the ageing of the leaves meant that photosynthesis rates then declined, especially during the last two months of bunch development. In a companion paper, Eckstein and Robinson (1995b) described the influence of weather variables on seasonal and diurnal changes in gas exchange processes. In the summer, photosynthesis peaked during the morning, and in the afternoon large saturation deficits of the air, up to 3.5 kPa, (and high leaf temperatures, 35–40 °C) resulted in partial stomatal closure and reduced rates of photosynthesis. In the winter, low night temperatures (as low as 6 °C), large saturation deficits (although very unlikely?) and a depleted root system were considered to be the main reasons for low physiological activity at this time of the year.

Eckstein and Robinson (1996) believed that reductions in photosynthesis, as a result of water stress, were the result of both stomatal limitations (84%) and a 'disturbed photosynthesis reaction' (13% higher internal concentration of CO_2 than in unstressed plants). The physiological responses of tissue-cultured plants (cv. Williams AAA Cavendish subgroup) to different irrigation treatments were monitored using measurements of leaf gas exchange. During the autumn, the first signs of water deficit appeared after four days of water shortage. After 12 days, there was a 79% reduction in photosynthesis relative to the irrigated (daily) control treatment. Physiological recovery of the plant from short-term water deficits took longer than three days. To maintain leaf gas exchange rates, Eckstein and Robinson (1996) recommended that the soil water potential (ψ_m) should be maintained above −15 to −20 kPa, and that irrigation intervals should not exceed three days. Water shortages in the winter had a less severe impact on assimilation than shortages during warmer seasons.

In Western Australia (32° S), Turner and Thomas (1998) measured rates of photosynthesis of pot-grown plants (cv. Grand Nain, AAA Cavendish subgroup) in a glasshouse over a five-day period. The photosynthetic photon flux density was typically below 800 μmol quanta m^{-2} s^{-1} (range 400–1400) and net photosynthesis rates, under relatively mild conditions, for irrigated plants were in the range 8–10 μmol quanta m^{-2} s^{-1}. Drought reduced net photosynthesis when ψ_m < −33 kPa and there was no leaf gas exchange at −62 kPa. This was despite there being few, small differences in the leaf water status between irrigated and droughted plants due to the high levels of hydration of banana leaves. Net photosynthesis was proportional to stomatal conductance, both variables being influenced by the photon flux density as well as by the drying soil. Raising the carbon dioxide concentration of the ambient air to 1600 μl l^{-1} increased net photosynthetic rates from 10 to 15 μmol CO_2 m^{-2} s^{-1} in irrigated plants and from 0.2 to 2.0 μmol CO_2 m^{-2} s^{-1} in droughted plants, indicating to the authors that non-stomatal inhibition of photosynthesis was not important. However, following a series of field and pot experiments conducted at two sites in Western Australia (16° S and 32° S), the same authors subsequently concluded, based on chlorophyll fluorescence and photosynthesis measurements, that it was a combination of stomatal and non-stomatal factors that reduced net photosynthesis in droughted plants (Thomas and Turner, 2001).

In a similar short-term field experiment to the one described above, but this time under hot, arid conditions, Thomas and Turner (1998) again compared leaf exchange processes of irrigated and droughted plants (cv. Williams, AAA

Cavendish subgroup). At this site in Western Australia (16°S; alt. 50 m), where air temperatures are commonly >35 °C and saturation deficits >4 kPa, they found that a single curve related photosynthetic rates to stomatal conductances for both droughted and frequently (daily) irrigated plants. With photosynthetic photon flux densities typically above 1000 μmol quanta m^{-2} s^{-1}, net photosynthesis rates, under these extreme conditions (E_{pan} = 10 mm d^{-1}), were in the range 15–20 μmol quanta m^{-2} s^{-1} for irrigated plants. Net photosynthesis was reduced when ψ_m fell below −30 kPa (at 0.30 m depth) and in irrigated plants when the saturation deficit of the air exceeded about 4.5 kPa.

Transpiration

In the Jordan Valley of Israel, Shmueli (1953) attempted to measure transpiration using a sensitive balance to record changes in weight of leaf sections taken from field-grown plants. This early work suggested that transpiration reached a peak in early to mid morning and then declined during the rest of the day. Sometimes there was a second peak later in the afternoon. Transpiration rates declined earlier in the day in plants where the available soil water content was below two thirds of the total. There was no relationship between transpiration and the degree of stomatal opening. High rates of transpiration were recorded even when the stomata were apparently partially closed.

In Taiwan, Chen (1971) also attempted to measure transpiration rates on a diurnal basis using a self-made automatic recorder. Rates tended to follow the same trends as those reported above for stomatal opening, with peak rates occurring around midday on sunny days.

In a carefully designed and monitored experiment, Robinson and Bower (1987) compared transpiration rates, measured with a continuous flow steady-state Licor porometer from plants (cv. Williams, AAA Cavendish subgroup) growing in containers in a fibreglass tunnel in the Eastern Transvaal in South Africa (25° 7′ S; alt. 720 m). They created two water regimes, soil rewetted to field capacity daily and continuous drying, and mild (0.4–0.6 kPa) and enhanced (1.5–2.0 kPa) saturation deficits of the air. The photosynthetically active radiation levels inside the shelter were about 45% of those in the open. With humid air there was little difference in transpiration rates between the two water regimes until the soil was relatively dry (ψ_m <−80 kPa at 0.2 m depth). By contrast, when the air was dry, transpiration rates from unirrigated plants fell below those from those watered daily when ψ_m reached about −20 kPa, two to three days after watering ceased, and then declined rapidly as the soil continued to dry. The coefficients of variation for measurements made on well-watered plants were always low (c. 10%) but very large (c. 100%) in droughted plants, with individual measurements showing much variability, despite precautions.

In an associated field experiment, diurnal measurements of transpiration were made on a plant crop of the same cultivar over a 14-d period without any rain, after the soil profile was wetted to field capacity. The control treatment was watered each evening. Peak transpiration rates were two to three times greater than those recorded in the shelter (26 and 10 $\mu g\,cm^{-2}\,s^{-1}$ respectively). Transpiration rates followed a diurnal trend, usually peaking in early or mid afternoon. Differences in transpiration between the two watering treatments first became evident when the saturation deficit of the air exceeded about 2.3 kPa. Transpiration from the unirrigated plants fell below those watered daily within three days of the treatments being imposed, when ψ_m, at a soil depth of 0.2 m, was between -20 to -25 kPa. For a clay loam soil with an available water capacity of about 150 mm m^{-1}, this corresponded to less than 25% depletion of the available water in the 0.3-m-deep root zone.

Stomatal conductance was found to be an even more sensitive indicator of plant water status than transpiration, with values declining rapidly as the soil dried below field capacity. There was a 50% reduction in conductance when ψ_m reached -65 kPa, which corresponded to a 54% depletion of the available water. The prevailing saturation deficit of the air at this time was 2.1 kPa. By comparison the corresponding reduction in transpiration rates was about 30%. When ψ_m fell to -525 kPa (73% depletion), stomatal conductances in the unwatered plants were reduced to 82% of those in plants watered daily, and transpiration to 70%. This was 14 d after differential treatments had been imposed. Leaf temperature was not a very sensitive indicator of water deficit, with a corresponding temperature difference of only 2°C.

As a result of these observations Robinson and Bower (1987) recommended that bananas, in this part of the subtropics, should be irrigated at intervals of one or two days only, depending on the prevailing evaporation rates, at a soil water deficit of only 10 mm. This is a value far less than those recommended for virtually any other field grown crop.

The above paper was exceptional in that it specified clearly the sampling procedures used for measurements. An associated paper by the same authors (Robinson and Bower, 1988) described in more detail how the measurement techniques for transpiration and stomatal conductance were standardised to minimise plant-to-plant variability (coefficient of variation only 5%). Measurements were then made on plants (cv. Williams, AAA Cavendish subgroup) of uniform size and height (1.5 m tall, plant crop), sequentially planted and watered daily. Diurnal trends in stomatal conductances and transpiration rates were monitored on the abaxial surface of the proximal third of the third, fourth, and fifth youngest leaf, angled towards the sun, on cloudless days, during different seasons in the Eastern Transvaal. Similar measurements (cv. Dwarf Cavendish, AAA Cavendish subgroup) were also made on one day at a site in south-eastern Transvaal (25° 30' S; alt. 369 m), where evaporation rates were higher than at the main research station. Seasonal trends in transpiration were also observed.

During the spring and summer the stomata were apparently wide open from 0800 to 1600 h, before starting to close as radiation levels declined. In the autumn there was a similar trend, but peak conductances were less than half of those recorded in the summer, and declined even further in the winter when plant-to-plant variability was also greatest.

By comparison, transpiration rates increased gradually from 0700 h and peaked at 1300 h in the autumn, or at 1600 h in the summer, before declining rapidly, especially in the summer. Under high evaporation rates, transpiration rates peaked earlier in the day (1100 h) before declining from 1400 h onwards. Cumulative transpiration from selected leaves ranged from the equivalent of 1.8 mm d^{-1} in the winter to 8.7 mm d^{-1} in the summer. Ambient air temperatures reached maximum values of 27.4, 33.4, 29.6 and 21.8°C in the spring, summer, autumn and winter respectively, and 37.2 °C at the second site, on the days that measurements were made. Leaf temperatures were close (within 1°C) to the ambient air temperature during the spring, summer and autumn, but exceeded air temperature by 3–4 °C in the winter, and in the summer at the site with high evaporation rates.

Diurnal cumulative transpiration rates expressed as a function of daily evaporation from a USWB Class A pan varied between seasons, from about 0.97 mm mm^{-1} in the summer down to 0.61 mm mm^{-1} in the winter. This difference was explained in part by variability in the responses of transpiration to diurnal changes in the saturation deficit of the air depending on the season. Thus in the spring, summer and autumn transpiration increased with the saturation deficit up to values of about 1.5 kPa. The response curves then diverged, and it was only on summer days that transpiration continued to increase as saturation deficits rose to about 3.5 kPa. In the winter, transpiration rates were low at all saturation deficits, but increased linearly over the range 0.5–1.7 kPa. These contrasting responses were explained in part by observed seasonal differences in root activity: root extension, for example, ceased in the winter months (see above). The results highlight the sensitivity of bananas to evaporative demand for water in relation to the rate of supply from the soil to the leaf, even when the soil water content is close to field capacity.

In a short-term study of leaf gas exchange processes in a hot arid site in Western Australia (16°S; alt. 50 m), Thomas and Turner (1998) found that transpiration, in contrast to photosynthesis, increased for well-watered plants of the same cultivar (Williams) with the saturation deficit of the air (range 1.8–5.5 kPa), but declined for 'droughted' plants when the saturation deficit exceeded about 3–4 kPa.

Drought tolerance

In a comparison of 17 *Musa* spp. genotypes in Ibadan, Nigeria (7°30′N; alt. 210 m), Ekanayake *et al.* (1994) identified, on the basis of leaf conductance measurements, two ABB cultivars ('Fougamou' and 'Bluggoe') that were potentially tolerant of transient dry conditions, and two others that were very sensitive to short dry spells ('Bobby Tannap' AAB and one of its hybrids TMPx582–4). Later, the same authors compared leaf stomatal conductances and morphology of 18 genotypes at two contrasting sites in Nigeria (Omme 4°43′ N; alt. 3 m; Ibadan 7°30′ N; alt. 210 m). Significant interactions, for example between clone and location, and between clone and sampling time, suggested that conductance and transpiration of a specific clone needed to be considered for a given environment for useful conclusions to be reached. In general, Ekanayake *et al.* (1998) concluded that ABB genotypes had higher conductances than other taxonomic groups (AAB, AA, AAA and AAAB), and that resistance to black sigatoka leaf spot disease was not due to differences in stomatal morphology or physiology.

In a laboratory-based pot study in Darwin, Australia, Thomas *et al.* (1998) compared the effects of environmental variables on leaf gas exchange processes (including transpiration) of three cultivars differing in their genomic constitution ('Williams', AAA; 'Lady Finger', AAB; 'Bluggoe', ABB). They found that, as the saturation deficit of the air was increased (from 1.5 to 5.7 kPa), both stomatal conductance and net photosynthesis declined linearly. Since increasing proportions of the B genome reduced this sensitivity to the dryness of the air and increased the instantaneous water-use efficiency of the leaf, Thomas *et al.* (1998) concluded that the B genome contributes to drought tolerance in *Musa* spp.

Summary: plant water relations

1. Stomata are present on both surfaces of the leaf: densities (and conductances) are up to four times greater on the lower surface than on the upper surface.
2. Stomata are closed at night; peak opening is reached by midday, sometimes followed by progressive closure.
3. The degree of stomatal opening is a very sensitive indicator of soil water availability and plant water status.
4. Reductions in photosynthesis as a result of water deficit are mainly due to stomatal limitations to gas exchange.
5. To maintain gas exchange rates the soil water potential should not fall below -15 to -33 kPa (at 0.2–0.3 m depth), leading to recommendations that irrigation intervals should not exceed three days.
6. Transpiration rates can be limited by dry air (e.g. when saturation deficits exceed 2.0–2.3 kPa).
7. In the subtropics, there are seasonal differences in the sensitivity of transpiration rates to the dryness of the air and evaporative demand, thought to be due, in part, to differences in root activity.
8. As a result, cumulative transpiration, expressed as a proportion of USWB E_{pan}, can vary from 0.61 in the winter to 0.97 in the summer.
9. There is some (limited) evidence that the presence of the B genome contributes to drought tolerance (as indicated by measurements of gas exchange).

Crop water requirements

Rainfall interception

Harris (1997) reported the results of measurements of rainfall partitioning by a banana canopy (cv. Robusta, AAA Cavendish subgroup, first ratoon) in St Lucia in the Windward Islands in the eastern Caribbean (*c.* 13° 54′ N). Over a three-and-a-half month period there were 28 rainfall events totalling 723 mm of rain. Both through-fall and stem-flow were highly correlated with rainfall amount and accounted, on average, for 80 and 10% respectively of incident rainfall. The balance of 10% was assumed to have been intercepted and to have been lost to the atmosphere by evaporation. The volume of water reaching the base of the pseudostem was considered to be about 13 times larger than would be expected on the basis of a proportional plan area. The spatial distribution of the water passing through the canopy changed with time and the age of the plants. These results are important when considering the water balance of a crop, including irrigation scheduling, and nutrient placement beneath the canopy, but rarely taken into account. Recently, this issue has started to be addressed in Guadeloupe. Based on detailed measurements of rainfall interception processes and the partitioning between splash and storage of raindrops differing in drop diameter, kinetic energy and fall height, empirical relationships were developed for the application of a rainfall interception model to simulate water flows beneath banana plants (Bassette and Bussière, 2008).

Evapotranspiration

Over the last 40 to 50 years, many different methods have been used to determine the actual water use of the banana plant, including those based on the water balance approach. These involve, for example, measurements of changes in soil water content and/or the use of lysimeters. Most recently, the sap flow and eddy correlation methods have also been used. Estimates of evapotranspiration, or transpiration alone, are then linked to direct measurements of evaporation from, commonly, a USWB Class A evaporation pan, or estimates of potential evaporation using weather data in order to develop a pan or crop factor (usually represented by K_c). All estimates are subject to large errors, and can be locally site specific. This topic is reviewed under three ecological headings: tropics, subtropics and Mediterranean climates.

Tropics

In the Upper Aguan Valley in Honduras ($c.$ 15° 22' N; alt. 250 m), an area where the banana is widely irrigated, Arscott et $al.$ (1965) attempted to develop a method of determining the water requirements of the cv. Giant Cavendish (AAA Cavendish subgroup). Using measurements of changes in soil water content (volumetric) over a one-year period they developed a multiple regression equation (without standard errors) relating daily crop water use to mean air temperature and relative humidity. Actual measured rates of water use for irrigated crops ranged from about 5.7 to 9.2 mm d^{-1}. Interestingly they reported that when 'hot dry winds' prevailed, rates of water use declined. This they associated with the dryness of the air (saturation deficit) causing transient wilting whenever potential transpiration rates exceeded the capacity of the plants to supply water from the soil to the leaf surface. This tended to occur in those months when the mean air temperature was about 28°C and the relative humidity of the air, recorded on five occasions during each day (0700, 0900, 1200, 1500 and 1730 h) with a sling psychrometer, averaged about 63%. This corresponds to a saturation deficit value for the air of about 1.5 kPa.

Subsequently Ghavami (1973), in the irrigated Sula Valley of Honduras (c. 15° 20′ N; alt. 30 m), used non-weighing, drainage lysimeters (2.1 m deep), with the soil surface protected from rainfall, to determine the actual water use of bananas (cv. Valery, AAA Cavendish subgroup). Cumulative monthly evapotranspiration (ET_c, mm), from a well-watered group of mature and immature plants, was linearly related ($r = 0.93$; $n = 12$) to cumulative evaporation from a USWB Class A pan sited (conventionally) in an adjacent area:

$$ET_c = 18 + 1.2E_{pan}.$$

Evapotranspiration was also linearly ($r = 0.87$) related to cumulative evaporation from a Class A pan placed on an elevated platform at crop canopy level (4.57 m). In this case the corresponding relationship was:

$$ET_c = -2.5 + 1.08E_{pan}.$$

In this location, the seasonal (annual) evapotranspiration totalled about 2200 mm. Cumulative relationships like these always need to be viewed with caution.

In the islands of Guadaloupe (16°N) and Martinique (14°N), six drainage lysimeters (2×2 m square $\times$ 1.5 m deep; single plants) were used to measure actual rates of water use over a 16-month period (1974/5). Peak rates of evapotranspiration (8–10 mm d^{-1}) were up to 1.6 times the calculated potential rates from a grass reference crop (Penman equation), or 1.4 times evaporation from a USWB Class A pan (Meyer and Schoch, 1976).

In Darwin, Australia (12.4°S; alt. 0 m), Lu et al. (2002) used the sap-flow technique to measure the diurnal water use of whole-plant, pot-grown and field-planted banana (cv. Williams, AAA Cavendish subgroup) during the dry season. 'Granier' sensor probes were inserted in the upper region (above-ground) of the central cylinder of the rhizome. Diurnal rates of water use were positively and linearly correlated with solar radiation (range 0–1.0 kW m^{-2}) and saturation deficit of the air (0–5 kPa). The daily water use of 2.5-m-tall (11 leaves, 3.8 m^2 total leaf area) field-grown plants was 9–10 l (equivalent to only 2 or 3 mm d^{-1} at plant densities of 2000 or 3000 ha^{-1} respectively), substantially less than the gross irrigation rate (42 l d^{-1}) then recommended for the Darwin region. The low sap flux densities recorded were attributed to the small leaf area/sapwood area ratio relative to those of other 'woody' species.

In semi-arid north-eastern Brazil (9°9′ S; alt. 365 m), Bassoi et al. (2004b) monitored crop water use over a 33-month period (from planting in 1999 to the end of the third harvest in 2001) for cv. Pacovan (AAB Prata subgroup). The soil was a low water-holding latosol and the crop was irrigated with microsprinklers at $\psi_m = -30$ kPa as recorded at 0.4 m depth by tensiometers. The total surface area between plants, spaced 3×3 m, was wetted. Actual crop water use (ET_c), excluding drainage, was determined using the water balance approach, and related to reference crop evapotranspiration (ET_0) as estimated by a USWB Class A pan (Doorenbos and Pruitt, 1977) and the FAO Penman–Monteith (P-M) method (Allen et al., 1998) to develop crop coefficients (K_c).

Values of K_c obtained for (1) the vegetative growth stage (from planting to the start of flowering, 211 d) were 0.7 ± 0.3 (pan), or 0.8 ± 0.3 (P-M), followed by (2) first flowering (121 d), 0.9 ± 0.3 (pan) and 1.1 ± 0.4 (P-M), and (3) from first harvest (plant crop) to flowering of the first ratoon (single sucker), 110 d, 1.0 ± 0.5 (pan) and 1.1 ± 0.5 (P-M). Values then declined (4) to 0.8 ± 0.4 (pan) and 1.0 ± 0.4 (P-M) from then until the start of the flowering of the second ratoon (217 d), and afterwards (5) up to and during the third harvest (with no suckers present) to 0.7 ± 0.4 (pan) and 0.8 ± 0.5 (P-M). Note that the standard errors presented are suspiciously large, and that the crop factors associated with the pan estimates of ET_0 are always greater than the Penman–Monteith values.

Recently, Montenegro et al. (2008) have reported the results of a similar study undertaken in a coastal region of Brazil ($3°28'$ S; alt. 31 m). Actual water use (ET_c) was measured over two production seasons (2004–05, 680 d) using the water balance approach (based on tensiometers) for an irrigated crop (microsprinklers; cv. Pacovan, AAB Prata subgroup; double row spacing $4.0 \times 2.0 \times 2.4$ m). These values were then compared with Penman–Monteith estimates of reference crop evapotranspiration (ET_0) for different growth stages to obtain the corresponding crop coefficients (K_c). Average ET_c values increased from 2.6 mm d^{-1} for the initial vegetative stage of the plant crop up to 4.3 mm d^{-1} for the flowering and fruit development stage for the ratoon crop. The corresponding average K_c values were 0.60 (0–170 d), 1.05 (plant crop, flowering and fruiting, 240–360 d), 0.86 (ratoon crop, vegetative, 415–560 d) and 1.05 (580–680 d). Although no standard errors are presented, these are clear and convincing estimates of both ET_c and K_c for an irrigated crop in this location.

Using the eddy correlation, and energy balance, micrometeorological methods, dos Santos et al. (2009) measured the actual water use (ET_c) of a drip-irrigated banana plantation (5 m tall; spaced 4 m $\times 2.4$ m) for nine months during the wet and dry seasons in a semi-arid area of Brazil ($5°08'$ S $38°06'$ W; alt. 147 m). The wind speed and saturation deficit of the air (range 3.2–4.3 kPa) had the largest influence on ET_c, which averaged 3.3 mm d^{-1} in the dry season (range 2.9–3.5) and 3.1 mm d^{-1} during the rains (range 1.8–4.3). These values are close to those described above by Montenegro et al. (2008). During the dry season, K_c ($= ET_c/ET_0$) averaged about 0.54, and 0.70 in the rains, less than those reported by Montenegro et al. (2008).

On the coastal plain in Surinam ($5°47'$ N; alt. 14 m), Van Vosselen et al. (2005) made a brave, but largely unsuccessful, attempt to calibrate and validate a model (SWAP) to assess the water requirements of banana. It was difficult to specify precisely enough the boundary conditions, especially those associated with a high water table and a swelling clay soil. The availability of reliable data with which to run the model was also a constraint compared with a conventional water balance approach.

Subtropics

Following the detailed studies by Robinson and Bower (1988) of transpiration from individual leaves of well-watered plants, Robinson and Alberts (1989) reported the seasonal variations in the crop coefficient (ET_c/E_{pan}, where E_{pan} is evaporation from a USWB Class A pan) for cv. Williams (AAA Cavendish subgroup), planted at a density of 1666 ha^{-1}. Soil samples were taken from positions midway between two plants, at 0.1 m depth increments down to 0.5 m, and the water contents were determined by drying to a constant weight. These values were converted to the corresponding volumetric water contents using the appropriate dry bulk density value. The soil was a sandy clay loam (density 1.68 g cm^{-3}) overlying a clay loam (1.60 g cm^{-3}). Samples were taken the morning following irrigation (by microsprinklers), and again immediately prior to the next irrigation at an adjacent point. Estimates of water loss were made in the four main seasons for the second ratoon crop, and only in the summer for the third ratoon, in selected treatments within an irrigation experiment. There were five replicate samples at each point.

Despite all these precautions variability between replicates was large (coefficient of variation 44%). The authors therefore recommended that the data should be treated with caution. In the spring and autumn ET_c/E_{pan} ratios were similar at about 0.6–0.7. In the summer the ratio increased to about 1.0, but was only 0.6 in the winter. Evapotranspiration rates in the summer reached 7.3 mm d^{-1}, but only 3.5 mm d^{-1} in the winter. Most (67–80%) of the water was extracted from the surface down to 0.3 m depth only across all treatments, and mainly from the 0–0.1 m layer (up to 48%). Some water (c. 15%) came from the depth increment 0.4–0.5 m in the drier treatments in the summer months.

In the subtropics of Australia and South Africa plants remain functional throughout the year, although low temperatures in the winter restrict growth rates. The seasonal variation in the ET_c/E_{pan} ratio observed in these studies by Robinson and Alberts (1989) confirms the findings of Robinson and Bower (1988) summarised above, namely that the transpiration rates from a banana plantation in the subtropics are not a fixed function of evaporation from an open water surface. Robinson and his co-authors explain this on the basis of the controlling influence of low soil temperatures on root activity during the year, which influences seasonal water use. In the winter months this is independent of the leaf area index. They concluded with the recommendation that 'evaporation pans and crop coefficients should be used with caution in the subtropics'. Data presented by Turner (1987) suggest that very little water is used by bananas during the winter in north-eastern New South Wales, Australia. This is despite the presence of large areas of leaves on the plants and suggests increased resistance to water flow through the soil/plant system. These observations are consistent with the work in South Africa.

In the Canary Islands (28° 22′ N; alt. 295 m), Santana et al. (1993) determined the water use of drip-irrigated banana (AAA Cavendish subgroup), and the associated crop coefficients (K_c), using two 60 m^3 drainage lysimeters (10 plants each). ET_c rates, averaged over ten-day intervals, varied from 1.5 to 4.6 mm d^{-1} in a 360-day period, depending on the season. The corresponding annual ET_c total was 1127 mm. Over the same time periods, K_c values ranged between 0.50 and 1.30 (ET_c/E_{pan} screened USWB Class A), and from 0.48 to 1.69 (ET_c/ET_0 Penman–Monteith), with the minimum values in the spring (May). There were very highly significant linear relations between K_c and the proportion of ground shaded by the canopy, as recorded at solar noon.

Mediterranean climates

In these locations the leaves of banana die back in the winter months, which explains the low ET_c/E_{pan} ratios at this time (Lahav and Kalmar 1988). At Tyr in the south of Lebanon (33° 16′ N), a large drainage lysimeter (7 × 7 m square × 1.2 m depth of soil) was used to measure the actual water use of five plants (a plant crop) spaced 3 m apart in a triangular array, over a 20-month period through 1971 and 1972. At this high-latitude location, evapotranspiration rates during the winter months averaged only 1–2 mm d^{-1} increasing through the spring and early summer before peaking in July and August at 5 mm d^{-1} in the plant crop, and 6 mm d^{-1} in the first ratoon, and then declining rapidly through the autumn. During the period covered, the K_c factor for a USWB Class A evaporation pan ranged, on a monthly basis, from 0.41 after planting to a peak of 1.13 late in the first ratoon. Over the 12-month, first ratoon, period K_c averaged 0.78, and 0.82 for the six months (May to October) when irrigation was applied. When compared with the Penman E_o (open water) estimate the corresponding K_c values were 0.90 and 0.85 respectively. Water was extracted from the soil to depths of 0.6–0.8 m, by plants both within and outside the lysimeter (Bovee, 1975).

Later, Israeli and Nameri (1987) used nine drainage lysimeters (about 1.2 m deep and 4.5 m^3 soil volume) to measure the water use of the banana (cv. Williams, AAA Cavendish subgroup) over a period of six years (1978–83) in the Jordan Valley of Israel (c. 32° N; alt. −210 m). The plants were irrigated through a drip system and drainage water was collected every 10 d. Seasonal water use for the ratoon crops increased from about 1.5 mm d^{-1} in April, at the start of the season, to peaks of about 7.9 mm d^{-1} in July and August before declining to 2 mm d^{-1} in November. A regression between water use (ET_c; mm d^{-1}) and evaporation from a USWB Class A pan (E_{pan}; mm d^{-1}) and the average leaf area (L; m^2) accounted for 92% of the variation over the 39 months of measurements ($n = 39$):

$$ET_c = -32.1 + 7.6E_{pan} + 0.75L.$$

Normalising for leaf area resulted in the following linear relationship ($r^2 = 88\%$):

$$ET_c/L^{2/3} = -0.58 + 0.75E_{pan}.$$

When an additional allowance was made for the age of the plants, there was a small improvement in the precision of this estimate ($r^2 = 92\%$).

In a short-term (one month) study Liu *et al.* (2008b) compared measured transpiration rates of well-watered bananas grown in a greenhouse in Israel in the winter with estimates of ET_0 calculated using a number of models, including the FAO Penman and FAO Penman–Monteith equations (Allen *et al.*, 1998), and a standard Chinese 200-mm diameter, 110-mm-deep evaporation pan. Linear relations were derived between daily transpiration and ET_0 for each method, with the pan method giving the best agreement ($\beta = 0.82$, $r^2 = 83\%$, $n = 21$).

Following the research by Lu *et al.* (2002) in Darwin (described above), Liu *et al.* (2008a) used the same sap flow method to estimate the transpiration of pot-grown banana plants in a greenhouse in Israel (32°0′ N; alt. 50 m) in winter months. When the effective radius for sap flow in the rhizome was assumed to be $0.63R$, rather than $0.60R$ (where R is the radius of the rhizome), there was a linear relationship ($\beta = 0.90$, $r^2 = 79\%$, $n = 16$) between daily sap flow estimates and the corresponding gravimetric measurements of transpiration. There was evidence that sap flow lagged 45 min behind transpiration for 4 h early in the morning, equivalent to 10.5% of daily transpiration, with gradual recovery in the afternoon. Actual daily water use averaged 1.46 l plant^{-1}, over the 16 days of measurement, substantially below the values reported by Lu *et al.* (2002) in northern Australia, due to the corresponding low radiation intensities in Israel in the winter.

FAO reconciliation

In an attempt to reconcile conflicting evidence and to produce standardised procedures for estimating the water requirements of the banana worldwide, Doorenbos and Kassam (1979) recommended the following values of the crop coefficient (K_c) based on the relationship given below:

$$ET_c = K_c \times ET_0$$

where ET_c is the maximum rate of evapotranspiration from a banana crop, ET_0 is the potential evapotranspiration from a reference crop, short grass or alfalfa, calculated using one of four methods proposed by Doorenbos and Pruitt (1977).

For a plant crop (first year) in the humid subtropics, K_c values ranged from 0.55 up to 1.0 depending on the stage of growth. In the dry subtropics with strong winds, they ranged from 0.45 up to a peak of 1.2. For the first ratoon (second year) the corresponding values were from 0.7 up to 1.05, and from 0.7 up to 1.25. For a tropical climate, K_c values increased from 0.4 in the first month after planting to 1.05 after 15 months.

In the subsequent FAO manual on crop evapotranspiration (Allen *et al.*, 1998), the tabulated K_c values for a plant crop (maximum height 3.0 m) ranged from 0.50 (early season), to 1.10 (mid season), to 1.00 (end of season). The corresponding values given for a first ratoon crop (4.0 m tall) were 1.00, 1.20 and 1.10. These values are for well-managed crops grown in a sub-humid climate (minimum relative humidity *c.* 45%). If, instead of a single crop coefficient, K_c is derived from its two constituent components, transpiration and bare soil evaporation, the corresponding values are given as 0.15, 1.05 and 0.90 (year 1) and 0.60, 1.10 and 1.05 (year 2), assuming a dry soil surface. In all these cases, the FAO version of the

Penman–Monteith equation must be used to estimate ET_0. The validity of these K_c values, which appear to be realistic, remains to be confirmed.

Summary: crop water requirements

In the tropics

1. Stem flow can account for 10% of incident rainfall.
2. Lysimeter studies have indicated that cumulative ET_c is linearly related to cumulative evaporation from a conventionally sited USWB Class A pan (E_{pan}) with slopes of 1.2–1.4 and peak rates of water use as high as 8–10 mm d^{-1}.
3. Sap flow measurements, by contrast, suggest transpiration rates as low as 2–3 mm d^{-1}, a suspiciously low value.
4. Water balance studies have identified K_c values for different growth stages for the plant crop and the first ratoon. These range from 0.6–0.7ET_0 (vegetative), and 1.0–1.1ET_0 (flowering/fruit development), with actual ET_c rates reaching 4.3 mm d^{-1}.
5. The eddy correlation method indicated similar ET_c rates, varying between 1.8 and 4.3 mm d^{-1}, but K_c values of 0.54 (dry season) to 0.70 (rains).

In the subtropics

6. Transpiration rates from a banana plantation in the subtropics are not a fixed function of evaporation from an open water surface. As a result, there are seasonal differences in K_c values ranging from 0.5–0.6E_{pan} in the winter months to about 1.0E_{pan} (or above) in the summer. The corresponding ET_c rates increase from 3.5 to 7.3 mm d^{-1} (based on a water balance approach). Caution is urged when using evaporation pans for scheduling irrigation.

In Mediterranean climates

7. In these locations, leaves of the banana die back in the winter. ET_c rates vary from 1–2 mm d^{-1} in the winter up to 5–8 mm d^{-1} in the summer. The corresponding K_c values range from 0.4E_{pan} to 1.1E_{pan} (using lysimeters).

Water productivity

The quality of papers reporting the results of irrigation experiments designed to quantify yield responses to water is very variable. Emphasis here is placed on those that provided information of more than local interest. Far too many experiments generate data that are site and time specific.

Tropics

Ghavami (1974) reported the results of an irrigation experiment conducted in the Sula Valley, Honduras (c. 15° 20' N; alt. 30 m) over a three-year period (1968–71) with cv. Valery (AAA Cavendish subgroup). Variable amounts of water were applied through under-tree sprinklers (from 25 to 83 mm weekly, inclusive of rainfall (average annual total 1140 mm), or half these quantities twice a week), but independent of evaporation rates. There was no unirrigated control treatment. They concluded from an analysis using third-order polynomial equations that a gross weekly water application need not exceed 44 mm, and that there were no benefits from irrigating more often than once a week. Yields of fresh fruit averaged across treatments were 55 t ha^{-1}.

The results of an experiment in St Lucia (14° N) on the effects of irrigation, at different levels of depletion of the available soil water, on the growth and yield in the plant and first ratoon crops (cv. Robusta, AAA Cavendish subgroup) were reported by Holder and Gumbs (1983a). The paper did not describe exactly how the water regimes were determined or how much water was applied, but harvested yields (see below for definition), 40 weeks after planting, were increased by 13% to 17% (from about 28 to 33 t ha^{-1}) with irrigation. The largest benefits were achieved in the most frequently irrigated treatments (soil returned to field capacity at 25% or 33% depletion of the available water). This followed an apparent increase in the proportion of harvested plants, from 82% to 91%, and a reduction in the number of broken pseudostems (from 10% to 0.9%).

For the first ratoon crop the irrigation treatments were modified, with either 50% replacement of the soil water deficit at eight- or four-day intervals, or 100% replacement at two-day intervals. The potential yield, defined as the product of the actual fresh weight of bunches with fruits within the specified size grade (23–32 mm) and the plant density, was increased from 46 (unirrigated) to 54 t ha^{-1} (wettest treatment). By contrast, the harvested yields, defined as the product of the potential yield and the proportion of harvested bunches 70 to 73 weeks after planting, were reduced from 38 to 35 t ha^{-1} by frequent irrigation. This time, irrigation reduced the proportion of harvested bunches (from 82% to 65%), and increased the number of broken pseudostems (from 16% to 31%) due to extra wind damage.

In the plant crop, irrigation had no effect on any of the vegetative characteristics recorded before bunch emergence, but there was a small increase (c. 1) in the number of leaves at harvest. In the first ratoon crop, irrigation increased the height and diameter of the pseudostem, but not the number of leaves at harvest. Irrigation advanced the time from planting to shooting by up to three weeks, and the time to harvest by 24 days.

In the plant crop, irrigation had no effect on the number of hands per bunch (7–8), but there was a small increase in the number of fingers per bunch (from 127 to 131) and in the fresh bunch weight (from 20 to 21 kg). In the ratoon crop, each of the components of yield was increased by frequent irrigation; the number of hands per bunch from nine to ten, fingers per bunch from 161 to 192, and bunch weight from 27 to 31 kg. It is difficult to draw clear conclusions on whether or not irrigation is worthwhile from this report.

Holder and Gumbs (1983b) also attempted to compare how two cultivars of the Cavendish subgroup responded to irrigation, 'Robusta' (AAA), which dominates trade in the Windward Islands, and 'Giant Cavendish' (AAA), which is considered to be susceptible to drought. 'Partial' (50% soil water replacement at 4 and 8 d intervals) was compared with 'full' irrigation (every 2 d) and an unirrigated control in the plant crop and the two following ratoon crops in a factorial experiment on a clay loam soil in St Lucia. No data on the depths of water applied are presented, nor indeed on the yield responses to irrigation as such. 'Giant Cavendish', the shorter of the two cultivars, outyielded 'Robusta' (by 13%) over the three cropping cycles. Irrigation increased the susceptibility to wind damage, especially of 'Robusta'. There were no differences in fruit yield per plant, pulp and peel consistency, and bunch uniformity between the two cultivars. 'Giant Cavendish' was more susceptible to bunch deformities such as 'choking' and 'openhandedness' when droughted, but despite this the authors supported its suitability for rain-fed and irrigated production in the Caribbean. The responses of the same two cultivars, with and without irrigation, to three (large) levels of nitrogen fertiliser (280, 560 and 840 kg N ha^{-1}) recorded in the same experiment have also been reported (Holder and Gumbs, 1983c). Most yield, and other, differences were small and insignificant. Wind damage restricted any commercial benefits.

Madramootoo and Jutras (1984) reported the use of drip irrigation on a heavy montmorillonitic clay soil, again in St Lucia in the eastern Caribbean (14°N). Supplementary watering was applied to the plant and first ratoon crops during the relatively dry, or less wet, period from January to May. The crop and the irrigation system were both managed by the farmer on 0.5 ha of land. Water was applied on a daily basis to the irrigated plots whenever rainfall was less than 5 mm, the assumed daily rate of evapotranspiration. In the two years, 635 and 720 mm of water were applied. Irrigation resulted in a small yield increase (7%) in the plant crop, on a base yield of 34 t ha^{-1}, and a small yield reduction in the ratoon crop, due it is claimed to water logging and a high water table. In the second ratoon, a hurricane severely damaged the crop. It was difficult to manage the irrigation system on this poorly draining soil.

In Bangalore, India (13°58' N; alt. 868 m), Hegde and Srinivas (1989a) compared the responses of cv. Robusta (AAA Cavendish subgroup) to irrigation at four different ψ_m values (from −25 to −85 kPa; sandy clay loam soil) as measured at depths of 0.15 m at 0.30 m from the stem. Fruit yields declined from 52 to 41 t ha^{-1} as the frequency of irrigation decreased. Actual evapotranspiration was measured using a water balance approach and annual totals ranged from 1100 to 1860 mm, giving corresponding water-use efficiencies of 37 to 28 kg ha^{-1} mm^{-1}. Although the experiment ran from 1984 to 1986, it is assumed that these values are annual average responses. Unfortunately, the amounts of irrigation water applied are not recorded. Mid-day leaf water potentials (measured with a Scholander pressure chamber in the pre-flowering period) fell to −0.20 MPa in the wettest treatment, and to −0.62 MPa in the driest. There was some evidence that osmotic adjustment maintained pressure potentials in the leaf. There was no evidence of an interaction between responses to irrigation and to nitrogen fertiliser.

Hegde and Srinivas (1989b) provided some additional information on the same experiment. Allowing the soil to dry beyond $\psi_m = -0.45$ kPa delayed flowering by up to 30 d and reduced finger and individual bunch weights (from 17 to 13 kg).

Hegde and Srinivas (1991) later compared the responses of cv. Robusta (AAA Cavendish subgroup) to supplementary irrigation, using drip and basin systems, during the plant and first ratoon crops (3086 plants ha^{-1}). Water was applied by the drip system (8 1 h^{-1}, single emitter), daily except at weekends, to replace evaporation (0.8 × USWB Class A pan regardless of growth stage) less 'effective' rainfall throughout the year. With basin irrigation, 30 mm of water was applied into 1.2 m square basins (plants spaced 1.8 m apart) whenever cumulative pan evaporation, less 'effective' rainfall, reached 38 mm. Comparisons of irrigation systems, as such, were therefore confounded with differences in irrigation intervals and amounts of irrigation water applied, which depended on rainfall frequencies and amounts.

There were yield advantages with the drip system in both years (from 73.5 to 83.9 t ha^{-1} in the plant crop, and from 65.7 to 72.5 t ha^{-1} in the first ratoon). These followed advancements in the date of flowering, by 14 and 15 d, increases in total dry matter production (13% in both years), and heavier individual bunches (+14% and +9%) resulting principally from heavier fingers (individual fruits) (+12% and +5%) especially in the plant crop. The efficiency of water use (total yield of fruit, fresh weight, divided by the sum of 'effective' rainfall, as defined by the authors, and irrigation water applied) was 13–17% higher with drip than with basin irrigation, from 43 to 49 (plant crop), and from 37 to 43 (first ratoon) kg ha^{-1} mm^{-1}. About 225 mm more irrigation water was applied with the drip system in both years than with the basin (1470 compared with 1245 mm). There was a corresponding reduction in the depth of 'effective' rainfall (from 475 to 275 mm in the plant crop and from 519 to 197 mm in the first ratoon). It is likely that more of the rainfall was 'effective' with drip irrigation than with basins, since not all the soil profile was wetted. Using a constant factor (0.8) to convert pan evaporation to ET_c was another weakness that reduces the value of this work. Comparing irrigation methods is not easy since many variables can be confounded.

Goenaga et al. (1993) and Goenaga and Irizarry (1995, 1998, 2000) reported the results of a series of related experiments in the Caribbean island of Puerto Rico. The yield responses of different cultivars to a range of levels of watering based on evaporation from a USWB Class A pan were compared on different soils (mollisol, oxisol and ultisol) in two contrasting locations (18° 02′ N; alt. 21 m; 18° 20′ N; alt. 185 m). One experiment ran for one year only (plant crop, cv. Maricongo, AAB), and the others for three years with either the plant crop and two ratoons, or three successive ratoons (cv. Grand Nain and cv. Johnson, both AAA Cavendish subgroup). Water was applied on alternate days through a drip system, and the irrigation treatments were based on pan factors, which increased in 0.25 increments from (variously) 0.0 (rain-fed) to 1.0 or 1.25. Plants were spaced at 1.8 × 1.8 m intervals (overall plant density 1990 ha^{-1}).

The amount of water to apply was calculated as the product of a pan coefficient (K_p = 0.7) and a crop coefficient (K_c = 0.88), to give an estimate of potential evapotranspiration each week. This was then multiplied by the area of the 'mat', and by the appropriate pan coefficient, from 0.0 to 1.25. Allowance was made for rainfall, and no irrigation was applied if rainfall in the previous week exceeded 19 mm. Fertiliser was either applied in the irrigation water or broadcast. All the yield data were plotted as functions of the pan factor, not the amount of water applied, although these two variables were directly related in individual years. For nearly all the years and sites, there were linear relations between the total bunch yield and the pan factor.

There were similar relationships for individual bunch weights and the number of marketable hands per bunch (sometimes quadratic). Various quality parameters were also often positively related to the pan factor in similar ways. These included hand weight, fruit length and fruit diameter. Irrigation reduced the number of days to flowering and, to a lesser extent, the number of days from flowering to harvest. The authors concluded that a pan factor of not less than 1.0 should be used when irrigating bananas. This ignores the fact that the response curves to water were nearly always linear throughout the range of irrigation treatments tested. It is a pity that the actual water applications in each year, or evapotranspiration, were not included as the independent variable. Although confirming the sensitivity of the banana to water availability, the authors unfortunately made no serious attempt to compare and contrast the results for each cultivar or site so that they could be interpreted and applied with precision to other situations. Some of the results were collated by Goenaga et al. (1995) to provide practical recommendations for drip irrigation on the semi-arid southern coast of Puerto Rico.

In Hawaii (21°21' N; alt. 6 m), Young et al. (1985), in a poorly reported paper, developed yield response functions to water using a design which allowed water to be applied, using a drip system, to individual rows at varying rates (from $0.2E_{pan}$ to $1.8E_{pan}$ in successive increments of 0.2). The volume of water to apply daily to each treatment was calculated using data from a screened USWB Class A pan, adjusted for the 'mat' area of individual groups of three plants (cv. Williams, AAA Cavendish subgroup) spaced at 2.44 × 3.05 m (1343 ha^{-1}). When rainfall was equal to or exceeded 38 mm in the previous week no irrigation water was applied in the following week to any of the treatments. Yearly totals of evapotranspiration were calculated using a water balance approach. When rainfall came in light showers it was assumed that there was no deep drainage, but if more than 38 mm fell in a week drainage followed. Changes in soil water content from one year to the next in the shallow root zone were assumed to be small (maximum 100 mm) compared with changes in other components of the water balance. The soil was kaolinite clay with weathered gravel.

The fruits were harvested weekly from ratoon crops for a period of two years. Since it took about 12 months for each sucker to develop and to produce fruit, evapotranspiration totals were calculated, for each treatment, for successive

52-week periods beginning 12 months before yields were recorded. This corresponded, approximately, to the period of production of each fruit bunch. In the first year the rainfall totalled 860 mm and in the second year 1390 mm; the corresponding estimates of deep drainage were 220 and 731 mm. Yields of fresh fruit (Y, t ha^{-1}) were linearly related to the estimated evapotranspiration total (ET, mm) during the previous 52 weeks. The slopes and the intercepts of the lines varied in each year.

Year 1	$Y = 49.1 + 0.0315ET$	($r = 0.88$)
Year 2	$Y = 39.7 + 0.0265ET$	($r = 0.70$)

It is important to note that, in both cases, the driest treatment ($0.2E_{pan}$) was excluded from the analysis (because there were no guard rows) and the results of one treatment from an adjacent experiment were added ($n = 9$).

Since the intercept on the yield axis is positive and large, the relationships are clearly limited to the ET range from which they were developed (about 1100 to 2700 mm, similar in both years). The corresponding annual yields (actual) were large and ranged from 67.5 to 131 t ha^{-1}.

Following Doorenbos and Kassam (1979 – see below) a single relationship was developed for both years relating the 'relative yield loss' ($1 - Y_a/Y_m$, where Y_a is the actual yield and Y_m is the maximum yield recorded in each experiment) to the so-called 'evapotranspiration deficit' ($1 - ET_a/ET_m$, where ET_a and ET_m represent the actual evapotranspiration totals corresponding to Y_a, for each treatment, and Y_m respectively). Thus:

$$1 - (Y_a/Y_m) = K_y(1 - ET_a/ET_m).$$

The slope of this linear relationship is known as the 'yield response factor' (K_y) for which Young et al. (1985) calculated a value of 0.63 ($r = 0.77$; $n = 18$). This compares with the range of values proposed by Doorenbos and Kassam (1979) of 1.2 to 1.35, for crops yielding 40–60 t ha^{-1}. Values of K_y less than 1.0 indicate relative insensitivity of yield to water deficit, and values above 1.0 relative sensitivity. Clearly there are big differences in the K_y values reported here and explanations are needed. Unfortunately it is not always obvious from the report by Young et al. (1985) how the variables in the water balance were actually calculated and evapotranspiration totals estimated.

In Brazil (9° 9′ S; alt. 365 m), Bassoi et al. (2004b) related the yield of fresh fruit over three harvesting seasons (totalling 409 t ha^{-1}) to measured crop water use over the same period (964 d, ΣET, c. 3950 mm) to obtain a water-use efficiency value of 10.4 kg ha^{-1} mm^{-1}.

Subtropics

Trochoulias (1973) studied the responses of cv. Williams (AAA Cavendish subgroup) to supplementary under-tree sprinkler irrigation over a three-year period

(1968–70) in New South Wales, Australia (28° 51′ S; alt. 148 m). Allowing the available soil water content to be depleted by 10%, 20%, 40% or 70% before irrigating the soil (clay loam) back to field capacity resulted in progressively larger reductions in yield. When averaged over all three years, maintaining the soil water content in excess of 90% of the calculated available water capacity, within an assumed root zone of 0.45 m, resulted in the highest yields (ratoon crops). This treatment, which yielded twice as much fruit as the rain-fed control, involved the application of 8 mm water every three to five days when there was no rain. This implies that evaporation rates were only about 1.5–3 mm d^{-1}, values that appear to be low. Total annual water applications were 400–500 mm. Frequent irrigation increased all the components of yield, principally the number of bunches and the number of fingers per bunch.

Later, again in eastern coastal Australia (28° 51′ S; alt. 148 m), Trochoulias and Murison (1981) compared different levels of replacement (deficit) irrigation with supplementary drip irrigation (single microtube at each plant station spaced 1.8 m apart) based on evaporation from a USWB Class A pan. Water was applied twice weekly from 0 to 1.20 × E_{pan} at 0.20 intervals. If rainfall during each interval exceeded evaporation no irrigation was applied. The experiment continued for six years (1971–76). The soil was a free-draining clay loam. The treatment effects on fruit yield and its components were small, since four of the six years were 'very wet'. Nevertheless the authors concluded that drip irrigation was suitable for bananas and that a useful working guide was to apply water at the rate of 0.60 × E_{pan}.

In subtropical north-eastern Transvaal in South Africa (23° 05′ S; alt. 750 m), Robinson and Alberts (1987) compared the responses of cv. Williams (AAA Cavendish subgroup), during the first and second ratoons, to under-canopy sprinkler and drip irrigation. Similar amounts of water were applied with both systems, daily in the case of drip (except at weekends!), and every two days with the under-canopy sprinklers in the summer or every 2.5 to 4 days in the winter. Applications were based on full replacement of the soil water deficit, based on the product of 0.75 × E_{pan}, where E_{pan} is evaporation from a conventionally sited USWB Class A pan, less any rainfall. There were two drippers per plant (spaced 3 × 2 m) each delivering 4 l h^{-1} in the first ratoon crop, and four drippers per plant in the second ratoon. The sprinklers were spaced at 6 m square with 100% water distribution overlap. This meant that the pseudostems and foliage were wetted to a height of 1.5 m whenever water was applied. Irrigation occurred throughout the year.

There were many interesting findings. Perhaps the most fundamental was that sprinkler irrigation reduced soil (at 0.2 m depth) and pseudostem temperatures by up to 3 °C. It was this effect that probably slowed the rate of emergence of leaves and, surprisingly (?) also increased, by two to three, the number of leaves produced before flower emergence. As a result flowering was delayed in the sprinkler-irrigated plants from early summer to the autumn (three to four months), which in turn extended the interval between one harvest and the next, by up to two months, compared with those irrigated by the drip system. In both ratoon crops, there were associated yield advantages with drip irrigation, about 30% per annum (on base yields of 40–50 t ha^{-1}), following the reduction in the duration of the crop cycle and the production of larger bunches (more hands per bunch, more fingers per hand and longer fingers), also linked to the time of the year when flowering occurred, compared with sprinkler-irrigated plants.

Assuming that plants in both treatments had similar amounts of 'available' water, evaporative cooling appears to have played an important role in differentiating how bananas respond to two contrasting methods of water application, in an area where pseudostem temperatures in the autumn and winter were close to or below about 22 °C during the daytime. Further analysis of the data showed that a 1 °C reduction in the daily maximum air temperature, over the range 23–32 °C, would reduce the number of leaves produced in a year by three to four. The effect of a reduction in temperature will however depend on its position on the temperature response curve. For the rate of appearance of new leaves the optimum temperature is near 30 °C (Turner and Lahav, 1983). We are dealing with interactions that are complex but that have a rational explanation.

Earlier the same two authors had reported the results of a trial in which the water was applied daily (from Monday to Friday) through a drip irrigation system to cv. Williams (AAA Cavendish subgroup) at the same site in north-eastern Transvaal. The soil was a sandy loam overlying a sandy clay loam, with a sandy clay below 0.4 m depth. In this subtropical location the annual effective precipitation, falling between October and April, is less than 1000 mm. Robinson and Alberts (1986) compared three different USWB Class A pan coefficients (ET/E_{pan}), namely 0.25, 0.5 and 0.75. Irrigation was applied throughout the year to supplement rainfall over three cropping cycles, a plant crop and two ratoon crops (1666 plants ha^{-1}). All the plants were irrigated to the same schedule (pan coefficient = 0.25) for the first seven months after planting until the crop cover had reached 40%. Afterwards differential watering was introduced. In the winter months during the second ratoon the three coefficients were reduced by 33% to 0.15, 0.35 and 0.50. Full randomisation of the treatments was not possible because of practical constraints with the irrigation equipment. Water was applied at the rate of 4 l h^{-1}, with two emitters per plant (spaced one metre apart) during the first two years, and four in the third and last year.

Applying insufficient water lengthened the period between planting and the first harvest by about three weeks, and between sucker selection and the harvest of the first ratoon crop by about six weeks. Applying the highest quantity of water (pan coefficient = 0.75) increased annual fruit yields by 33%, 51% and 8% in the plant, first ratoon and second ratoon crops respectively, compared with the treatment receiving least water (pan coefficient = 0.25). The second year of the experiment was considered to be an atypical drought year, when excessive evaporation rates linked to high temperatures (>29 °C) and dry air (>2.5–3.0 kPa) during part of the season (December to February) contributed to the plant responses observed in the driest treatment. Water deficits also reduced some aspects of vegetative growth including the rate of unfolding of leaves, and the leaf area index at flower emergence (as defined by the authors). The yield benefits recorded in the first ratoon (from 55 to 74 to 83 t ha^{-1} y^{-1} from relatively dry to wet, respectively) were associated with an increase in bunch mass (from 32 to 45 kg) resulting from more hands per bunch, more fingers per hand and longer fingers. These benefits were independent of the month of flowering. The total depths of irrigation water applied in that year were 488, 969 and 1453 mm, respectively. These equate to irrigation water-use efficiencies of about 40 kg ha^{-1} mm^{-1} Similar quantities of water were applied during the second ratoon, a so-called 'normal year' in terms of rainfall, but without the same yield differences between the three treatments. The soil water status in all three treatments was similar.

The authors concluded that bananas are sensitive to water deficits at ψ_m values as high as -20 kPa (recorded at 0.2 m depth, and 0.2 m from the emitter). They were unable to recommend a standard pan coefficient because of the variability in response between the two years, and therefore supported the use of tensiometers together with an evaporation pan for irrigation scheduling.

Mediterranean climates

In the northern coastal plain of Israel (c. 33°N), Lahav and Kalmar (1981) compared the responses of cv. Williams (AAA, Cavendish subgroup), grown on a brown grummisol, to two levels of water application and two frequencies of irrigation over three years (1975–77). Water quantities were based on evaporation from a USWB Class A pan ($\times$ 0.9 during the mid-summer months, but adjusted for leaf area at other times in the growing season) with full (660 mm) or partial (75%, 500 mm) replacement of the estimated soil water deficit. Water was applied, through a drip system (4 l h^{-1}), at intervals of three days, commercial practice at that time, or 'pulsed' at hourly intervals over five hours on a daily basis. Fertilisers were injected into the irrigation system at constant rates (30 mg N l^{-1} and 90 mg K l^{-1}). 'Pulsing' with the regular quantity of water always resulted in the highest yield (44 t ha^{-1} averaged over the plant crop and two ratoons, compared with 41 t ha^{-1} from the pulsed crop with partial replacement, and 42 and 40 t ha^{-1} from the two treatments irrigated at three-day intervals respectively). The corresponding yield responses to the water applied were equivalent to 66, 82, 64 and 80 kg ha^{-1} mm^{-1}.

Differences in yield, which were not always consistent between years, were associated with variation in individual bunch weights (averaging 29 kg over three years) and the number of bunches per hectare (1700), associated in part with the effects of the irrigation treatments on the time of flowering. Yield differences between watering treatments were related to the high ψ_m values maintained in the frequently irrigated treatments, always above -12 kPa at 0.225 m depth, and to the volume of wetted soil beneath the drippers, which ranged from 44% in the three-day interval treatment, with a reduced quantity of water, 60% for the corresponding 'pulsed' treatment, up to about 77 and 72% respectively for the two treatments irrigated with the regular amounts. Pulsing also cooled the soil in mid-summer (by 2–3 °C below 25–26 °C, at a depth of 0.1 m, which could be important in areas where temperatures can exceed the optimum), reduced the number of roots in the 1.0 m deep profile by nearly 18%, and reduced nitrate leaching. The authors concluded that shortening the irrigation interval was an effective way of saving water. Applying less water also improved water use efficiencies.

In experiments lasting five years (1979–83) in western Galilee in northern Israel, an area of winter rainfall, Lahav and Kalmar (1988) later studied the responses of Dwarf Cavendish bananas (AAA) to a range of watering regimes using drip irrigation, combined with two fertigation (fertiliser plus irrigation) treatments. In this high-latitude (c. 33° N) area, the leaves are killed during the winter but leaf areas increase from March through to August and September before starting to decline. Seasonal irrigation water requirements (evapotranspiration) in this region average about 1100 mm, with peak rates in mid-summer of 6–7 mm d^{-1}. In this trial, on a grummisol soil, there were two lateral pipes per plant row (double rows, 3×3 m with 6 m path) with drippers (output $4 l h^{-1}$) spaced at 1-m intervals. The irrigation treatments consisted of applying different quantities of water during the season based on variable pan factors and evaporation from a USWB Class A pan. During the mid-summer months (July, August and September) these reached peak values of 0.8, 1.0, 1.3 and 1.4 before declining. Similarly values from the beginning of the season in April until July were proportional to the peak values. Irrigation of the control treatment was based on a constant pan coefficient of 1.0 throughout the season regardless of the stage of growth. Fertiliser was applied in the irrigation water either continuously at a constant concentration or at weekly intervals.

Total water applications varied from 845 mm (pan coefficient = 0.8) to 1447 mm (pan coefficient = 1.4) averaged over the five years. The corresponding range of nitrogen applications in the continuous treatment was 170 to 295 kg N ha^{-1}. Irrigation treatment effects were much larger than any influence of fertiliser although weekly applications resulted in small, but non-significant, benefits. Aver-aged over the five years, irrigating at a pan coefficient of 0.8 reduced the height of the suckers (by 0.1 m), as measured in the autumn, compared with every other treatment, and also reduced fruit (bunch) yields from 68 to 65 t ha^{-1}. None of the treatments influenced the dates of flowering. This yield difference was attributed to a reduction in the corresponding numbers of bunches, from 1865 to 1775 ha^{-1}, as individual bunch weights were similar at about 32 kg.

The most efficient treatment, in terms of the fresh weight of fruit produced per unit of irrigation water applied, was that in which the least amount of water was applied (pan coefficient = 0.8) at about 76 kg ha^{-1} mm^{-1}. The lowest value was 48 kg ha^{-1} mm^{-1} (pan coefficient = 1.4).

Israeli and Nameri (1986) reported the relationships between water use, as determined using drainage lysimeters (Israeli and Nameri, 1987), and production, including leaching requirements. These studies were conducted in the Jordan Valley of Israel (c. 32° N; alt. −210 m). There were linear relations between total dry matter production from each 'mat' and the actual total water use (evaporation plus transpiration), in each of the three years of the experiment. The slopes of the regression lines were 1.71 (plant crop), 1.60 (first ratoon) and 2.35 g l^{-1} (second ratoon); the corresponding values for the dry weight of the fruit were 0.78, 0.77 and 0.86 g l^{-1}. The dry weight of fruit was greatest in the first ratoon and least in the plant crop. Since the total plant dry weights were similar in the first and second ratoons the harvest index was greatest in the second year after planting.

FAO reconciliation

Doorenbos and Kassam (1979) concluded from their review that the banana requires an ample and frequent supply of water. They quoted annual water requirement totals from 1200 mm in the humid tropics to 2200 mm in the dry tropics. The relationships that they developed from the literature between relative yield loss and relative transpiration deficit, as defined above, were similar for all four growth stages that were rather arbitrarily specified: (1) Establishment, and early vegetative growth; (2) Vegetative when the leaves are developing; (3) Flowering, from the start of flower bud differentiation, and (4) Yield formation, as the fruits develop. Water deficit during any of these stages was considered to be equally damaging to yield, through effects on either quantity or quality of the fruit. The value of K_y proposed was 1.2–1.35, indicating great sensitivity to water deficit. The irrigation interval, depending on the soil type, rooting depth and evaporation rates, would range from as little as three days up to 15 d. They suggested that the allowable soil water deficit should not exceed 35% of the available water in the root zone. Most roots they stated were to be found in the top 0.75 m. of soil. Good commercial yields are in the range 40–60 t ha^{-1}, and water-utilisation efficiencies (fruit 70% water content) vary from 2.5–4.0 for the plant crop to 3.5–6.0 kg m^{-3} for the ratoon crops. These values equate to 25–40 and 35–60 kg ha^{-1} mm^{-1}, which are within the range cited in this text.

Summary: water productivity

It has not always been easy to draw generic conclusions with wide applicability from the irrigation experiments reported. Poor experimental designs together with lack of information and poor/incomplete analysis have limited their value. Rarely have authors considered the weaknesses/limitations in the experimental techniques employed before presenting the results. The need for an integrated approach to irrigation research is highlighted, linking empirical field experiments (but well designed and implemented) with supporting environmental physiology measurements, so that the mechanisms responsible for determining the observed yield responses can be fully understood, and the results applied elsewhere (time and space) with confidence (Carr, 2000a).

In the tropics

1. Although there were often, usually small, ancillary benefits from irrigation, including advancements in flowering and harvesting dates and improvements in

many aspects of fruit quality (including heavier bunches, more hands per bunch, greater fruit length and diameter), yield increases between sites were variable and inconsistent.

2. In Puerto Rico, yields of fresh fruit increased linearly with the quantity of water applied from, for example, 35 to 65 t ha^{-1} (more in the ratoon crops than the plant crop) based on the range of pan coefficients tested.

3. In Hawaii, annual fresh fruit yields increased linearly as ET increased by an average of about 30 kg ha^{-1} mm^{-1} on base yields of 40–50 t ha^{-1}. This compares with water-use efficiencies of 40–50 kg ha^{-1} mm^{-1}, based on the sum of 'effective' rainfall and irrigation, reported in South India, and 10 kg ha^{-1} mm^{-1}, based on ET, in Brazil.

4. In Hawaii, the corresponding yield response factor (K_y) was 0.63, which is substantially below the range of values (1.2–1.35) proposed by Doorenbos and Kassam (1979).

In the subtropics

5. To ensure large yields of banana, soil water deficits should be kept low (by maintaining $\psi_m > -20$ kPa at 0.2 m depth). This means irrigation intervals should not exceed 2–3 d during the summer.

6. Although annual yield responses to irrigation are variable, water-use efficiencies of 40 kg ha^{-1} mm^{-1} have been achieved, a value similar to that obtained in the tropics.

7. All the components of yield are enhanced by irrigation. Applying insufficient water delays crop development.

8. The cooling effect of irrigation with microsprinklers on the soil and pseudostem temperatures, compared with drippers, delays crop development and can reduce annual yields by 30%.

9. It has not been possible to specify a standard pan coefficient because of variability in yield responses to irrigation between years.

Mediterranean climates

10. Using drip irrigation, water-use efficiencies of up to 80 kg ha^{-1} mm^{-1} (water applied) have been obtained with 'partial' replacement (75%) of the soil water deficit, compared with 50–60 kg ha^{-1} mm^{-1} with 'full' replacement (100%).

11. Yield responses are not consistent between years.

Irrigation systems

The choice of irrigation system depends on so many factors that are site and context specific. All can be made to work effectively given appropriate technical skills, financial resources, and awareness. Robinson (1996) and Daniells (2004) have reviewed these in the context of banana plantations, while Eckstein *et al.* (1998) reported the preliminary results of a comparison of various irrigation systems at two sites in South Africa. Bananas differ from most other crops in that they are 'nomadic'. Plant positions change after the plant crop depending on which sucker is selected for the following ratoon crops. In the context of this review, the cooling effects of under-tree microsprinklers have to be taken into account when assessing their relative merits. Well-designed and well-managed microsprinklers and drip-irrigation systems are likely to be the most efficient and effective methods of applying water to banana.

Conclusions

With the increasing need to use water effectively, and to maximise water productivity (yield per unit of water consumed) in irrigated agriculture worldwide, there is a continuing need to identify the minimum amount of water required to optimise farm incomes rather than always seeking to aim for maximum marketable yields per unit area. Fereres and Soriano (2006) have recently reviewed the role that so-called 'deficit irrigation' (the application of water below ET_{max} levels) can play in improving the water productivity of field crops; in particular they emphasised how 'regulated' deficit irrigation can increase farm profits, particularly in the case of fruit trees and vines. With the banana, few experiments have been reported in which the aim was to identify at what stages in the development of the crop water applications can be reduced below the maximum without a proportional loss in marketable yields (this assumes that there are yield-determining processes that are differentially sensitive to water deficits). As Fereres and Soriano (2006) stated, research linking the physiological basis of these responses (generally well understood for the banana) to the design of practical 'regulated deficit irrigation strategies' could have a significant impact in water-limited areas (or where it is expensive to deliver water to the field).[3] A tall, aerodynamically rough crop like banana is also better coupled to the atmosphere than a short field crop and a decline in stomatal conductance will be translated into a corresponding reduction in transpiration (and to a large extent photosynthesis). In addition, micro-irrigation systems are ideally suited for controlling water applications and therefore for this form of stress management. Their design and operating criteria, usually the preserve of engineers, need to be specified with appropriate levels of precision (for specific farming systems) in order to maximise (marketable) crop water productivity, while ensuring minimum adverse effects on the water environment (Molden, 2007). These are some of the challenges that face researchers and managers working with banana.

Summary

The results of research on the water relations and irrigation need of banana are collated and summarised in an attempt to link fundamental studies on crop physiology to irrigation practices. Background information on the ecology of the banana and crop development processes, with emphasis on root growth and water uptake, is presented, followed by reviews of the influence of water stress on gas exchange (stomatal conductance, photosynthesis and transpiration), crop water use, and yield. Emphasis is placed on research that has international relevance and, where appropriate, three geographical areas (the tropics, subtropics and Mediterranean climates) are considered. Although banana roots can extend to depths of 1.0–1.5 m, the 'effective' depth of rooting is usually taken to be 0–0.40 m, sometimes extending to 0.60 m. Stomatal conductance is a sensitive measure of soil water availability and plant water status, while evapotranspiration rates (ET_c) can be limited by dry air (saturation deficits >2.0 kPa). In the tropics, typical ET_c rates are 3–4 mm d^{-1}, but there is no agreed value for the crop coefficient (K_c). In the subtropics, there are seasonal differences in K_c, with values ranging from 0.6 in the winter months to about $1.0E_{pan}$ in the summer. It is difficult to draw generic conclusions with wide applicability from the irrigation experiments as they were reported. All the components of marketable yield can be enhanced by irrigation, while applying insufficient water delays crop development. Annual yield responses to irrigation are variable, but water-use efficiencies of 40 kg ha^{-1} mm^{-1} (fresh fruit/ water applied) have been achieved in the tropics and subtropics (and elsewhere up to 80 kg ha^{-1} mm^{-1} with 'partial' replacement of the soil water deficit). To ensure large yields of (marketable) fruit, soil water deficits must be kept low ($\psi_m > -20$ kPa at 0.2 m depth). In the subtropics, this means irrigation intervals should not exceed 2–3 d during the summer. The cooling effect of irrigation with microsprinklers on the soil and pseudostem temperatures, compared with drippers, can delay crop development and reduce annual yields by 30%. There is some (limited) evidence that the presence of the B genome contributes to drought tolerance. Yield response factors to irrigation for different growth stages have yet to be confirmed. Opportunities to improve the water productivity of the many and diverse banana cultivars need to be explored further.

Endnotes

1 This conventional view is now being challenged, for example by Lejju *et al.* (2006) who have identified archaeological evidence of the presence of bananas at Munsa in Uganda during the fourth millennium BC.
2 Banana root systems formed the subject of a recent symposium held in Costa Rica and the reader is referred to the proceedings for detailed reviews on specific aspects of the factors that influence root development and function (Turner and Rosales, 2005). Those properties influencing plant responses to drought and affecting irrigation need are considered here.
3 However, a difficulty for banana is that in ratoon crops numerous shoots at different stages of development are present on the mat at any one time. A straightforward regulated deficit irrigation strategy may therefore not be so useful unless the crop is grown as an annual. This highlights the need to know what effect a water deficit at a certain stage of development of the parent has on the next ratoon crop (D.W. Turner, 2009, personal communication).

3   Cocoa[1]

Introduction

The centre of diversity of the cocoa tree (*Theobroma cacao* L.) is believed to be within the rainforests of lowland northern South America where the greatest range of variation in natural populations exists (Cheesman, 1944; Simmonds, 1998). The plant is grown for its fruits known as cocoa pods (botanically indehiscent drupes). The pods contain seeds, which are fermented with the mucilage surrounding them and then dried to give fermented dried cocoa, the raw material used in the food industry for the production of chocolate and powder (for drinking, baking and ice cream manufacture). A small proportion is also sold as cocoa butter, which is used in the pharmaceutical and cosmetic industries (Wood, 1985a).

In its natural habitat cocoa is a small tree in the lower storey of the evergreen rainforest. Cocoa has been cultivated since ancient times (at least 2000 years) in Central America, from Mexico to southern Costa Rica. After the arrival of the Spaniards in the sixteenth century, cocoa spread rapidly in the New World. It was introduced into South-east Asia in the seventeenth century and to West Africa in the nineteenth century. By 2008, the total planted area of cocoa in the world was estimated to be about 8.6 million ha, of which 67% is in West Africa and only 17% in the Americas and the Caribbean. It is nearly all grown within 20°N and 20°S of the equator (World Cocoa Foundation, 2010). The largest individual producer is Côte d'Ivoire (1.22 million tonnes; 2.3 million ha in 2008), followed by Ghana (662 000 t; 1.75 million ha), Indonesia (490 000 t; 990 000 ha), Nigeria (250 000 t; 1.15 million ha), Cameroon (227 000 t; 500 000 ha) and Brazil (157 000 t; 641 000 ha) (areas harvested from FAO, 2010a: production from ICCO, 2010). West Africa now produces 69% of the world's annual total of 3.6 million tonnes of cocoa. The market value of the cocoa crop is US$5.1 billion (World Cocoa Foundation, 2010).

Cocoa is of international importance as a smallholder crop (5–6 million farmers) with only about 5% (our best estimate) of the world crop (annual total = 3.6 million t) produced on plantations. Historically, most cocoa was planted in

---

[1] By M.K.V. Carr and G. Lockwood.

Figure 3.1 Cocoa. Trees growing under the shade of forest trees – near Itajuipe, Bahia, Brazil (AJD).

newly thinned forest (Figure 3.1) with little investment of monetary capital during the establishment or maintenance phases (Ruf, 1995). Nowadays, little forest remains in the traditional growing areas, and its conversion to cocoa is not acceptable. Replanting former cocoa land is costlier and, as it is usually lower yielding, it is less profitable. In West Africa, where cocoa is usually grown under shade, there are long-standing problems over the tenure of the cocoa trees and shared cropping is widespread. As an internationally traded commodity, cocoa contributes to the livelihoods of an estimated 40–50 million people (World Cocoa Foundation, 2010).

The climates of the major cocoa growing countries have been described by Wood (1985b). The average annual rainfall totals range from 1300 mm (Gagnoa, Côte d'Ivoire) to 2800 mm (Keravat, Papua New Guinea), but it is not the total alone that matters but rather the distribution of rain through the year. In Malaysia (Perak, 4°N) and Papua New Guinea (Keravat, 4°S) rainfall is uniformly distributed, but in Ecuador (Pichilingue, 1°S) and South India (8–10°N), for example, there is a single, long dry season. In West Africa (5–8°N) there are generally two dry, or less wet, seasons. Only a very small proportion (c. 0.5%) of the area planted with cocoa is thought to be irrigated. In a comprehensive analysis of growth processes of cocoa in different regions of the world, Hardy (1958) identified the following critical air temperatures for commercial cocoa production: the mean annual temperature should not be less than 22 °C, the mean daily minimum not less than 5 °C, and the absolute minimum not below 10 °C.

A great deal of research has been reported on the ecophysiology and water relations of cocoa, and various aspects of this topic have previously been reviewed (e.g. Fordham, 1972a; Alvim, 1977; Balasimha, 1999) including, most recently, a paper by Almeida and Valle (2007) with a focus on factors influencing the growth and development of the cocoa tree.

Crop development

Traditionally, cocoa was classified according to trade perceptions of its physical and sensory quality and associated botanical traits, albeit imperfectly as Criollo, Forastero or Trinitaro (Cheesman, 1944).

- *Criollo*: cultivated on a small scale because of its lack of vigour, although the area is currently expanding due to renewed interest in speciality cocoa.
- *Forastero*: the bulk of the world's cocoa production is from this type.
- *Trinitaro*: usually considered to be the result of crosses between the other two types and is not found in the wild. Material designated as Trinitario is grown commercially in the Caribbean and Papua New Guinea.

Recently, Motamayor *et al.* (2008) have proposed a new classification of cocoa germplasm. Based on an analysis of over 1000 individual trees from different geographic areas within the Amazon basin, and using microsatellite markers, they identified 10 genetic groups or clusters, rather than the three listed above, namely: Marañón, Curaray, Criollo, Iquitos, Nanay, Contamana, Amelonado, Purús, Nacional and Guiana. The authors believe that this new classification reflects more accurately the genetic diversity available to breeders.

Traditionally, cocoa has always been propagated from seed although the potential for clones (with a *plagiotropic* – at an angle to the vertical – growth habit, see below) has long been recognised (G. Lockwood and K. Boamah Adomako, unpublished data). The breeding and adoption of new seedling cultivars has led to large increases in productivity, for example in Ghana (Edwin and Masters, 2005).

Cocoa seedlings exhibit two growth forms: a vertical (*orthotropic*) stem, known as a *chupon*, with spiral phyllotaxi, together with three to five lateral (*plagiotropic*) fan branches (⅝ phyllotaxi) (Figure 3.2). The point from which the buds grow out sideways from the terminal end of the main stem is known as a *jorquette*. Clones propagated from *chupon* buddings have the same growth habit as seedlings, while material propagated from fan buds has the same morphology, growth and flowering habit as fan branches (and very rarely produce an *orthotropic* shoot).

The phenological growth stages of cocoa plants have been described in detail by Niemenak *et al.* (2009) using the 'extended BBCH-scale'. In summary, development can be divided into the following principal growth stages:

- growth stage 0: seed germination or vegetative propagation
- growth stage 1: leaf development on the main (vertical) stem, and on the fan branches

Figure 3.2 Cocoa. Mature trees with orthotropic (vertical) main stem and plagiotropic (fan) branches – Bonto Murso district, near Kumasi, Ghana (AJD).

- growth stage 2: main stem elongation, formation of a *jorquette* of fan branches and another *chupon*
- growth stage 3: fan branch elongation
- (no growth stage 4 – this applies only to cereals)
- growth stage 5: inflorescence emergence
- growth stage 6: flowering
- growth stage 7: development of fruit
- growth stage 8: ripening of fruit and seed
- growth stage 9: senescence.

These growth stages are further subdivided into secondary and tertiary growth stages. For the purposes of this review we are principally concerned with the development of the leaf canopy (growth stages 1–3), flowering (growth stages 5–6) and fruit development and ripening (growth stages 7–8).

Stem and leaf canopy

Leaf development occurs on the main stem (*orthotropic*) as well as on the fan (*plagiotropic*) branches. Initially about 10 leaves begin to develop and expand at the same time. This is known as a leaf flush (Greenwood and Posnette, 1950).

After about 40 days a second (and third and so on) flush occurs. Although this rhythmic process is activated by changes in the environment, there is also believed to be some form of endogenous control (Almeida and Valle, 2007). Leaves differ in appearance depending on the type of stem on which they arise (the petioles are about 60 mm longer on the main stem or *chupon*, allowing the leaf to be orientated in relation to the light) (Burle, 1961).

An individual shoot passes through alternate periods of growth and 'dormancy', during which leaf primordia/small leaves are developing in the apical bud (Hardwick *et al.*, 1988a), and follows a similar pattern to the leaf flush cycle (Greathouse *et al.*, 1971). The growth period is characterised by the expansion of leaves and elongation of the shoot. During dormancy the length of the shoot remains constant and no new leaves expand. The main stem reaches physiological maturity after one or two years from planting (at a height of 1.0–1.2 m) when the apical meristem stops growing and a *jorquette* of *plagiotropic* branches (3–5) begins to develop, and these in turn eventually branch. At the same time a second vertical stem (*second chupon*) appears on the main stem (*first chupon*). In the course of time, a second whorl of fan branches develops. This process may be repeated several times such that the canopy increases in height (up to 10 m when cultivated and even 20 m when uncultivated and under heavy shade). Growing cocoa under shade is customary practice in West Africa. Cocoa may be pruned to limit its height to 3–5 m, when only the initial *jorquette* and subsequent branching growth are retained, further *chupons* are continually removed to restrict the eventual vertical growth, and fan branches may be pruned in order to develop an efficient plant structure and for cultural management.

The importance of light interception and distribution within the canopy in influencing productivity in cocoa germplasm was highlighted by a field study in Malaysia (Yapp and Hadley, 1994). For cocoa with a full canopy (48 months after transplanting), planted at a conventional density (1096 trees ha^{-1}), precocity and yield were a simple function of intercepted radiation (f). In a comparison of 12 genotypes, there were large variations in the value of f (from 0.50 to 0.82), and in the cocoa bean yield/intercepted radiation conversion efficiency (0.085 to 0.133 g MJ^{-1}). In contrast, for cocoa planted at high density (3333 ha^{-1}), with full canopy cover, yield was related to light distribution in the canopy and the degree of attenuation represented by the light extinction coefficient (k). Shaded leaves low in the canopy can contribute more to dry matter production when the value of k is small and sunlight penetrates into the canopy (Hadley and Yapp, 1993). The effect of the age of a leaf and its position within a canopy on its productivity (rates of photosynthesis and respiration) has been described by Miyaji *et al.* (1997).

Subsequently, the importance of canopy structure in cocoa was confirmed by Daymond *et al.* (2002a). In a comparison of 10 clones in Brazil, the leaf area index (L), for example, reached values ranging from 2.8 to 4.5 (plants spaced 3 m × 3 m square). There were associated differences in the relationship between L and the fractional light interception. In addition, clones vary in the base temperature for main stem growth, which ranges, for example, from 18.6 °C (genotype AMAZ

15/15) to 20.8 °C (SPEC 54/1) (Daymond and Hadley, 2004). For comparison, the minimum duration of a leaf flush cycle occurs at a daily mean air temperature of about 26 °C (Hadley *et al.*, 1994).

Analysing the results of a large trial in Sabah, Malaysia, Lockwood and Pang (1996) concluded that the optimum plant density can vary between clones (the densities compared were 1096 and 3333 plants ha^{-1}). Later, Pang (2006) confirmed that, at the current stage of cocoa breeding, selection for adaptation to planting density was of higher priority than selection for yield efficiency (the ratio of cumulative yield over a period of time to the increment in the cross-sectional area of the trunk over the same time period).

Drought symptoms. Water availability, by influencing the rate of canopy expansion, and hence light interception, can be expected to influence the productivity of young cocoa. When droughted, seedlings quickly reach a 'point of no return' from which they cannot recover. Visible symptoms of drought include premature leaf fall (progressively younger leaves absciss), the yellowing of basal leaves, wilting, small leaves and slow trunk growth. Provided plants survive the drought, the relief of water stress is followed by a large, synchronised flush of leaf growth (Hutcheon, 1977b), but the flushing cycle that follows is independent of plant water status (at least until the next dry season) (Hardwick *et al.*, 1988b). Synchronised leaf growth can also be triggered by an increase in temperature (Hutcheon, 1981a).

Flowering

Flowers form in meristematic tissues located above leaf scars on the main stem and woody branches (*cauliflorus*). Seedling trees do not flower until after the second flush has hardened on *jorquette* branches, but plants budded from *plagiotropic* growth can flower when the second growth flush has hardened. An individual flowering site (commonly known as a 'cushion', or strictly as a 'compressed cincinnal cyme'; Purseglove, 1968), can have flowers at different stages of development, while up to 120 000 flowers can be produced each year on a single tree (Figure 3.3). The floral meristem produces flowers over the tree's entire lifespan. There is marked seasonal variation in flower production (Edwards, 1973). Flower opening is synchronised and the anthers release their pollen early in the morning. Various small insects, including midges (Ceratopogonidae: *Forcipomyia*), are responsible for pollination. Unfertilised flowers absciss from the stem about 24–36 h after anthesis. Only 0.5–5% of flowers develop into mature pods.

In Ghana, Smith (1964) found that irrigation of young cocoa increased growth rates, advanced flowering and increased the number of flowers (in the second year after planting) but did not affect the percentage of fruit set compared with the unirrigated control treatment. In his rainfall-only treatment, flushing (leaf production) within a population was synchronised by the onset of the rains, while in the irrigated treatments trees responded as individuals. Sale (1970a) found that each time potted cocoa plants were watered following drought vigorous flushing (and flowering) occurred, beginning about 10 days after watering. Similarly, field

Figure 3.3 Cocoa. Flowers form in meristematic tissues located above leaf scars on the main stem and woody branches – Bahia, Brazil (AJD). (See also colour plate.)

observations in Brazil indicated that after a period of water stress, during which flowering was inhibited, the start of the rains triggered synchronous flushing (and flowering) (Alvim and Alvim, 1977). By transferring plants from low (50–60%) or medium (70–80%) relative humidity to high humidity (90–95%), Sale (1970b) was able to show that flowering could also be induced by the relief from atmospheric-induced water stress. These and other factors influencing seasonal periodicity of flushing and flowering in Ghana have been described by Hutcheon (1977b).

Fruit

From anthesis to the maturation of the fruit takes 150 days or more. In Papua New Guinea, Bridgland (1953) reported actual total gestation periods (in a progeny trial) lasting from 163 to 200 days (mean 182). These were similar durations to those found later in Cameroon (between 167 and 200 days) (Braudeau, 1969). The fruit remains attached to the tree until harvested (Figure 3.4). There are two critical points during the development of the small fruits (known as cherelles) when they may stop growing (known as 'cherelle wilt'). The first, at about 40 days after fertilisation of the flowers, coincides with the first division of the fertilised egg. The second, after about 75 days, coincides with greatly increased fat and starch metabolism and the onset of rapid growth of the fruits (McKelvie, 1956). At either susceptible stage, more of the cherelles, even all of them, may wilt if the tree is flushing heavily. However, if the tree is not flushing and all the cherelles are at

Figure 3.4 Cocoa. Fruits remain attached to the tree until harvested – Bahia, Brazil (AJD). (See also colour plate.)

the same developmental stage following synchronised manual pollination, few if any wilt and if the crop is large, flushing is suppressed. If, when the pods are harvested, replacements are set by synchronised manual pollination of freshly opened flowers, they develop normally, without wilting. The trees do not flush, and they may die back.

When fruit set is not synchronised artificially, cherelle wilt is considered to be a fruit-thinning mechanism, which occurs in response to a limited carbohydrate supply and competition between pods for carbohydrates, whereby younger pods are preferentially 'wilted'. In this respect, water stress has an indirect impact on the 'wilting' process as carbohydrate assimilation is reduced. Whether water stress also has a direct impact on another growth process, for example cell division in the cherelle, is not known. Rather than being a yield-limiting factor itself, cherelle wilt therefore reflects any carbohydrate limitation to pod growth. The number of seeds per fruit is normally in the range 20–50. The rate of increase in pod dry weight is slow for the first 60 days after pollination and peaks after about 100 days before declining.

The timing of harvests varies from region to region depending on temperature and rainfall distribution (Alvim, 1988). For example, in the state of Bahia, Brazil (14°S), where there is no clearly defined dry season, low mean air temperatures (<23 °C) in June, July and August restrict flowering, which results in low yields seven months later. In warmer locations, harvests are mainly influenced by rainfall distribution, again with a six to seven month delay between cause (the start of the rains) and effect (mature fruit).

Genotypes differ in the sensitivity of fruit growth to changes in air temperature, which can affect time to fruit ripening, fruit losses from cherelle wilt, final pod size, bean size and lipid content (Daymond and Hadley, 2008). There are also differences between cultivars in the base temperature for fruit growth (range: 7.5 °C for genotypes Amelonado and AMAZ 15/15 to 12.9 °C for SPEC 54/1).

Potential yield

The potential annual biomass (above ground) productivity of cocoa was estimated by Corley (1983) to be 56 t ha^{-1}. Based on an assumed harvest index of 0.20 (perhaps an overestimate) the maximum seed yield was predicted to be 11 t ha^{-1}. This is considerably more than the best recorded yield of 4.4 t ha^{-1} (without shade) and the best commercial yields of 1.5 to 2.5 t ha^{-1} (J. T.-Y. Pang and G. Lockwood, personal communication). Limiting factors, in addition to water stress and pests and diseases, included shade trees (limiting photosynthesis; competition for water), excessive leaf temperatures (especially when grown without shade), low air temperatures (e.g. in Bahia, Brazil) and foliage susceptible to wind damage (benefits from shelter, but reduced leaf cooling because of less wind).

Daymond et al. (2002a, 2002b) have confirmed the potential for yield improvement in cocoa by selectively breeding for efficient partitioning of biomass to the yield component. In a field experiment in Bahia, Brazil, they found, over an 18-month period, a sevenfold difference in dry bean yield between the 12 genotypes tested, ranging from the equivalent of 200 to 1400 kg ha^{-1}. During the same interval, the increase in trunk cross-sectional area ranged from 11.1 cm^2 to 27.6 cm^2, with yield efficiencies varying between 0.008 kg cm^{-2} and 0.08 kg cm^{-2}. Of the seven clones compared, beans accounted for between 32% and 45% of the pod biomass.

Roots

According to Kummerow et al. (1981), summarising the work of others, cocoa seedlings are characterised by a fast-growing tap root on which are formed a limited number of rootlets. After three or four months a first ring of lateral roots has been established. These grow vigorously, always close to the soil surface (<100 mm depth), and branch at their distal parts into dense clusters of fine roots. After four to six years these laterals have spread 4–6 m from the stem forming a mat of fine roots, which extend into the decomposing litter layer. Kummerow et al. (1981) took detailed measurements in Brazil, and estimated the total length of these roots for an 11-year-old cocoa tree to be 1200 m m^{-2} or the equivalent of 1.2 m^2 of root surface for every 1 m^2 soil surface. The tap root continues to grow in thickness and length, reaching depths of 1.0–1.5 m depending on soil conditions. Lateral roots develop on the tap root at depth, some of which grow upwards. By combining fine root production with reasonable estimates of the dry mass of other roots, including the tap root, the total dry mass of roots was estimated to be at least 10 t ha^{-1}.

Similar descriptions of the root systems of cocoa have been provided by Wood (1985b) and Toxopeus (1985). Both refer to the detailed work in Trinidad summarised by McCreary *et al.* (1943), while van Himme (1959) has described with illustrations the results of a major study of the root system of cocoa in the Democratic Republic of the Congo. The proportion by dry mass of the tap root to the total root system was found by Thong and Ng (1980) on an inland soil in Malaysia to be constant at 0.84 almost independent of the plant age (up to seven years). By contrast, in a similar study but this time on a coastal clay soil where a water table limited the development of the root system, the tap root represented only 0.15–0.30 of the total dry mass of the root system (Teoh *et al.*, 1986).

By observing roots in a glass-sided box filled with soil, Vogel (1975), working in Cameroon, found that roots of cocoa seedlings grew in length in a rhythmic pattern having the same period as the leaf flush. The greatest root growth rate occurred *before and during* each flush. Subsequently, Sleigh *et al.* (1981), using a similar method in the UK, confirmed the cyclic nature of root extension, but with maximum root development *alternating* with active leaf expansion. By contrast, Taylor and Hadley (1988), also in the UK, found that periods of rapid root growth *coincided* (within three or four days) with maximum rates of shoot growth (and vice versa). During the periods between successive leaf flushes, root growth continued but at a reduced rate. These observations were made in a controlled environment greenhouse with a nutrient film system.

So roots of cocoa appear to grow in a rhythmic pattern but its timing in relation to a leaf flush is uncertain (possibly this is a result of the different methods of root measurement used). There are practical implications in terms of the sensitivity of cocoa to water stress at different stages in the flush cycle, and the best time to transplant seedlings to the field, so the subject merits further research.

In a remarkably detailed experiment, Moser *et al.* (2010) simulated an El Niño drought in central Sulawesi, Indonesia (1.55°S 120.02°E; alt. 585 m) by reducing rain reaching the soil by 70–80% (with plastic roofs placed beneath the cocoa canopy) for 13 months in a six-year-old cocoa plantation. The cocoa was shaded by *Gliricidia sepium*. All aspects of dry matter production were monitored, including roots, relative to the unprotected control area. Excavations before the drought revealed a superficial distribution of roots in the profile with more than 83% of the fine roots (<2 mm diameter) and more than 86% of coarse roots (2–50 mm), both by dry mass, being found in the top 0.40 m. There were significant quantities of fine roots to a depth of 1.0 m, the deepest being found at 2.0 m. Coarse roots extended to 1.5 m. Surprisingly, there were no differences in root distribution or mass between the control and the sheltered cocoa at the end of the drought 13 months later. A few roots extended laterally up to >1.0 m, although coarse and large roots (2–150 mm) were concentrated within 0.5 m of the stem. The roots of *Gliricidia* penetrated much deeper than those of cocoa with considerable numbers of fine roots present at 2.5 m depth.

Little work appears to have been done on genotypic variation in root: shoot ratios.

Summary: crop development

1. Cocoa has a dimorphic growth habit.
2. Flushes of leaf and shoot growth, synchronised by the commencement of the rains after a dry season (or an increase in temperature), alternate with periods of 'dormancy'.
3. Light interception by the leaf canopy, and light distribution within the canopy, are important determinants of yield.
4. Visible symptoms of drought include premature leaf fall (progressively younger leaves absciss), the yellowing of basal leaves, wilting, small leaves and slow trunk growth.
5. Flowering is inhibited during periods of water stress (and by low temperatures) but the start of the rains results in synchronous flowering.
6. Water stress during fruit development results in small pods and may be an indirect contributory factor to 'cherelle wilt'.
7. Roots grow in a rhythmic pattern but there is conflicting evidence of its timing in relation to shoot growth flushes.
8. Although cocoa is considered to be shallow rooting, roots can extend to depths of 1.5–2.0 m. A large proportion of the roots (80%) are located in the top 0.2–0.4 m.
9. Roots can extend laterally considerable distances (>5 m), but the majority are within 0.5 m of the stem.
10. Only a few of these processes have been quantified in practically useful ways.

Plant water relations

Stomata

Stomatal size and frequency as well as the structure and distribution of leaf waxes of two seedling cocoa varieties (Catongo and Catongo/SIAL) were described by Gomes and Kozlowski (1988). In a greenhouse study in Wisconsin, USA, the average density of stomata, which are found only on the abaxial (lower) surface of the leaf, was about 700 mm^{-2}. Both leaf surfaces were covered with heavy deposits of amorphous wax, except near the stomatal pores. By comparison, in Malaysia, stomatal densities in mixed hybrid seedlings averaged 820 mm^{-2} in irrigated seedlings but 1110 mm^{-2} in unirrigated plants (increase in density associated with smaller leaves) (Huan et al., 1986). In a comparison of eight contrasting genotypes in a greenhouse environment in the UK, Daymond et al. (2009) recorded stomatal densities ranging from 788 to 1081 mm^{-2} (mean 960). In contrast, the densities recorded by Balasimha et al. (1985) in a comparison of 40 accessions in South India averaged only about 100 mm^{-2} (range 80–122). Such a low density compared with other research suggests an error in measurement or in reporting.

In pioneering research, Alvim (1958) used the infiltration technique to monitor the degree of stomatal opening in cocoa, and showed, with cuttings and young plants, that the method could be used as a practical indicator of water stress in cocoa. In a field study with five-year-old plants, which were not suffering from water stress, stomatal opening increased as the light intensity increased, with maximum opening in strong sunlight.

Using a diffusion porometer in Ghana, Hutcheon (1977a) found that stomatal conductance was controlled mainly by the leaf water status. Partial stomatal opening at low light intensities is a feature of shade-tolerant species like cocoa, as is the capacity of the stomata to respond quickly to changing light intensity. The stomata of well-watered field-grown cocoa plants generally remained open in full sunlight.

Leaf water status

In a review of early research on the water relations of cocoa, Fordham (1972a) emphasised the importance of recording the internal water status of a plant if the results of research on crop water use were to have application at sites away from where the research was undertaken. Attempts to relate rainfall amount and distribution to yields in Ghana, Nigeria and Trinidad had, for example, merely emphasised that the relationships are not simple and have little generic value. Similarly, attempts to quantify the role of irrigation in crop productivity in Ghana and Trinidad had failed to establish causal relationships as they did not always take into account either the influence of pore size distribution on soil water availability, or the impact of the aerial environment (e.g. effects of solar radiation/shade, humidity and wind) on evaporative demand at the crop surface. Since preliminary results had been encouraging, Fordham (1972a) advocated the use of the pressure chamber technique for determining the water status of field-grown cocoa plants, a method subsequently confirmed, in a comparative test, as being suitable for cocoa by Yegappan and Mainstone (1981) in Malaysia.

In Ghana, Hutcheon (1977b) described the sequential changes in leaf water potential (and its components) as water stress increased, as recorded in a diverse range of experiments. Initially in a wet soil, values were high early in the morning, declining to about -1.2 MPa by the middle of the day before recovering in the afternoon. As the soil dried, midday values declined, reaching lows of -1.5 MPa when midday stomatal closure restricted further water loss. As the soil dried out, recovery from midday closure became weaker until leaf water potentials remained low even at night. There was no evidence of diurnal changes in osmotic potential but some evidence of considerable seasonal changes (down to -2.4 MPa).

In Ecuador, by contrast, low evaporation rates, due to dull, cool conditions, at the start of the long (six months) dry season prevented the midday leaf water potential, measured in the upper canopy of unirrigated, shaded 12-year-old trees, from declining below that recorded in irrigated trees (about -0.8 MPa) until there had been three to four months without rain. By the end of the dry season, the leaf water potential had reached -1.6 MPa. There was no similar reduction in stomatal conductance, which remained relatively constant throughout the dry season. Osmotic adjustment may have occurred. Similar measurements made in semi-arid north-east Brazil, where irrigation is also necessary and where evaporation rates are higher than in Ecuador, indicated how drought stress occurred earlier in the dry season (leaf water potential reached −1.4 MPa within two months) with a large differential in stomatal conductance between irrigated and unirrigated trees (Orchard, 1985).

Gas exchange

In Ghana, Hutcheon (1977a) found a close relationship between photosynthesis rate (measured with the $^{14}CO_2$ method on a wide range of cocoa genotypes and growing conditions) and stomatal conductance, indicating that stomata are an important controlling factor in photosynthesis. Stomata are sensitive to the dryness of the air. For example, in a greenhouse experiment in the UK (under low light conditions), stomatal conductance fell (cv. West African Amelonado) when the saturation deficit of the air exceeded about 1.0 kPa, and continued to decline over the range tested (up to 3.0 kPa). Rates of photosynthesis (measured with an infrared gas analyser) also declined over the same range of values. In contrast, transpiration at first increased with the saturation deficit but, at values above 1.0 kPa, remained constant (Raja Harun and Hardwick, 1988). In another greenhouse experiment, this time in Indiana, USA, Joly (1988) judged that net photosynthesis of seedlings declined once the leaf water potential fell below about −0.8 to −1.0 MPa. The relationship (with pooled data) between net photosynthesis and stomatal conductance was again (Hutcheon, 1977a) best described by an exponential asymptotic equation. Assimilation rates varied linearly with transpiration. The three cultivars tested (EET399 and EET400, half siblings of Ecuadorian Oriente origin, and one Trinitario type, UF613) differed in their instantaneous water-use efficiencies, that is net photosynthesis divided by transpiration (Joly and Hahn, 1989).

By contrast, in a greenhouse/controlled environment study in Maryland, USA, three different genotypes (CCN 51, LCT EEN 37/A and VB 1117 representing different types) responded in similar ways to changes in ambient conditions (light intensity, carbon dioxide concentration and dryness of the air). As an example, the short-term effects of increasing the saturation deficit of the air (from 0.9 to 2.2 kPa) were to reduce the rate of net photosynthesis slightly, to increase transpiration but to have minimal effect on stomatal conductance. Cocoa was considered to be unusually ineffective in limiting transpiration in dry air compared with other rainforest tree species (Baligar et al., 2008).

The sensitivity of the stomata of cocoa to the saturation deficit of the air was demonstrated further in Colombia by Hernandez et al. (1989). Stomatal conductance declined rapidly in shaded plants (cv. ICA4 × IMC67) as the saturation deficit was increased from 0.5 to 3.5 kPa, whilst apparent carbon dioxide uptake was reduced at values above about 2.0 kPa. Transpiration at first increased as the saturation deficit was increased from 0.5 to 1.0 kPa, but then declined, reaching very low values at 4.0 kPa. The implications of these observations to the sustainable management of cocoa under field conditions were considered, in particular the value of shade (water-use efficiencies are likely to be higher under shade, although in practice this will depend on such things as the amount of light intercepted by the shade canopy, the leaf area index of the cocoa canopy and the ambient light conditions) and irrigation (responses will be limited during periods of hot, dry air).

In a field study in India, Balasimha *et al.* (1991) also found that photosynthesis declined as the saturation deficit of the air increased above about 2 kPa. Drought-tolerant accessions maintained higher leaf water potentials during dry months than drought-susceptible ones.

In Malaysia, container-grown plants were subjected to water stress of different durations. Leaf water potentials fell to minimum values of about −3.0 MPa in droughted plants with a suggestion that the critical value below which stomatal conductance and photosynthesis both declined was about −1.5 MPa. There was an indication that one genotype (KKM25) could withstand water stress better than the other two (KKM4, KKM5) tested (Razi *et al.*, 1992).

A similar study was undertaken in a semi-arid region of Venezuela (08° 31′N 71° 71′W; alt. 1100 m) this time with field-grown, shaded, four-year-old plants (cv. Guasare, Criollo type) irrigated every three, 12 or 25 days. Minimum leaf water potentials were in the range −1.4 to −1.7 MPa, and there was evidence of osmotic adjustment, but only in the 3- and 12-day treatments. In severely stressed plants, daily photosynthesis was reduced by 25% and transpiration by 39% compared with the 3-day treatment, which implies an increase in water-use efficiency (Rada *et al.*, 2005). In a greenhouse experiment in the UK, significant differences in the instantaneous water-use efficiencies were recorded in the eight genotypes compared, ranging from 3.1 (IMC47) to 4.2 mmol mol^{-1} water (ICS1) (Daymond *et al.*, 2009).

Shading, as well as water, also influences gas exchange processes. In a nursery experiment in Ghana, young clonal plants were shaded with netting at three different levels (32.5%, 55% and 76% shade). The plants were kept well watered. During the two rainy seasons, stomatal conductance, photosynthesis and transpiration all declined as the level of shade increased, but during the dry season the situation was reversed (Acheampong *et al.*, 2009). This difference between the seasons was explained by the dryness of the air. Saturation deficits averaged about 1.4 kPa in the rains and 2.7 kPa in the dry season.

That wind can physically damage cocoa was confirmed in a wind tunnel experiment in Wisconsin, USA. Of particular interest was the observation that an increase in wind speed from 1.5–3.0 m s^{-1} to 6.0 m s^{-1} increased stomatal conductance, but *reduced* transpiration. This was apparently the result of a reduction in the leaf:air saturation deficit associated with cooling of the leaves by the wind (Gomes and Kozlowski, 1989).

Drought tolerance

In South India, where there is an extended dry season, Balasimha *et al.* (1988) developed a procedure for determining the relative drought tolerance of cocoa accessions. Based on an assessment of a number of plant parameters, including effective stomatal control of transpiration when under water stress, five accessions (NC23, NC29, NC31, NC39 and NC42) were identified as being drought tolerant. Later, Balasimha *et al.* (1999) evaluated the drought tolerance of the progeny of crosses between four high-yielding and three drought-tolerant pollen parents (NC23/43, NC29/66 and NC42/94). Using the rapid excised leaf screening method (based on changes in leaf water potential of an excised leaf in 90 minutes (Balasimha and Daniel, 1988), and follow-up field measurements during the dry season over a five-year period, they identified two hybrids (1–21 × NC42/94; 1–29 × NC23/43) exhibiting drought tolerance characteristics on the basis of their higher leaf water potentials and stomatal resistances during dry weather.

In a comparison of eight clones (all grafted on to local Comum variety rootstocks) in a greenhouse experiment in South Bahia, Brazil, three were identified as being drought tolerant on the basis of the degree of osmotic adjustment recorded, but only when dehydration was imposed rapidly (Almeida *et al.*, 2001). The osmotic adjustment was associated with the accumulation of potassium and phosphorus ions in the leaf. These three clones (SPA5, SIAL70 and TSH516) were recommended for growing on drought-prone shallow soils in the region. By contrast, Premachandra and Joly (1994) could find no evidence of osmotic adjustment in cocoa seedlings subjected to water stress over a 22-day period in a glasshouse experiment in Indiana, USA.

According to Bae *et al.* (2008), it may be possible to enhance drought tolerance in cocoa by altering polyamine levels, which are associated with the response of plants to drought, either by selection or by genetic manipulation. Although transgenic techniques in cocoa are at present discouraged because of consumer concerns, the genes involved in polyamine biosynthesis have been identified.

Bae *et al.* (2009) have also identified another possible way of enhancing drought tolerance in cocoa. They found that colonisation of young cocoa seedlings by the endophytic fungus *Trichoderma hamatum* (isolate DIS 219b) promoted growth and delayed the onset of drought symptoms through changes in gene expression. The primary direct effect of colonisation was to promote root growth (dry weight and fresh weight) regardless of the plant water status. This delayed drought-induced changes in stomatal conductance, net photosynthesis and green fluorescence emissions as well as wilting.

Summary: plant water relations

1. Stomata are found on the abaxial surface of the leaf at densities of 700 to 1100 mm^{-2}.
2. Stomata open in low light intensities and remain fully open in full sunlight on well-watered plants.

3. Leaf water potentials in well-watered plants can decline to -1.2 MPa at midday, before recovering.
4. As the soil dries, minimum values of leaf water potential can reach -3.0 MPa in droughted plants.
5. Partial stomatal closure begins at a leaf water potential of about -1.5 MPa.
6. Stomatal conductance is sensitive to dry air, declining as the saturation deficit increases from about 1.0 up to 3.5 kPa.
7. Net photosynthesis declines over a similar range of values.
8. Transpiration initially increases but when the saturation deficit exceeds about 1.0 kPa it begins to decline (or remains constant), reaching very low values at 4 kPa.
9. Cultivars differ in their instantaneous water-use efficiencies.
10. Leaf water potential and stomatal conductance measurements have been used to identify drought-resistant cultivars.
11. There is conflicting evidence about the contribution of osmotic adjustment to drought tolerance.

Crop water requirements

Few attempts have been made to quantify the actual water use of cocoa in the field. In Côte d'Ivoire, Jadin *et al.* (1976) used a neutron probe to monitor changes in the soil water profile beneath young cocoa plants irrigated with sprinklers or with drip. Using a similar set of measurements, Jadin and Snoeck (1981) developed linear relations between actual water use for irrigated and unirrigated crops and a Penman equation (date unspecified) estimate of evapotranspiration, and also between water-use and evaporation from a pan (Colorado). It is not easy to apply these findings with confidence elsewhere.

Using the Penman–Monteith equation (Allen *et al.*, 1998) and published values of the key parameters (including crop, aerodynamic and surface resistances) controlling transpiration and evaporation from crop and soil surfaces, Radersma and Ridder (1996) computed evapotranspiration by cocoa (and three other crops) for La Mé, Côte d'Ivoire ($c.$ 5° 20′ N 4° 02′ W; alt. 35 m). Assuming an annual rainfall total of 1500 mm, daily transpiration rates (T) were estimated to be between 3.0 and 6.1 mm d^{-1} during the rains, depending on net radiation levels and the saturation deficit of the air, and from 1.0 to 1.9 mm d^{-1} during the dry season. The corresponding seasonal and annual totals for evapotranspiration (ET_c) were 584 mm (wet season), 294 mm (dry season) and 878 mm (total).

The sap-flow technique was used by Colas *et al.* (1999) in Indonesia (40° 38′ S 105° 15′ E) to measure transpiration of individual cocoa trees (eight years old) grown alone (1333 ha^{-1}) or in association with coconut (1186 and 87 trees ha^{-1} respectively). On dry days, transpiration increased rapidly early in the morning, reaching maximum values of 2.5 to 3.0 l dm^{-2} h^{-1} before declining from about 1100 h onwards, when the saturation deficit of the air reached about 2 kPa, as a result of stomatal closure. On wet days, transpiration rates remained relatively constant during the middle of the day. Over a consecutive 18-day period, transpiration averaged the equivalent of about 1.31 mm d^{-1} (or 10 l $tree^{-1}$ d^{-1}), compared with a Penman (1948 version) potential ET_0 estimate of 3–5 mm d^{-1}. This equates to a crop factor (K_c) value of about 0.3 (see below). For comparison, transpiration by cocoa in the crop combination averaged 1.19 mm d^{-1}.

In the simulated El Niño drought experiment reported by Moser *et al.* (2010) in central Sulawesi, Indonesia (described above in the section 'Roots'), the combined average rate of water use by both species, cocoa and *Gliricidia sepium* (as measured with heat dissipation sap flux sensors), was only 1.3 mm d^{-1} in the protected plots and 1.5 mm d^{-1} in the control (70% of which was from the cocoa trees), values similar to those reported above, but still surprisingly low (Köhler *et al.*, 2010). Drainage represented 55% of the annual rainfall in the control plots and 11% in the protected plots.

Water requirements (ET_c) derived from estimates of potential evapotranspiration by a reference crop (ET_0) require a crop factor ($K_c = ET_c/ET_0$). Allen *et al.* (1998) suggested a K_c value of 1.0–1.05 for a cocoa crop with a complete canopy. The K_c value is based on a theoretical understanding of the processes of transpiration and evaporation from a tall crop, and assumes full crop cover or frequent wetting of the soil surface. The evidence on evapotranspiration presented above suggests this K_c value is too high, but it depends on the actual crop cover in the two field experiments described.

Summary: crop water requirements

1. Little research has been reported on the water use of cocoa.
2. Field data (based on the sap flow method) suggest ET_c rates of <2 mm d^{-1} which, for a crop with a complete canopy, would appear to be low.
3. The corresponding K_c value is only 0.3, much less than the theoretical value for a cocoa crop or any crop with complete ground cover.

Water productivity

Yield forecasting

A number of attempts have been made to relate cocoa yields to rainfall, for example in Trinidad (Dunlop, 1925), Papua New Guinea (Bridgland, 1953), Ghana (Maidment, 1928; Skidmore, 1929; Ali, 1969) and Nigeria (Toxopeus and Wessel, 1970), also reviewed by Fordham (1972a). But, as the role of rainfall in crop growth is not a simple one, these attempts have had limited success. Recently, a physiological growth and production model for cocoa has been developed and described by Zuidema *et al.* (2005). Known as SUCROS-Cocoa, it is based on the SUCROS-family of models with parameter values taken from the literature. It simulates biomass production and bean yield for different situations and locations. Lack of good independent data has limited the validation of the model. Regression analysis showed that over 70% of the variation in simulated bean yield could be explained by a combination of annual solar radiation and rainfall in the two driest months. Yield losses due to drought of up to 50% depending on location and soil type were predicted. Heavy shading (>60%) reduced yields by more than 30%. Opportunities to improve the model were identified.

Responses to irrigation

In view of the apparent sensitivity of cocoa to drought, there have been surprisingly few irrigation experiments, but perhaps less surprising given the limited likelihood of commercial scale irrigation. Murray (1961) reported the results of one such experiment in Trinidad, but problems in applying and monitoring the amount of water applied in the dry season (acknowledged by the author) meant that the results obtained over five years (no consistent yield advantage from irrigation) had little value. Low atmospheric humidity in the dry season was suggested as a possible limiting factor. An irrigation experiment in Ghana was also compromised by poor design (recognised as a weakness) that meant the data could not be analysed statistically. Yield increases of 12%, 17% and 40% were obtained from mature Amelonado trees by keeping the soil close to field capacity in the three years 1960–62. These increases were less than anticipated at the time, owing, it was again thought, to 'dry air' constraints. No absolute yields were presented (Hutcheon et al., 1973).

Another irrigation experiment was undertaken in Malawi (16° 31′ S 35° 10′ E; alt. 52 m) by Lee (1975). The depth of water applied was based on three evaporation pan factors (0.6, 0.8 and 1.00 × E_{pan} (Kenya type)), with two irrigation intervals based on calculated soil water deficits (52 or 104 mm). Detailed records were kept for two years (1971/72 and 1972/73) when the average annual rainfall was 710 mm and the depths of water applied over the long dry season were 920–1650 mm. Yields of wet beans from a four- to five-year-old Amazon cultivar (T76) were similar for all three water applications, averaging about 2200 kg ha^{-1}. More frequent irrigation out-yielded less frequent irrigation by 285 kg ha^{-1} (2318 and 2032 kg ha^{-1}), although plot to plot variability was large (CV >30%). Hot dry winds damaged the crop towards the end of the dry season. The alluvial soil had a water-holding capacity of 230 mm m^{-1}. The seasonality of harvest was not affected by any of the treatments.

In Côte d'Ivoire, Jadin and Jacquemart (1978) compared two methods of irrigation (sprinklers and drip) with an unirrigated control treatment on the development and yield of young cocoa over a two-year period. Sprinkler irrigation was applied when the measured soil water deficit reached 20 mm, and drip irrigation was applied when the soil water tension reached 20 kPa at 0.20 m depth. As a result in part of these differences in scheduling irrigation, considerably more water was applied with the sprinklers in the dry season (535 mm) than with drip

(224 mm). Irrigation, particularly drip, speeded up the rate of development, increased the number of flowers, and increased yields but did not affect the periodicity of the growth cycle.

The results of an unreplicated field-scale irrigation trial in Peninsular Malaysia were later reported by Huan *et al.* (1986). Supplementary irrigation (drip) was applied daily to a 0.5-ha block of mixed hybrid seedlings on a coastal estate (marine clay) after a dry period (no rain for two weeks), except on days after there had been 5 mm or more rain, or when it was actually raining. A similar 0.5-ha block acted as the unirrigated control. The trial lasted nearly three years (1981–83). Annual dry bean yields were increased by irrigation from 1500 to 2400 kg ha^{-1} (+60%) in 1982 and from 1150 to 1450 kg ha^{-1} (+28%) in 1983. This followed an increase in pod number (averaging +39%) and in bean weight (+7%). The quantities of water applied were not specified. These results can only be considered indicative of responses to irrigation at this site, and have limited generic value.

In the simulated El Niño drought experiment reported by Moser *et al.* (2010) in Indonesia (described above under 'Roots' and 'Crop water requirements') there were no significant differences in cocoa leaf, stem and branch wood, or fine root biomass production (above and below ground) between the control treatment and the one in which rain through-fall was reduced by 70–80%, even though the soil profile dried to permanent wilting point during the year. By contrast, there was a reduction in dry bean production over the year as a whole from 740 ± 180 to 670 ± 30 kg ha^{-1} (both low yields when compared with the best commercial yields), with the later harvests more affected than the early ones. Possible causes of the limited response to drought in terms of net biomass production were proposed. These included active osmotic adjustment in the roots (measured), high atmospheric humidity in both treatments, and drought mitigation through shading by *Gliricidia*.

In contrast to the trials described above, the principal practical objective of an experiment summarised by Hutcheon (1981b) was to find out if irrigation would induce Amelonado trees to flower throughout the dry season in order to produce pollen for use in manual pollination of a seed orchard in Ghana. When the *cherelles* were continuously removed, irrigated trees produced 30% more flowers than those that were unirrigated, although the flowering patterns were the same. There was no benefit from using over-tree sprinklers (to reduce internal water stress by raising the humidity) rather than microsprinklers under the trees.

Drought mitigation

Repeated mulching (with fresh, moist plantain pseudostems) improved the establishment of cocoa seedlings during the extended six-month dry season in Ecuador (Orchard and Saltos, 1988). Mulched plants maintained stomatal opening at similar levels to irrigated plants and, at the end of the first year of establishment, there were substantial increases in the dry weight of shoots and roots, and in leaf area, over and above those recorded in the irrigated treatments. The benefits of applying a mulch during the first three to four years after establishing cocoa seedlings in the field can be large. But, the cost of growing, cutting, transporting and spreading the mulch can be prohibitive unless it is available on the spot as, for example, when leguminous shrubs such as *Flemingia macrophylla* are used as temporary shade (Wood, 1985c). The role of shade trees in drought mitigation (or enhancement) is complex and variable depending on many factors and is worthy of a separate review.

Specific advice to growers on irrigation practices is even harder to find but the Central Plantation Crops Research Institute, Kerala, South India gives this (edited) advice to cocoa producers in some areas of southern India, where long periods of dry weather lasting three to six months can occur (CPCRI, 2010):

Irrigate at weekly intervals during the summer [this presumably refers to a sole crop]. When cocoa is grown as a mixed crop with arecanut, irrigate once a week during November–December, once every six days during January–March and once in four to five days during April–May with 175 l water tree^{-1}. Maximum yields are obtained when cocoa is drip irrigated with 20 l day^{-1} tree^{-1}.

Assuming a planting density of 1600 trees ha^{-1} (2.5 m × 2.5 m) these figures equate to rates of water use equivalent to 5.6–7.0 mm d^{-1} or at 1100 trees ha^{-1} (3 m × 3 m) 3.9–4.8 mm d^{-1}, and 2.2 or 3.2 mm d^{-1} for drip irrigation. No estimates of the yield benefits are given nor is the total quantity of water to be applied over a season. Presumably this is not known.

Summary: water productivity

1. There is a paucity of reliable published data quantifying the yield and other benefits that could result from well-managed irrigation of cocoa in different locations.
2. It remains to be seen under what conditions irrigation (and/or drought mitigation, including mulching of young trees) is financially worthwhile.
3. Dry air can be expected to limit responses to irrigation.
4. It is not possible to specify yield response functions to water with the limited information available.
5. Shade trees add a further complication.

Conclusions

Since cocoa is a drought-sensitive crop, and a large proportion of the world's cocoa is grown in parts of the tropics having a distinct alternation between wet and dry seasons, it is to be expected that the water relations of cocoa would have been the subject of research. What is surprising is the limited amount of work done in the field with mature crops as compared with research on immature plants in relatively controlled conditions. Although research on immature plants has led to a good understanding of aspects of cocoa physiology, there is a paucity of information of direct practical value. For example, the lack of data on crop water use and water productivity means it is impossible to quantify yield losses due to drought or yield benefits from irrigation. With the threat of climate change leading to less, or more erratic, rainfall in the tropics, and higher temperatures and drier air, uncertainty in yield forecasting will increase and yields will decrease on average. In particular, there is evidence of a decline in the annual rainfall in the agro-ecological area where cocoa is concentrated in Ghana (Owusu and Waylen, 2009). This may be of particular significance when the existing cocoa is uprooted and the land replanted on soils that have already deteriorated in terms of organic matter and nutrient content. Taken together this could result in crop establishment problems. The vulnerability of the cocoa industry to climate change in Ghana, which employs over 800 000 smallholder farm families on farm sizes ranging from 0.4 to 4.0 ha, has been the subject of a study by Anim-Kwapong and Frimpong (2006). They proposed a number of policy options that needed to be put in place to enable the industry to adapt to these anticipated changes. These included a drought management policy and, specifically, the promotion of irrigation.

Why has there been this emphasis on fundamental research and less on its practical application? The answer must be in part due to the structure and nature of the industry, and the way research is organised and funded. Often the 'drivers for change' in agriculture are the more progressive farmers who are already industry leaders, but believe they can do it even better given additional information. They will also probably be those with access to funds. Cocoa producers are, however, predominantly (95%) smallholders with very low shadow wages spread over protracted time spans, little if any access to capital and all too often low profitability. Cocoa suffers because the barriers to smallholder entry are very low and people who have few alternative sources of income are prepared to do a lot of work for very little (Figure 3.5). Most world cocoa production is organic by default – pesticides might be used if insect pests in particular get out of hand – but otherwise there are few purchased inputs, so irrigation has never seemed an option worth pursuing vigorously by researchers (despite the sensitivity of cocoa to water stress).

But perhaps in future the cocoa industry will develop in a different, more intensive way. Following an eight-year research project, a fledgling modern, mechanised, clonal, cocoa industry is being created in northern Queensland, Australia, with supplementary irrigation. Estimates from field trials give the annual irrigation requirement to be up to 470 mm, depending on the site, with peak weekly requirements of about 200 l tree^{-1} (1250 trees ha^{-1}). Dry bean yields of between 1.5 and 2.7 t ha^{-1} have been achieved from young trees (Diczbalis *et al.*, 2010).

Research on the water relations and irrigation of cocoa is obviously complicated by such issues as shade/shelter (for relative advantages and disadvantages see Beer, 1987, and Obiri *et al.*, 2007), mixed cropping (e.g. with coconut, rubber, banana), diverse scales of production (from smallholders to large estates) and interactions with plant density. All of these variables affect crop water status and the yield potential, and hence the justification (in financial and livelihood terms),

Figure 3.5 Cocoa. Beans drying on mats erected in a 'cocoa village' – Bonto Murso district, near Kumasi, Ghana (AJD).

or otherwise, for irrigation or other forms of drought mitigation, for example mulching at the establishment phase, appropriate shade usage, and planting drought-tolerant cultivars (Acheampong, 2010). These are the issues that need to be addressed, preferably in a coordinated international research programme.

Whether managing water stress is justified in a given situation will depend on the growth stage of the cocoa, the local climate and predicted climate change for the region. But when there are still so many other limiting factors to yield, including generally low levels of field maintenance, pests and diseases, lack or inbalance of nutrients and, on occasion, competition from shade trees, water management will not always be the priority issue of concern. Irrigation is a luxury for many farmers. It should only be considered when other limiting factors have been addressed. In view of cocoa's international importance as a traded commodity, it is difficult to understand why cocoa production at a field level is so under-resourced.

Summary

The results of research into the water relations of cocoa are reviewed in the context of drought mitigation and irrigation need. Background information on the centres of production of the cocoa tree, and the role of water in crop development and growth processes, is followed by reviews of the effects of water stress on stomatal conductance, leaf water status and gas exchange, together with drought tolerance, crop water use and water productivity. Leaf and shoot growth occur in a series of flushes, which are synchronised by the start of the rains following a dry season (or an increase in temperature), alternating with periods of 'dormancy'. Flowering is inhibited by water stress but synchronous flowering occurs soon after the dry season ends. Roots too grow in a rhythmic pattern similar to that of leaf flushes. Roots can reach depths of 1.5–2.0 m, with the main mass of roots found in the top 0.2 to 0.4 m, and can spread laterally greater than 5 m from the stem. Stomata open in low light intensities and remain fully open in high-intensity sunlight in well-watered plants. Partial stomatal closure begins at a leaf water potential of about -1.5 MPa. Stomatal conductance is sensitive to dry air, declining as the saturation deficit increases from about 1.0 up to 3.5 kPa. Net photosynthesis and transpiration both consequently decline over a similar range of values. Little has been published on the actual water use of cocoa in the field. Measured ET_c values equate to <2 mm d^{-1} only, whereas computed ET_c rates of 3–6 mm d^{-1} in the rains and <2 mm d^{-1} in the dry season have also been reported. Despite the sensitivity of cocoa to water stress, there is a paucity of reliable, field-based published data of practical value on the yield responses of cocoa to drought or to irrigation. With the threat of climate change leading to less, or more erratic, rainfall in the tropics, uncertainty in yield forecasting as a result of water stress will increase. Social, technical and economic issues influencing the research agenda are discussed.

4   Coconut

Introduction

The coconut (*Cocos nucifera* L.) contributes to the livelihoods of millions of people in the developing world, not only through its production but also through employment generated by the many associated industries. It is the most wide-spread, economically useful palm of the wet tropics, being found mainly in coastal areas between latitudes 20°N and 20°S of the equator.

Although there still remains some uncertainty about the coconut's centre of origin (it may be South America), it is generally accepted that the coconut originated in the south-west Pacific or the Indian Ocean and that it became domesticated in Malaysia and on the coasts and islands between South-east Asia and the western Pacific. It has been present in the Pacific islands (a centre of genetic diversity) for millions of years, long before their settlement by Polynesians, where it is an indigenous species. It is believed that coconut was distributed widely by nuts floating in ocean currents and germinating after they were washed ashore in new locations. In this way it probably reached southern India and Sri Lanka, where the coconut has a recorded history of 2000–3000 years, and possibly Madagascar and eastern Africa. The nuts were probably also carried by humans as a source of food and drink on long sea voyages, reaching West Africa and the Americas about 500 years ago when the coconut became pantropical (Purseglove, 1972; Harries, 1978, 1995; Persley, 1992).

The coconut is often referred to as the 'tree of life' because of the diverse range of products, more than 100, which are derived from all parts of the plant (Persley, 1992). The oil is used for both edible and industrial purposes. The edible uses include the production of margarine and cooking oil, and the industrial uses include the manufacture of detergents, soap and fatty acids. The residue left after oil extraction from the copra (the dried endosperm of the nut from which oil is extracted) is used as animal feed. The balance of world production (about 50%) is consumed fresh in producing countries. Other components of the nut, as well as the rest of the palm, provide such diverse products as food, drinks, medicines, building materials, furniture, ornaments and household goods (Figure 4.1).

As already stated, coconut is a crop of the humid tropics. It grows best at altitudes below 1000 m and near the coast where the mean air temperature is in the range 25–29 °C (Persley, 1992). In their review paper, Gomes and Prado (2007)

Figure 4.1 Coconut. The 'tree of life' is the source of many different products – Kerala, India (MKVC). (See also colour plate.)

specify a slightly broader optimum temperature range of 23–34 °C, with the absence of temperatures below 15 °C, and a relative humidity between 60% and 90%. Annual rainfall should exceed 1500 mm, preferably evenly distributed throughout the year. Coconut palms can be grown on a wide range of soils, providing they are free draining (Purseglove, 1972; Murray, 1977). As Murray (1977) explains, deficiencies in rainfall can sometimes be compensated for by access to groundwater by roots especially, for example, where coastal beach plantations are backed by rain-fed freshwater swamps and lagoons that drain towards the sea.

In 2007, the total planted area was estimated by FAO (2009a) to be about 11.2 million hectares, spread across more than 90 countries, and producing 10 million tonnes of copra equivalent and 6 million tonnes of oil equivalent. The main producing countries are Indonesia, the Philippines (both with more than three million hectares) and India (>one million ha). For many small-island and coastal communities the coconut is a major source of wealth. In such places, apart from fishing, there are often few other viable enterprises. Internationally it is a smallholder crop, with planted areas typically in the range 0.5 to 4.0 ha per household (Purseglove, 1972; Persley, 1992). Brazil, the fourth largest producer, has, for example, 220 000 small-scale producers occupying 300 000 ha mostly in the coastal zone. In addition, Brazil has about 80 000 ha of dwarf cultivars grown commercially in non-traditional areas (with irrigation; Figure 4.2) (Gomes and Prado, 2007).

Figure 4.2 Coconut. Irrigated Green Dwarfs – Brazil (LM). (See also colour plate.)

A great deal of research has been reported on the ecophysiology and water relations of coconut and the adverse effects of water stress on yield are well established. Various aspects of this topic have previously been reviewed (e.g. Murray, 1977), most recently in a paper by Gomes and Prado (2007), which focuses on the physiological responses of coconut to water stress.

Crop development

There are two naturally occurring types of coconut, the tall and the dwarf. In addition, hybrids (mainly tall × dwarf) have been bred in many countries. But, as Harries (1995) pointed out, although this is a convenient distinction it is misleading as even dwarf types can reach a considerable height given time. Dwarfness in coconut really means precocity as the first fruit is set close to the ground.

The coconut has a single, unbranched trunk, which in the tall type (var. *typica*) grows up to 30 m in height (Figure 4.3). The production of fruit begins five to seven years after planting, and reaches a peak at 15–20 years. The economic life is considered to be 60 years but plants can live to 90 years. Under favourable conditions they produce 60–70 nuts a year. By contrast, the dwarf type (var. *nana*) starts bearing after three years and peaks after six years. Hybrids are early bearing, after four years, with peak production within 20 years when they can produce 160 nuts a year (30 kg copra) (Purseglove, 1972; Persley, 1992).

As tall palms are largely cross-pollinated (by insects and wind), coconut populations are very variable. Dwarf palms in contrast are generally self-fertilising. A study of genetic diversity by Lebrun *et al.* (1998), using restriction fragment length polymorphism (RFLP) analysis, identified two major groups.

Figure 4.3 Coconut. 'Talls' can reach a height of 30 m, and live for up to 90 years – Dares Salaam, Tanzania (MKVC).

One included ecotypes from the Far East and the South Pacific while the other comprised ecotypes from India, Sri Lanka and West Africa. All dwarf types belonged to the first group, even those collected in West Africa. Tall ecotypes were generally more polymorphic than dwarfs. Where they are sufficiently distinctive, local and regional populations receive names that reflect their location (e.g. Malayan Dwarf; West African Tall). Colour variants and local ecotypes within such populations have also been given names (Harries, 1995).

Stem and leaves

As Tomlinson (2006) highlights, palms as a family are unique, possessing distinctive features of leaf development, vascular structure and anatomical properties of the stem. The coconut palm is conventionally propagated by seed. After germination, no visible trunk (stem) is formed in the tall ecotype for several years as there is no cambium (therefore no secondary thickening), and the trunk can only be formed when the apical meristem has attained its full diameter. The single stem develops entirely from cells derived from the apical meristem. Its structure means that it not only is a conductor of water to the leaves but also, because of its volume, acts as an important water store, or capacitor (Gomes and Prado, 2007). For example, Passos and Da Silva (1991) have described the diurnal variation in the circumference of the stem that can occur in response to changes in the water balance of the palm, a response that is enhanced when the fruits are removed. The carbohydrate reserves of coconut are mainly stored in the stem in soluble form (primarily sucrose). By contrast, the roots do not appear to have a storage function (Mialet-Serra *et al.*, 2008).

The terminal bud differentiates a single leaf at a time (it takes about 30 months from differentiation to the emergence of the leaf), an internode (the growth of which constitutes the incremental increase in height of the stem) and an inflorescence, in regular succession (Mialet-Serra *et al.*, 2008). The pinnate leaves are borne in a terminal radiating crown, which in an adult palm consists of 25–35 opened leaves and a central bud. In tall cultivars one leaf opens at intervals of about one month and in dwarfs about every three weeks. A fully mature leaf remains on the (tall) palm for two and a half to three years before falling. There is a waxy cuticle on the upper epidermis beneath which are two layers of large hypodermal cells, which serve as water-storage tissue. These, and other, leaf anatomical adaptations to drought tolerance have been described by Naresh Kumar *et al.* (2000) and cited by Gomes and Prado (2007).

As Huxley (1999) points out, the canopy of the young palm is dense and spreading but as it grows older it becomes sparser and, as it is also well above ground, allows light penetration. So although young palms in a plantation cover the ground well, older palms do not. Using a mobile sampling method in Thailand, Moss (1992) measured the amount of photosynthetically active radiation (PAR) intercepted by the leaf canopy at monthly intervals throughout one year. Seasonal reductions in the amount of light intercepted were related to water stress, its most immediate effect being to increase the rate of frond shedding and to slow the production of new fronds. As coconuts take about one month to produce a new frond, it can take many months for lost leaf area to be replaced, with long-term effects on yield. For crops well supplied with potassium, 60–70% of the incident light was intercepted by the canopy of an eight-year-old dwarf × tall hybrid (Malayan Yellow Dwarf × West African Tall) over the year (plant density 156 ha^{-1}). Increasing the plant density from 142 to 205 ha^{-1} (cv. Thai Tall; 13 years old) increased light interception over the same year from 55% to 73%.

Using these measurements, Moss (1992) was able to demonstrate a close linear relationship between yield of copra (and the whole fruit) and cumulative light interception ($n = 6$; $r^2 = 90\%$). As the line(s) did not pass through the origin, it was suggested that there is a minimum level of light interception below which no yield is produced.

In the Solomon Islands, Friend and Corley (1994) showed how the maximum leaf area was reached about five years after planting in all three (tall, dwarf and hybrid) of the cultivars compared. The 'dwarf' had the smallest leaf area (*c*. 4 m^2) whereas the 'tall' and the 'hybrid' had similar values (*c*. 6–7 m^2). The leaf area indices were in the range 4–5 (modified by differences in plant density; average density = 217 ha^{-1}).

In a comparison of five cultivars in Zanzibar, Juma and Fordham (1998) found that increases in the saturation deficit of the air had a negative effect on the rate of leaf expansion, while an increase in temperature within the range 24–30 °C showed a positive effect.

Figure 4.4 Coconut. A single inflorescence is borne in the axil of each leaf – Zanzibar, Tanzania (MKVC).

Flowering

When flowering commences (after about three years for dwarfs and five years for talls), a single inflorescence (botanically a spadix) is borne in the axil of each leaf (Figure 4.4) with numerous male and a few (up to 30–40) female flowers (subject to seasonal variation). The fruit (botanically a drupe) develops its full size about six months after fertilisation and matures within 12 months (Purseglove, 1972; Murray, 1977; Gomes and Prado, 2007). The inflorescence is in fact initiated up to 32 months before the spathe (a large bract enclosing the spadix) opens, or 44 months before the fruit is harvested (Figure 4.5) (Wickramaratne, 1987).

Fruit

Botanically the fruit consists of (1) the exocarp, or outer skin, (2) the mesocarp, or fibrous layer (source of coir), (3) the endocarp, or shell, and (4) a single seed. The 'nut' of commerce consists of the seed and endocarp (Purseglove, 1972). According to Murray (1977), variations in monthly and annual yields are determined largely by rainfall distribution. During the first few months after the inflorescence appears a considerable number of immature fruits are shed, and later-maturing nuts may be shed at the end of a long dry spell.

Figure 4.5 Coconut. An inflorescence is initiated up to 44 months before the fruit is harvested – Tanzania (MKVC). (See also colour plate.)

Roots

Roots are adventitious, being formed on the lower part of the stem (the bole), which is usually beneath the soil surface. They are produced throughout the life of the plant; like the stem, they are also without cambium. The major roots (up to 4 mm in diameter) produce numerous secondary and tertiary roots; there are no root hairs (Murray, 1977). A marked capacity to generate new roots allows palms to be transplanted easily. Different measurement techniques have been used to study the root systems of the coconut in several countries, which makes direct comparisons difficult.

In a detailed study in the Philippines, Magnaye (1969) recorded roots extending laterally 5 m from the base of the trunk, and to a depth of 2 m. On a free-draining, sandy loam soil, the greatest concentration of roots was within 0.3–1.0 m from the stem, and at 0.6–1.2 m depth. In a similar study in Kerala, again on a sandy loam soil, roots of mature palms extended laterally about 3.5 m from the trunk (spacing 7.5 × 7.5 m), although the majority were within 2.0 m, and reached depths of 1.5 m (Kushwah et al., 1973).

In northern Venezuela, Avilan et al. (1984) compared the root systems of tall and dwarf cultivars of different ages (four and 12 years, and five and 11 years, respectively). Root distribution was markedly influenced by the physical properties of the soil and by fertiliser application and irrigation. Roots of both cultivars reached depths of at least 0.8 m, regardless of age, with the majority (77–94%, by number) in the top 0.5 m.

In the Côte d'Ivoire, Pomier and Bonneau (1987) completed a detailed study of factors influencing the depth and distribution of roots (dry mass per unit volume of soil). On free-draining sandy soils, roots of mature palms (hybrid PB-121) reached 3.5–4 m depths and extended laterally at least 4 m from the stem. The greatest density of roots was within a 1-m depth and 1.2–1.5-m spread. Root growth beyond 1.5 m was restricted when the water table was within 3 m of the soil surface.

In north-east Brazil, on a sandy soil, Cintra *et al.* (1992) compared the root distribution (dry mass) of six dwarf cultivars (six years old) at the end of the rainy season and again at the end of the dry season. Roots reached a depth of 1.0 m (water table at 1.10 m), but with the highest root density in the 0.20–0.60-m layer. About 70% of the roots were within a 1.0-m radius of the trunk and 90% within 1.5 m. Cultivars Malayan and Gramame Yellow Dwarfs had better vertical and horizontal root distribution than, in particular, Jiqui Green Dwarf and Cameroon Red Dwarf. In a parallel study, Cintra *et al.* (1993) compared the root distribution of six tall cultivars (also six years old). The maximum depth of rooting was 0.8 m, with 70% of the roots in the 0.1–0.5-m layer and within a 1.0-m radius from the trunk. Cultivars Polynesia Tall and Praia de Forte Brazil had 'better' root distributions (in terms of total root production and fine root density, and more roots at depth when subjected to water stress) than the others. It is not clear whether differences in the surface area of circles with different radii from the trunk were taken into account in the comparisons of total root mass at increasing distance from the trunk.

Also in north-east Brazil (10° 17′ S 37° 35′ W; alt. 120 m), Azevedo *et al.* (2006) used digital image analysis to record root distribution of six-year-old dwarf green palms growing on a sandy soil (irrigated). All the roots were found in the top 1.0 m of soil with the following distribution: 0–0.2 m 8%; 0.2–0.4 m 24%; 0.4–0.6 m 24%; 0.6–0.8 m 15%; 0.8–1.0 m 8%.

As part of a study to specify the design criteria for drip irrigation in Sri Lanka, Arachchi (1998) excavated roots of 15-year-old palms (cv. CRIC 60) to a depth of 1.5 m. In a Xanthic Ferralsol soil, root distribution by mass declined exponentially from 0.3 m downwards and spread laterally 2.0 m from the base of the palm (spaced 7.7 m square).

Summary: crop development

1. The single stem of the coconut palm is an important water store (or capacitor).
2. It takes about 30 months from differentiation of a leaf to its emergence. New leaves open at intervals of about four weeks (talls) or three weeks (dwarfs).
3. Water stress increases the rate of frond shedding and slows the emergence of new leaves.
4. There is a close linear relation between yield and cumulative light interception. Seasonal changes in light interception are related to water stress.

5. An inflorescence is initiated up to 44 months before the fruit is harvested.
6. Variations in monthly and annual yields are determined largely by rainfall distribution.
7. Roots are adventitious.
8. Root systems have been described in different ways, which makes direct comparisons difficult.
9. Roots can reach depths of >2 m (down to 4 m), but the root density is usually greatest in the top 0.5–1.0 m.
10. Roots can extend laterally >3 m (up to 10 m), but the root density is usually greatest within 1.0–1.5 m of the trunk.
11. Tall and dwarf cultivars appear to have similar root distributions, but there are apparent differences in root distribution between cultivars within each ecotype.

Plant water relations

Perhaps the first detailed review of the water relations of the coconut was that by Copeland (1906) who described in great detail the results of his experiences in the Philippines. Because of the simplicity of the apparatus available little of what he reported is of value today, although he did describe the structure and function of the root and the leaf, including the location (lower surface), size and density (144 mm^{-2}) of the stomata. He also attempted to measure transpiration by recording the loss in weight of individual severed leaves and by the cobalt-chloride colour test. By extrapolation, he derived daily water use figures for a single palm tree ranging from 25 to 45 l. He also recorded his observations of the effects of drought (including: the folding of the pinnae, ageing and loss of leaves, check to the growth of young leaves and flowering branches, and premature falling of nuts). He concluded that, as recovery after the onset of rain was a slow process, a 'dry season' occurring every other year would reduce the yield of nuts by half.

In more recent research, measurements of stomatal conductance, leaf water status and gas exchange (photosynthesis and transpiration) have been used to monitor the effects of water stress on the coconut. As one output from this process, procedures for screening cultivars for drought tolerance have been proposed.

In Sri Lanka, Manthriratna and Sambasivam (1974) compared stomatal densities of different cultivars and forms of the coconut palm. The results, expressed on an unspecified unit area basis only, suggested that there may be varietal differences as all three colour forms of the self-pollinating cv. *nana* had higher densities than cv. *typica* (+12%). In Kerala, Mathew (1981) reported stomatal densities in the range $170–180 \text{ mm}^{-2}$ for healthy palms but more in root (wilt) diseased plants ($220–230 \text{ mm}^{-2}$). Again in Kerala, Rajagopal *et al.* (1990), in a comparison of 23 cultivars, recorded stomatal densities averaging 208 mm^{-2} for talls ($n = 10$), 232 mm^{-2} for dwarfs ($n = 6$) and 216 mm^{-2} for hybrids ($n = 7$).

In healthy palms, partial stomatal closure was observed from mid-morning (cv. West Coast Tall) in Kerala. By contrast, stomata of palms suffering from root (wilt) disease remained open throughout the day regardless of the season (Rajagopal et al., 1986). Leaf water potentials were also less in diseased plants (down to -1.99 MPa) leading to flaccidity (Rajagopal et al., 1987). Progressive stomatal closure (from 1000 to 1200 h) was also observed by Kasturi Bai et al. (1988) alongside a decline in the leaf water potential (from -0.6 to -1.55 MPa). Stomatal conductances began to fall when the saturation deficit of the air exceeded about 2.4 kPa. During dry weather, conductances were always greater in a hybrid (dwarf $\times$ tall) cultivar than in cv. West Coast Tall.

Using both a diffusion porometer and the infiltration technique, diurnal changes in stomatal opening were also monitored in Brazil (Passos and Da Silva, 1990). Both methods showed the stomata (leaf 14, mature palm, Géant de Brésil) to be wide open during the middle of the day (from 0800 h to 1600 h) before closing rapidly as solar radiation levels declined. In this example, the temperature (range 25 to 34 °C) and saturation deficit (up to 2.8 kPa) of the air, and leaf water potential (pressure bomb; -0.5 to -2.5 MPa) appeared to have little effect on stomatal opening.

The pressure bomb technique was first used on coconut in Florida by Milburn and Zimmerman (1977) to record diurnal and seasonal changes in leaf water potential. The base of the 5 m tall tree (cv. Malayan Dwarf) was only 0.5–1.5 m above sea level at high tide. Using carefully prepared leaflet samples (rolled and stored in situ in a sealed plastic bag), leaf water potentials as low as -1.3 MPa were reached during the day in the rainy season before increasing to -0.2 MPa at night. Surprisingly, the minimum values were larger during the dry season (-0.9 MPa) than during the rains. This was attributed to partial stomatal closure (measured with a viscous flow porometer) from mid-morning onwards. In Kerala, Voleti et al. (1993b) compared water potential measurements made on different leaves within the canopy. There was a vertical profile, with the spindle (unfolding) leaf having the highest (least negative) leaf water potential throughout the day in both irrigated and rain-fed plants.

Changes in leaf water potential (cv. West Coast Tall) were monitored over a six-month period in Kerala, South India (12°30′ N 70°00′ E; alt. 11 m) by Shiva-shankar et al. (1991). Daytime values declined as the dry season progressed, from about -1.0 MPa to -2.0 MPa. By contrast, values for irrigated palms remained above -1.3 MPa. These values were linked to the time when stress-induced changes in the activities of three enzymes were recorded.

In an irrigation trial in Kerala, there was a progressive reduction in stomatal conductance and leaf water potential (from -0.9 to -1.4 MPa), as the irrigation interval was extended from 12 to 16 and 24 days. The epicuticular wax content increased over the same range. Irrigation treatments were applied from December to May over two consecutive years. At this time of the year, there was both soil- and atmospheric- (saturation deficits >3 kPa, as recorded from 1000 to 1200 h) induced stress (Rajagopal et al., 1989).

The leaf gas exchange processes and water relations of six tall genotypes were compared by Prado *et al.* (2001) in Sergipe State, north-east Brazil (10° 26' S 36° 32' W; alt. 26 m). During the wet season, the controlling factor for photosynthesis and transpiration was solar radiation, whereas in the dry season it was stomatal conductance. One cultivar (Brazilian Tall) maintained gas exchange at a higher level than the other five during the dry season, despite low daytime leaf water potentials (−1.9 MPa).

Similarly, Gomes *et al.* (2002b) compared gas exchange processes in two dwarf coconut genotypes (Malayan Yellow Dwarf, MYD, and Brazilian Green Dwarf, BGD) over four days in Bahia State, Brazil (15° 16' S 39° 06' W; alt. 105 m). The two cultivars differed in rates of photosynthesis and transpiration, and in stomatal conductance, with values for BGD exceeding those of MYD in all three processes. For both genotypes, stomatal conductance (and photoynthesis) was negatively correlated with the leaf-to-air saturation deficit (within the chamber): BGD was more sensitive than MYD (but the evidence presented is not very convincing due to large variability). In a parallel study, Gomes *et al.* (2002a) confirmed (based on a comparison of two contrasting sites; see below for details) that the dryness of the air (saturation deficit between 0.7 and 3.5 kPa) influenced stomatal control of gas exchange in Brazilian Green Dwarf.

More recently, Passos *et al.* (2009) compared the responses of cv. Jiqui Green Dwarf to atmospheric water stress at the same two sites in north-east Brazil. One, described as wet tropical, was near the coast (10° 17' S 36° 30' W; alt. 75 m), while the other was at an inland site in a semi-arid area (09° 09' S 42° 22' W; alt. 387 m). Both crops were irrigated (150 l palm^{-1} d^{-1}). A portable infrared gas analyser was used to measure the diurnal and seasonal changes in stomatal conductance, photosynthesis and transpiration. Leaf water potentials were measured with a pressure chamber. Measurements were made on leaf 14 counting from the top. Although several correlations between these variables are presented, they are not all statistically convincing. Leaf water potentials declined linearly with increases in the saturation deficit of the air at both sites. In general, stomatal conductance was less at the semi-arid site where the saturation deficit was larger (monthly mean values reaching 1.6 kPa; daytime summer maximum *c.* 2.8 kPa) than at the coastal site (1.0 kPa; 1.8 kPa). Instantaneous water-use efficiencies *increased* linearly with increases in the saturation deficit, but only at the coastal site. The authors concluded that, as global warming would lead to increased saturation deficits, coconut plantations could not be justified in semi-arid areas even with irrigation. They did not, however, consider the effects of increased carbon dioxide levels on photosynthesis and water-use efficiency.

In Brazil, Gomes *et al.* (2009) monitored abscisic acid (ABA) accumulation in the leaves of droughted coconut palms (Brazilian Green Dwarf ecotypes) grown in a greenhouse. ABA is produced in the root tips in response to dry soil conditions and is carried to the leaves in the transpiration stream. ABA accumulation in the leaflets occurred before there were significant changes in pre-dawn leaf water potentials and remained high even after eight days of rewatering. Under mild stress stomatal conductance was 'controlled' by ABA accumulated in the leaflets, and at greater stress levels by the leaf water status (pre-dawn leaf water potential down to -1.2 MPa, when photosynthesis ceased). Stomatal conductances (and photosynthetic rates) were slow to recover on rewatering but sufficient to sustain photosynthesis and to allow rapid recovery of transpiration. Intrinsic water-use efficiency was improved at mild stress levels without impairment of the photosynthetic rate. This suggested that 'regulated deficit irrigation' may have a role in increasing the water productivity of irrigated palms.

In a related paper on the same experiment, Gomes *et al.* (2008) described in more detail the photosynthetic limitations encountered during the recovery phase after the relief of water stress. Non-stomatal factors were identified as contributing to the delayed and incomplete recovery of photosynthesis following rewatering. The two ecotypes tested differed in their responses. The one normally cultivated in a hot dry environment (Una Green Dwarf) recovered more quickly than the one cultivated under hot humid conditions (Jiqui Green Dwarf).

The influence of canopy shape on gas exchange processes of mature coconut palms (cv. West Coast Tall) has recently been described by Naresh Kumar and Kasturi Bai (2009). An oval-shaped canopy was considered to be superior to X-shaped or semi-circle-shaped canopies in terms of photosynthetic and water-use efficiencies under both irrigated and rain-fed conditions (in Kerala). Nut productivity was, however, only influenced by canopy shape in rain-fed palms. Gas exchange properties varied with the position of the leaf within the canopy, and with the stage of development of the subtended fruit.

In most species, chloride ions (Cl^-), and potassium ions (K^+), increase in concentration in the guard cells during stomatal opening with a corresponding reduction in their concentration in the adjacent subsidiary cells. In coconut the important role of Cl^- in regulating stomatal opening has been demonstrated by Braconnier and d'Auzac (1990) in a greenhouse study in which plantlets were exposed to osmotic stress in a hydroponic medium. Chloride deficiency resulted in delayed stomatal opening at the start of the day, and a reduction in the capacity for osmoregulation when plants were stressed. Subsequently, Braconnier and Bonneau (1998) showed clearly in a field study in Sumatra (cv. PB-121) how stomatal conductance was reduced in chloride-deficient palms, but only in the dry season *not* in the rains. This observation has implications in terms of gas exchange (net assimilation and transpiration were also reduced) and drought mitigation (see below). The role of the chloride ion (and other biochemical mechanisms) in water regulation at the cell level, including osmotic adjustment to maintain leaf turgor when plants are under water stress, has been reviewed in detail by Gomes and Prado (2007).

Screening for drought tolerance

Being able to identify drought-resistant cultivars early in the selection process using physiological traits would be a great asset to breeders of coconuts. Many attempts have been made using a range of techniques. Some of these are summarised below.

In Kerala, Rajagopal *et al.* (1988) described how the relative rate of decline in leaf water potential in dehydrating excised leaflets could act as an index of drought tolerance. Despite a lack of statistical analysis, three cultivars were identified as being drought tolerant in this way.

In a comparison of the response to drought of 23 cultivars, again in Kerala, Rajagopal *et al.* (1990) recorded stomatal conductance (steady state porometer), leaf water potential (pressure bomb) and epicuticular wax content of physiologic-ally mature leaves (the eleventh counting from the top) on 22-year-old palms. Measurements were made between 1000 h and 1200 h on successive days before there was water stress (November) and when the plants were experiencing stress (March). Stomatal conductances declined over this period in all cases but the dwarf types showed the least change (that is, the stomata remained open longer), together with one hybrid (Chowghat Orange Dwarf × West Coast Tall). In general, leaf water potentials of all types fell to similar mean values in March, -1.27 MPa talls, -1.36 MPa dwarfs; -1.24 MPa hybrids. The epicuticular wax content was consistently least in the dwarf types. Ranking drought tolerance on the basis of all four indicators (including stomatal frequency) suggested that a hybrid (West Coast Tall × West Coast Tall) was the most drought tolerant cultivar, followed by six of the tall genotypes. The dwarf types were, with one exception, the most drought susceptible because of limited stomatal and epicuti-cular control of water loss by transpiration. Some of the hybrids had desirable drought-resistance characteristics, including Laccadive Ordinary × Gangabon-dam, LO × Chowghat Orange Dwarf and West Coast Tall × COD. A selection of these indices allowed a rapid screening method to be developed (Rajagopal *et al.*, 1993).

In a similar follow-up study, Voleti *et al.* (1993a) compared the responses of three genotypes (WCT, WCT × COD and COD × WCT) to drought stress on two contrasting soils (laterite and sandy loam). Cultivar West Coast Tall again showed effective stomatal regulation of water loss, while there was some evidence (not very clear) that the three cultivars differed in their responses (in terms of stomatal conductance and the components of leaf water potential) depending on the soil type.

In Sri Lanka, Jayasekara *et al.* (1993) used similar physiological criteria to screen for drought tolerance amongst a selection of 32–35-year-old individual tall × tall hybrid palms. Their criteria were based on the relative sensitivity of stomatal conductance, transpiration rates and leaf water potentials of individual palms thought to be drought tolerant compared with the environmental mean for each variable measured over a four-year period. It is not easy to follow the details of the methodology as described in the paper, but genotypes demonstrating stability in two of the three variables (after screening twice) were considered to be drought tolerant.

In the Côte d'Ivoire, Repellin *et al.* (1993, 1997) compared several physiological tests for characterising the response to drought (by withholding water for 29 days) of young (two years old) coconut palms grown in plastic containers. The dwarf varieties tested (Malayan Yellow Dwarf and Cameroon Red Dwarf, both known to be susceptible to drought) dehydrated (as measured by the rate of decline in leaf water potential and relative water content) more quickly than West African Tall (moderately drought resistant), while the hybrid PB-121 was intermediate between its two parents (WAT × MYD). These differences between cultivars were only observed under severe stress when stomatal closure was complete. The reduction in gas exchange (transpiration and carbon assimilation) with drought, primarily due to stomatal closure, was similar in all the cultivars tested. (This is in contrast to the findings of Rajagopal *et al.* (1990) who reported that stomatal conductance discriminated between adult palms.) As this response was virtually independent of the plant water status, it suggested that a chemical signal (possibly abscisic acid) from the root may be the cause of stomatal closure. The activity of two hydrolytic enzymes (lipase and protease) increased in response to drought and the research suggested that their microsequencing could provide molecular tools for selecting drought-tolerant coconut parents.

Analysing research done in Kerala, Rajagopal and Kasturi Bai (2002) summarised the many possible drought-tolerance mechanisms in coconut. These included the maintenance of high leaf water status through effective stomatal regulation, deposition of wax on the leaf surface, and the accumulation of organic solutes aided by anatomical adaptations. They believed that genotypes with these characteristics could be used in breeding programmes. Rajagopal *et al.* (2007) have since reported on their attempt, with some success, to understand the genetics of drought resistance, while Kasturi Bai *et al.* (2008) have suggested that the chlorophyll fluorescence technique (in combination with measurements of leaf water potential) may have a role in screening coconut seedlings for adaptation to water stress.

Summary: plant water relations

1. Stomata are confined mainly to the abaxial surface of the leaf at a density of about 200 mm^{-2}. Dwarf ecotypes may have a higher density than talls (+ 12%).
2. Depending on the weather conditions, partial closure of the stomata usually occurs from mid-morning onwards.
3. Stomata are slow to reopen fully after the relief of water stress.
4. Stomatal conductances decline as the saturation deficit of the air increases (atmospheric drought).
5. As water stress levels increase, the stomata of adult dwarf ecotypes remain open for longer than those of tall ones.
6. Abscisic acid may be involved in controlling stomatal closure, at least under mild stress conditions.
7. Leaf water potential is a sensitive indicator of plant water status, declining to values as low as -1.3 MPa in the middle of the day even when the soil is wet, or -2.0 MPa if the soil is dry (unless stomatal conductance is reduced by dry air).
8. Leaf water potentials decline linearly with increases in the saturation deficit of the air (at least until the stomata begin to close).

9. Pre-dawn leaf water potentials of -1.2 MPa correspond to complete stomatal closure during the day.
10. Instantaneous water-use efficiencies appear to increase in plants experiencing moderate water stress.
11. Chloride ions play an important role in the stomatal opening process, in osmoregulation and in drought tolerance.
12. Epicuticular wax content increases as the soil water deficit increases. It is normally higher in dwarf ecotypes than talls.
13. The activities of certain hydrolytic enzymes increase in response to drought.
14. Various physiological traits have been identified that can assist in the screening for drought tolerance.
15. In general, talls are considered to be more resistant to drought than dwarfs, and than most dwarf $\times$ tall hybrids. West Coast Tall in particular is recognised for its drought resistance.

Crop water requirements

Several techniques have been used in different regions of the world to quantify (or predict) the water use of the coconut palm. They include lysimeters, the soil water balance, sap flow and the eddy-flux methods. Results are reported in terms of litres palm^{-1} d^{-1} and/or as mm d^{-1}, depending on whether the plant population is specified.

In Kerala, Jayakumar *et al.* (1988) used a pair of drainage lysimeters to measure the actual water use of two, six-year-old, irrigated palms (cv. West Coast Tall; leaf area index 2.4) over a six-month period (May to November, the dry season). Actual water use (ET_c), averaged over 5-day intervals, ranged from 2.7 to 4.1 mm d^{-1} (mean 3.3). By comparison, reference crop evapotranspiration (ET_0) averaged 6.2 mm d^{-1} (Penman equation), 4.6 mm d^{-1} (Blaney Criddle), and 5.3 mm d^{-1} (USWB Class A pan) (Doorenbos and Pruitt, 1977). The corresponding crop coefficients (K_c) were 0.54, 0.73 and 0.65 respectively.

A similar study was reported by Rao (1989), also in Kerala (11° 13′ N 75° 52′ E; alt. 70 m). A five-year-old palm (cv. West Coast Tall; 3 m tall with six or seven functional leaves) was transplanted into each of a pair of drainage lysimeters (3.5 $\times$ 3.5 m square). Monthly evapotranspiration rates (ET_c) were recorded over a year. For irrigated palms, ET_c varied from 3.3 mm d^{-1} in June to 7.8 mm d^{-1} in April, with an annual mean of 5.1 mm d^{-1}. By comparison, evaporation from a USWB Class A pan averaged 4.4 mm d^{-1} over the year. Depending on the season, K_c varied from 0.60–0.68 in the rains to 0.87–0.96 in the summer, with an annual mean value of 0.82. An attempt was made to allow for the wider spacing (7.0 $\times$ 7.0 m) of palms in a plantation on crop water use, based on changes in soil water content, but it is not easy to evaluate how well this was done.

In Sri Lanka (08° 02′ N 79° E; alt. 35 m), on a gravelly soil, water use of 15-year-old palms (CRIC 60, spaced 7.7 $\times$ 7.7 m, 170 plants ha^{-1}) was monitored with a neutron probe over four consecutive dry periods. Water use (ET_c) averaged about

3.8 mm d^{-1} for the first eight days before declining as the soil dried. This equates to about 220 l palm^{-1} d^{-1}. The average ET over the whole 45-day dry period was 2.5 mm d^{-1}, or 150 l palm^{-1} d^{-1} (Arachchi, 1998).

Using a soil water balance approach, Azevedo *et al.* (2006) estimated actual evapotranspiration (ET) for six-year-old dwarf green palms in north-east Brazil. Depending on the irrigation treatment, mean values were 2.5, 2.9 and 3.2 mm d^{-1}, with cumulative annual totals of 900–1100 mm. At a planting density of 205 ha^{-1} (7.5 $\times$ 7.5 $\times$ 7.5 m) these equate to 120–160 l palm^{-1} d^{-1}. By comparison, the Penman–Monteith estimate of reference crop evapotranspiration (ET_0; Allen *et al.*, 1998) over the year averaged 4.6 mm d^{-1}, giving a peak crop coefficient (K_c) of 0.7. For comparison, in Brazil, Miranda *et al.* (2007) derived K_c values for an irrigated (microsprinklers) dwarf green cultivar (Jiqui) over a 32-month period, beginning 11 months from planting (spacing 7.5 m triangular; 205 palms ha^{-1}). Using the water balance approach (based on tensiometers), ET_c increased from a minimum 0.5 mm d^{-1} (25 l palm^{-1} d^{-1}) up to a maximum of 5 mm d^{-1} (244 l palm^{-1} d^{-1}) at a coastal site in Ceará State (3° 17′ S 39° 15′ W; alt. 30 m). Over the same period, ET_0 (Penman–Monteith) varied between 3 and 6 mm d^{-1}. During the canopy development phase, K_c increased linearly from 0.63 (11 months after planting) to 1.0 (23 months, when the palms were flowering). Thereafter it remained constant, with an average value of 1.02.

In a detailed experiment in Vanuatu (15° 26′ S 167° 11′ E; alt. 80 m), Roupsard *et al.* (2006) monitored water use of a coconut plantation (Vanuatu Red Dwarf $\times$ Vanuatu Tall Hybrid) over a three-year period. The eddy-flux method was used to estimate actual evapotranspiration (ET_c) from the palms and grass understorey, and the sap flow method to measure transpiration (T) from the palms alone. Water was freely available in the soil throughout the experimental period. The leaf area index was constant ($L = 3$), and the crop cover averaged 75% (both values for palms only). Transpiration (annual total 640 mm) represented 68% of ET_c (950 mm). ET_c rates varied seasonally between 1.8 and 3.4 mm d^{-1}, T from 1.3 to 2.3 mm d^{-1}, and ET_0 (Penman–Monteith) from 2.4 to 5.8 mm d^{-1}. At a density of 144 palms ha^{-1}, these ET_c values equate to 93 to 160 l palm^{-1} d^{-1}. The crop coefficient K_c ($= ET_c/ET_0$) values averaged 0.79 and 0.59 in the cool and warm seasons respectively. Canopy transpiration during the warm season was apparently limited by partial stomatal closure linked to the saturation deficit of the air, although the maximum daytime value did not exceed 1.2 kPa.

Recently, Madurapperuma *et al.* (2009a) used the 'compensation heat pulse method' to measure actual water use (T) of two cultivars of mature palms (20 years) grown on two contrasting soils in Sri Lanka (7° 35′ N 80° 57′ E; alt. 100 m; square spacing 8.3 $\times$ 8.3 m, 145 palms ha^{-1}). Diurnal patterns of sap flow were clearly discernable on successive days, averaging about 3 l h^{-1} at night before, on the water-retentive soil, typically rising rapidly from about 0600 h until 1000 h. Sap flow then remained high until about 1600 h before declining in the late afternoon. Peak rates of water use differed between the two cultivars, reaching 13–14 l palm^{-1} h^{-1} for CRIC 60 (a tall $\times$ tall hybrid) but only 9–10 l palm^{-1} h^{-1} for CRIC 65 (a dwarf $\times$ tall hybrid). Total daily water use averaged 120 l palm^{-1} d^{-1}

Table 4.1 Coconut. Summary of measured crop water use (ET_c, mature palms) and corresponding estimates of the crop coefficient ($K_c = ET_c/ET_0$, mean and seasonal range in brackets). T = transpiration only

Site	Method	Type	ET_0 method	ET_c (mm d^{-1})	K_c	Reference
Kerala 1	Lysimeter	Tall	Class A pan	2.7–4.1	0.65	Jayakumar *et al.* (1988)
Kerala 2	Lysimeter	Tall	Class A pan	3.3–7.8	0.82 (0.65–0.91)	Rao (1989)
Sri Lanka 1	Water balance	Tall		3.8		Arachchi, 1998
Brazil 1	Water balance	Dwarf	Penman–Monteith	3.2	0.7	Azevedo *et al.* (2006)
Brazil 2	Water balance	Dwarf	Penman–Monteith	5.0	1.02	Miranda *et al.* (2007)
Vanuatu	Eddy-flux + sap flow	Dwarf	Penman–Monteith	1.8–3.4	0.69 (0.59–0.79)	Roupsard *et al.* (2006)
Sri Lanka 2	Sap flow	Tall	Not specified	1.7 (= T)	0.5 (T/ET_0)	Madurapperuma *et al.* (2009a)
		Dwarf		1.3 (= T)	0.37 (T/ET_0)	

(range 105–135 l palm^{-1} d^{-1}) or 1.74 mm d^{-1} for CRIC 60, and 25% less at 90 l palm^{-1} d^{-1} (range 75–97 l palm^{-1} d^{-1}) or 1.31 mm d^{-1} for CRIC 65. By comparison, daily water use on the second, less water retentive soil was lower, averaging 92 and 79 l palm^{-1} d^{-1} (1.33 and 1.15 mm d^{-1}) for each of the two cultivars respectively. The mean daily ET rate (method not specified) over the period of measurement was stated as 3.5 mm giving a K_c value (strictly, for T) of 0.37–0.50. Concurrent measurements over the study period indicated that stomatal conductance was substantially greater in CRIC 60 (tall) than in CRIC 65 (dwarf). Palms growing on the water-retentive soil had larger leaf areas and trunk diameters (and hence more stem water storage) than the corresponding palms grown on the second soil. In Sri Lanka, CRIC 65 is known for its sensitivity to water stress, whilst CRIC 60 is recognised as being drought tolerant. The 'compensation heat pulse method' had previously been successfully evaluated (except at very low flow rates) on palms in Australia by Madurapperuma *et al.* (2009b).

Summary: crop water requirements

1. A diverse selection of methods has been used to measure the water use of coconut palm so that direct comparisons of the results are not easy (Table 4.1).
2. Across all the sites, ET_c for mature palms ranged from 1.2 to 7.8 mm d^{-1}. A 'typical' rate of water use is probably 3.0–3.5 mm d^{-1}.
3. Corresponding values of K_c for mature palms range from 0.5 to 1.02, with some evidence of seasonal variability. A 'working' value for irrigation planning purposes is probably 0.7, but this needs to be confirmed. For immature palms the value is proportionally less.

Water productivity

This section covers, first, the attempts made to forecast yields based on statistical correlation techniques and, more recently, the development of a process-based simulation model. Second, the results of irrigation experiments in which yield responses to water are quantified are reviewed. Third, some of the recommended drought-mitigation practices are summarised.

Yield forecasting

In an analysis of the effects of the 1931 drought (80 mm rain in five months) on the commercial yield of coconuts in Sri Lanka, Park (1934) observed that the minimum nut yield occurred 13 months after the drought ended. The spathes that opened when the drought was most severe were more affected by drought than the flowering and fruiting branches at other stages of development. Yields of nuts did not recover fully until two years after the conclusion of the drought, but the yield of copra per nut had recovered within 12 months. This analysis was followed up by Abeywardena (1968) who attempted to develop a yield forecasting model using long-term rainfall and yield data (1935–66) in Sri Lanka. But because of the great time lapse between the initiation of leaf and flower primordia and flowering, and with many other inflorescences at various stages of development present at the same time, it is difficult to relate yield responses to any particular climatic condition. It is not surprising therefore that Abeywardena (1968), using multiple regression techniques, was unable to establish a causal link between rainfall over the 12 months prior to harvest and yield. However, by breaking the year into different periods, when other external factors were similar, he developed an equation (with 12 variables) that explained 86% of the yield variation. Because of the complexity, it is doubtful if this analysis has generic value. Later, Peiris et al. (1995) reviewed the many attempts to develop statistical relationships between climatic factors and nut yield (including button shedding and premature nut fall). This highlighted the difficulties (and weaknesses) in this approach to the development of yield forecasting models. Nevertheless, Peiris and Thattil (1998) subsequently used multivariate analysis in an attempt to explain within and between year variation in nut yield in Sri Lanka. They believed that their 'parsimonious' approach (three key variables were identified as important determinants: maximum air temperature, afternoon relative humidity and pan evaporation) could be used to develop meaningful models in other locations. It has not been reported whether or not this is the case.

Of greater potential generic value is the simulation model (InfoCrop-coconut) developed, calibrated and validated in India by Naresh Kumar et al. (2008). The model simulates development stages of growth (based on thermal time), dry matter production (solar radiation interception × radiation conversion efficiency) and dry matter partitioning. There were good linear relationships between simulated dry matter production and measured values ($r^2 = 0.95$) and between the corresponding nut yields ($r^2 = 0.86$) across a range of sites and experimental treatments.

Nut productivity under rain-fed and irrigated conditions at four sites within India was simulated. In north Kerala, for example, irrigation ($200 \, l \, palm^{-1}$ every four days plus fertiliser) was predicted to increase the (simulated) average annual nut yield from about 4000 to 15 500 kg ha^{-1}. Further evaluation of this model is justified.

The long-term effect (up to four years) of an extended dry period on nut yields was confirmed by Naresh Kumar *et al.* (2007) in an analysis of the impact of rainfall amount and distribution on yields from commercial plantations in the different agro-ecological zones in India. Whether the simulation model takes this into account is not clear.

Yield responses to irrigation

Actual yield responses to irrigation have been recorded in field experiments in India (Kerala, Karnataka), Sri Lanka and Brazil.

The irrigation requirement of immature palms (from five to seven years in the field; cv. West Coast Tall) was reported by Nelliat and Padmaja (1978) in Kerala. The 'best' combination of treatments in terms of yield of nuts and water-use efficiency was the application of 40 mm of irrigation water (I) when the cumulative potential evaporation (CPE) total reached 53 mm (I/CPE ratio 0.75, which is equivalent to a K_c value of 0.75). In this way, an average total of 680 mm of water was applied in the summer months, yielding a total of 157 nuts palm^{-1} over the three years after the palms started to come into bearing. By comparison, when the I/CPE ratio was 0.5 the total yield was significantly less, at 126 nuts palm^{-1}. There was no unirrigated/rain-fed control treatment.

In Kerala, Nair (1989) reported the results of an irrigation trial in which 500 litres of water per palm (at a plant density of 178 ha^{-1} this is equivalent to 9 mm) were applied at different intervals (cv. West Coast Tall) during the summer months (December to May) over a five-year period. Water was applied in 1.8-m-radius basins to a sandy clay loam soil. Compared with the control rain-fed treatment (average yield *c.* 90 nuts palm^{-1}), significant increases (range $+15$–39 nuts palm^{-1}) in yield were obtained in the third and subsequent years from irrigation applied when the 'cumulative potential evaporation' totalled 50 or 25 mm. No indication is given on how much water in total was applied in each year.

Previously, Bhaskaran and Leela (1978) had described a similar trial in Kerala lasting 12 years with the same cultivar (WCT). Water was applied at a rate of 800 l palm^{-1} every seven days (equivalent to 2 mm d^{-1} only) in 2-m-radius basins during the summer months. The soil was a red sandy loam. It took three years before the full yield benefits (averaging $+30$ nuts palm^{-1} y^{-1} compared with pre-irrigation yields of a variable 40) were realised. Before that, in the 'transition period', the yield increase was about half this. The largest increase ($+39$ nuts palm^{-1}) came from palms initially classified as 'low' yielding (20–40 nuts palm^{-1}). Yield increases followed an increase in female flower production and setting percentage. This and other work on water management of coconut undertaken in India is summarised in Yusuf and Varadan (1993).

In Sri Lanka, Nainanayake *et al.* (2008) evaluated the responses of mature palms (20 years old; cultivar not named), growing on a shallow (0.6 m) sandy clay loam soil, to drip irrigation over a two-year period, two to four years after irrigation began. During this period there were three dry spells lasting 48, 78 and 83 days. Irrigation during these dry periods (80 l palm^{-1} d^{-1}, or 1.3 mm d^{-1}) reduced the afternoon soil temperature (from 31 to 27 °C), nut surface temperature (seventh bunch from the top, by up to 2.5 °C) and air temperature (by up to 2.0 °C) compared with the control (unirrigated) treatment. Irrigation therefore ameliorated the temperature regime and created conditions close to the optimum (27 °C) for coconut. Even with irrigation, stomatal conductance declined during the dry periods but applications of 80 l palm^{-1} d^{-1} (1.3 mm d^{-1}) maintained transpiration at rates similar to those recorded in the rainy season. Irrigation also increased female flower production and reduced premature nut fall. Over the two-year period, applications of 80 l palm^{-1} d^{-1} resulted in a 45% yield increase over the control. Halving the amount of water applied (to 40 l palm^{-1}; 0.65 mm d^{-1}) halved the yield benefit (to 20%). The absolute yields were not clearly specified.

In north-east Brazil, Azevedo *et al.* (2006) applied 50, 100 or 150 l palm^{-1} d^{-1} (equivalent to 1.0, 2.0 and 3.0 mm d^{-1} respectively) to six-year-old dwarf green palms on a sandy soil over a two-year period. It appears that these daily applications were made regardless of rainfall. There was no rainfall-only control treatment. The whole experimental area had previously been irrigated. Using a water balance approach, actual evapotranspiration was estimated, from which water-use efficiencies were calculated (i.e. apparently based on total *ET* not on the depth of irrigation water applied). There were no yield differences between treatments in terms of the number of bunches per palm or the number of fruits per bunch but extra irrigation water increased the volume of water per fruit by about 16%. When yield was expressed as the number of fruits per hectare there was a significant 12% yield loss from applying 1.0 mm d^{-1} compared with 2 mm d^{-1} (equivalent to a reduction in the number of nuts per palm from 93 to 82). Water-use efficiencies (as inadequately defined) were essentially the same for all three irrigation treatments.

In Kerala, irrigation was recently reported to have increased annual yields from mature coconut palms (cv. West Coast Tall) over a six-year period by, on average, an estimated 30–40 nuts palm^{-1} (from 50–60 to 90 nuts palm^{-1}; Naresh Kumar and Kasturi Bai, 2009). Similar yield increases (cv. Tiptur Tall) were also reported from Karnataka (13° 15' N 76° 15' E; alt. 800 m). Over a five-year period, drip irrigation at 33%, 66% and 100% replacement of evaporation (E_{pan}) resulted in average annual yields of 78, 87 and 98 nuts palm^{-1} respectively compared with 52 nuts palm^{-1} from the rain-only control (Basavaraju and Hanumanthappa, 2009). It is not clear what allowance was made for rainfall (average annual total 650 mm) when assessing irrigation need or if the 25-year old palms had previously been irrigated. The peak monthly water requirement was 75 l palm^{-1} d^{-1} and the least 45 l palm^{-1} d^{-1} (planting density = 100 palms ha^{-1}; clay loam soil).

Figure 4.6 Coconut. Drip irrigation experiment – Coconut Research Institute, Sri Lanka (MKVC).

In Sri Lanka, Arachchi (1998) developed criteria for the design of a drip irrigation system for coconuts grown on a gravelly soil (Figure 4.6). The maximum flow rate he recommended was $30\,l\,h^{-1}$ for 2.5 h from each of four drippers spaced equidistant around, and 1.0 m from, the base of the trunk. This regime wetted a large volume of soil within the effective root zone of 15-year-old palms (cv. CRIC 60; 170 plants ha^{-1}) and equates to 5.1 mm d^{-1} ($300\,l\,palm^{-1}\,d^{-1}$). By comparison, actual crop water use during the first eight days of the dry period averaged about 3.8 mm d^{-1} ($220\,l\,palm^{-1}\,d^{-1}$) before declining (see *Crop water requirements* above). Eight days became the recommended irrigation interval. Such an analysis does not allow for the fact that drip irrigation enables small quantities of water to be applied frequently, rather than requiring the whole root zone to be wetted at extended intervals. The productive and economic advantages of designing drip systems for deficit (under) irrigation as compared with standard drip systems designed for full irrigation to meet potential evapotranspiration have been described by Keller *et al.* (1992) in Kerala.

However, the limited size of the wetted soil volume under drip irrigation was identified as a cause for concern during an on-farm evaluation of drip irrigation in South India, although that view may have been influenced by the fact that only very small quantities of water ($32\,l\,palm^{-1}\,d^{-1}$; 0.6 mm d^{-1}) were being applied (Thamban *et al.*, 2006). What also appear to be very small amounts of water (30–$40\,l\,palm^{-1}\,d^{-1}$) are recommended for drip (and basin) irrigation on sandy soils in the Konkan region of Maharashtra (India) (Nagwekar *et al.*, 2006). This advice is inconsistent with the recommendation of the Coconut Development Board of India, which states

Figure 4.7 Coconut. Drought mitigation; husk pits increase the retention and availability of water, especially in sandy soils – Sri Lanka (MKVC).

'generally, an adult palm requires 600 to 800 litres of water once in four to seven days' (CDB, 2010). This apparent inconsistency may be due to whether irrigation is for 'survival' or for nut production. The amount of water that it is worth applying will in turn be influenced by the market value of the nut at the time.

Drought mitigation

In view of the sensitivity of palms to water stress, water conservation is strongly recommended (Mahindapala and Pinto, 1991). In Sri Lanka, this means: *mulching* (to restrict evaporation from the soil surface) by placing a layer of vegetation (such as coconut husks) in a 1.75-m-radius circle around the trunk. *Husk* or *coir dust pits* or *trenches* increase the retention and availability of water especially in sandy soils (Figure 4.7). They should be at least 0.6 m deep and 1.2–1.5 m wide and situated within reach of the roots of each palm. They are filled with alternating layers of husk (or coir dust) and soil. The results of original studies undertaken in India by Marar and Kunhiraman (1957) and by Balasubramanian *et al.* (1985) on husk burial as a drought-mitigation measure have been summarised by Yusuf and Varadan (1993). For young palms in South India, Shanmugam (1973) describes other drought-mitigation practices including deep planting (with the

Figure 4.8 Coconut. Drought symptoms; petiole breakage of the lower fronds of a mother palm in a seed garden (cv. Malayan Yellow Dwarf) – Gunung Batin, Lampung, Indonesia (XB). (See also colour plate.)

bole 0.6–1.2 m below ground level), the placement of porous earthen pots regularly filled with water (pitcher irrigation) adjacent to the stem, in addition to husk burial (500–1000 husks per palm) and husk mulch. Xavier Bonneau (personal communication) confirms the benefits that can result from these interventions.

In Indonesia (Gunung Patan, Sumatra), hybrid palms, four to six years old, were observed to be more susceptible to drought than younger or older palms because of an imbalance between the relative sizes of the foliage and the root systems (Bonneau and Subagio, 1999). Other things being equal, dwarfs (e.g. Malayan Yellow Dwarf and Cameroon Red Dwarf) were more susceptible to drought than talls or hybrids (Figures 4.8–4.11). The researchers also found that the mortality of commercially grown palms during the long dry season was negatively correlated with the chloride status of the palm (as measured in leaf 14 at the beginning of the dry season). Applying common salt (sodium chloride) at annual rates of up to 4.5 kg palm^{-1} reduced mortality, reduced defoliation (Figure 4.12), advanced recovery after the start of the rains, and increased the yield of copra from 10 to 15 kg palm^{-1} (averaged over eight seasons). Previously, Braconnier and Bonneau (1998) had confirmed the important role that the chloride ion plays in maintaining gas exchange (net assimilation and transpiration) in dry weather through stomatal regulation. Indeed, sodium chloride is recommended as a cheaper alternative to potassium chloride as a fertiliser in Indonesia, particularly in dry areas, where it contributes to drought mitigation (Bonneau *et al.*, 1997).

Figure 4.9 Coconut: severe drought symptoms; toppling over of a Malayan Yellow Dwarf ×
West African Tall hybrid – Gunung Batin, Lampung, Indonesia (XB).

Figure 4.10 Coconut. Extreme water stress; total collapse of a Malayan Yellow Dwarf ×
West African Tall hybrid – Gunung Batin, Lampung, Indonesia (XB).

Figure 4.11 Coconut. View of a Malayan Yellow Dwarf × West African Tall hybrid block during a long dry season – Gunung Batin, Lampung, Indonesia (XB).

Figure 4.12 Coconut. Drought mitigation; palms without salt fertilisation in the foreground, with salt in the background – Gunung Batin, Lampung, Indonesia (XB)

For example, in a field experiment in Indonesia, in which different levels of common salt (from 0 to 4.5 kg palm^{-1} y^{-1}) were compared, mature hybrid palms (cv. PB-121) responded as follows to these two extreme treatments, averaged over the period 1989–96: an increase from 55 to 76 nuts palm^{-1} y^{-1}; from 184 to 201 g copra nut^{-1}; from 10.7 to 15.8 kg copra palm^{-1} y^{-1}; from 6.4 to 14.9 green leaves palm^{-1} (at the end of the 1991 drought); from 78 to 148 palms ha^{-1} (at the end of the 1996 drought, the original plant population was 152 ha^{-1}); and from 1.47 to 2.38 t copra ha^{-1} y^{-1}. The critical chloride concentration in the leaf (fourteenth) of mature palms is considered to be 0.5% (Bonneau *et al.*, 1997).

The role of chloride in the nutrition of palms and its contribution to drought resistance in coconut has been well reported. Only in plantations close to the sea, where salt spray occurs naturally, is there apparently no benefit from its application (Ollagnier *et al.*, 1983; Von Uexhull, 1985; Bonneau *et al.*, 1993).

In this context, Yusuf and Varadan (1993) cited a study by Shanmugam (1973) which demonstrated that coconut can withstand irrigation with sea water (salt content 0.6–1.0%; 90 l palm^{-1} twice a week; sandy or sandy loam soils; 1.5-m-radius basins). The monsoon rains leached out any residual salt. In Brazil (5° 46′ S 35° 12′ W; alt. 18 m), Marinho *et al.* (2006) evaluated the viability of using saline water for irrigation. In a two-year-study with (initially) three-and-a-half-year-old green dwarf coconut (cv. Anão Verde), the number of female flowers was increased when salt was added to the irrigation water, but at an electrical conductivity (ECw) greater or equal to 5 dS m^{-1} the mean weight of a fruit (beginning at the eleventh harvest) and the number of fruits (fourteenth harvest) were both reduced compared with the control treatment (ECw = 0.1 dS m^{-1}). However, even when the ECw was 10 dS m^{-1}, acceptable yields were still achieved.

Summary: water productivity

1. Partly because of the long time interval (44 months) between flower initiation and harvest of the mature nut, it has not been possible to establish a direct causal link between yield and rainfall for coconut.
2. Full responses to irrigation are only obtained in the third and subsequent years after irrigation is introduced.
3. There is a limited amount of reliable and complete field data on actual yield responses to irrigation/drought. In general, it appears that relatively small quantities of water have been applied in the irrigation trials compared with ET_c.
4. In Kerala, yield increases of 20–40 nuts palm^{-1} (mature cv. West Coast Tall) have been recorded after the application of the equivalent of about 2 mm d^{-1} during the summer months. This represents a 50% yield increase on base yields averaging about 60 nuts palm^{-1}.

5. In Sri Lanka, applications of 1.3 mm d^{-1} during the dry season ameliorated the microclimate, maintained transpiration and increased yields by 45%.
6. In north-east Brazil, yields were reduced by 12% if only 1.0 mm d^{-1}, rather than 2.0 mm d^{-1}, was applied during the dry season to dwarf green cultivars.
7. Irrigation increases female flower production and reduces premature nut fall.
8. It has not been possible to derive yield/water-use (water productivity) response functions here because of incomplete data.
9. Drought-mitigation practices include husk burial and mulching, and the application of common salt.
10. Mature palms can withstand irrigation with sea water.

Conclusions

Until relatively recently much of the research reported was empirical, so that the results were only of value in the immediate location of the experiments. They were time and space limited. This is understandable and is due, in part, to the difficulty of undertaking research on this fascinating crop. It is also due in part to limited funding at the relatively small research institutes with the mandate to undertake this research. There has also been, with some exceptions, a notable lack of international collaboration in research (coconuts are outside the CGIAR system) for a crop on which millions of people depend for their livelihoods (Carr and Punchihewa, 2002).

Potential and actual yields are determined mainly by climate, and its day-to-day variability which is known as weather. In the case of coconuts, this simple statement is complicated by the long time interval between the initiation of the inflorescence and the harvesting of the mature nut (44 months). Changing weather conditions during this period will influence yield development in different ways.

In 1992, following a visit to Sri Lanka, I wrote 'to the best of my knowledge only a limited amount of research on the physiological basis of yield development in coconuts has been reported. For example, how do environmental (and agronomic) factors influence such variables as fractional light interception by the canopy; dry matter production and partitioning and the harvest index; conversion efficiency for solar radiation; crop water use and water-use efficiency? What are the potential yields of copra (and other commercially valuable products) in given locations with existing cultivars? Why do actual yields differ from potential yields? How important is water stress as a limiting factor? Is irrigation justified economically? What are the other principal limiting factors? If we can begin to quantify some of these variables and the relationships between them in systematic ways, it should be possible to develop procedures for yield forecasting (with and without water stress as a limiting factor) and even perhaps to identify selection criteria for new cultivars. To do this successfully though will require a coordinated team approach within an agreed framework for analysis.

Experiments must be designed so that key measurements are taken to provide basic data that can be used to develop and validate a yield forecasting model' (Carr, 1992).

Substantial progress has been made since 1992 in many of these areas, including the development of a yield forecasting model, although this needs to be developed and validated further. In the immediate context of this review, there is still a lack of knowledge on the actual water use of coconut and of yield responses to water. Indicators of drought tolerance have been identified but it is not clear whether these have resulted in new genotypes. The challenge that remains is to quantify how little rather than how much water is needed to produce an economically viable crop. A coordinated international approach to addressing this issue is recommended.

Summary

The results of research on the water relations and irrigation need of coconut are collated and summarised in an attempt to link fundamental studies on crop physiology to drought-mitigation and irrigation practices. Background information on the centres of origin and production of coconut and on crop development processes is followed by reviews of plant water relations, crop water use and water productivity, including drought mitigation. The majority of the recent research published in the international literature has been conducted in Brazil, Kerala (South India) and Sri Lanka, and by CIRAD (France) in association with local research organisations in a number of countries, including the Côte d'Ivoire. The unique vegetative structure of the palm (stem and leaves) together with the long interval between flower initiation and the harvesting of the mature fruit (44 months) mean that causal links between environmental factors (especially water) are difficult to establish. The stomata play an important role in controlling water loss, while the leaf water potential is a sensitive indicator of plant water status. Both stomatal conductance and leaf water potential are negatively correlated with the saturation deficit of the air. Although roots extend to depths >2 m and laterally >3 m, the density of roots is greatest in the top 0–1.0 m soil, and laterally within 1.0–1.5 m of the trunk. In general, dwarf cultivars are more susceptible to drought than tall ones. Methods of screening for drought tolerance based on physiological traits have been proposed. The best estimates of the actual water use (ET_c) of mature palms indicate representative rates of about 3 mm d^{-1}. Reported values for the crop coefficient (K_c) are variable but suggest that 0.7 is a reasonable estimate. Although the sensitivity of coconut to drought is well recognised, there is a limited amount of reliable data on actual yield responses to irrigation, although annual yield increases (50%) of 20–40 nuts palm^{-1} (4–12 kg copra, cultivar dependent) have been reported. These are only realised in the third and subsequent years after the introduction of irrigation applied at a rate equivalent to about 2 mm d^{-1} (or 100 l palm^{-1} d^{-1}) at intervals of up to one week. Irrigation increases female flower production and reduces premature nut fall. Basin irrigation, microsprink-lers and drip irrigation are all suitable methods of applying water. Recommended methods of drought mitigation include the burial of husks in trenches adjacent to the plant, mulching and the application of common salt (chloride ions). An international approach to addressing the need for more information on water productivity is recommended.

5   Coffee

Introduction

There are two principal types of coffee grown commercially: *Coffea arabica* L., commonly known as 'arabica' coffee, and *Coffea canephora* Pierre ex Froehner, commonly known as 'robusta' coffee. A third (*Coffea liberca* Bull ex Hiern) is grown only on a very small scale. The useful product is the bean, a relatively heavy seed that ripens within a sweet red fruit. The seeds, rich in caffeine, form the basis of a beverage widely traded and consumed throughout the world.

Many of the physiological characters of coffee can best be understood by recalling the conditions under which these species evolved, particularly the responses to water. The centre of origin of *C. arabica* is considered to be the cool, shady environment in the understorey of forests in the Ethiopian highlands. At these latitudes (6–9°N) and altitudes (1600–2000 m) the mean average air temperature is in the range 15–20°C, the annual rainfall is 1600–2000 mm and there is a single dry season lasting three to four months. Only relatively recently has arabica coffee been grown in open sunny situations (Figure 5.1), including regions closer to the equator where there are two annual dry seasons (Cannell, 1985). According to Van der Vossen (2001, 2005), virtually all current cultivars of *C. arabica* are descendants of early introductions of coffee from Ethiopia to Arabia (Yemen), where they were subjected to a relatively dry ecosystem without shade for a thousand or more years, before being introduced into Asia and South America around AD 1700 and two centuries later into East Africa.

By comparison, *C. canephora* occurs wild in the lowland equatorial forest extending from West Africa to Lake Victoria. Robusta coffee (Figure 5.2) grows best in areas where the mean annual temperature is around 26°C (Wrigley, 1988). It represents about 20% of the world trade in coffee and is an important crop in Cameroon, Indonesia, Ivory Coast, Uganda and Vietnam. *Coffea liberca*, another lowland species, contributes less than one per cent of the total. Unless otherwise stated this review concentrates on *C. arabica*.

Commercial plantations of coffee are now distributed from Hawaii (20–25°N) and Cuba (22°N) in the north to Parana State, Brazil (22–26°S), in the south. The total planted area of coffee in the world (arabica and robusta) is about 8 million ha producing (in 2009) 9.75 million tonnes of green beans (but annual

Figure 5.1 Coffee. New estate planted on the contour; note the run-off control measures –
Roscommon Estate, Zimbabwe (MKVC). (See also colour plate.)

Figure 5.2 Coffee. 'Robusta' is an important smallholder crop in western Tanzania (MKVC).

production figures are notoriously variable). Brazil is by far the largest producer with 2.1 million ha (arabica and robusta) producing 2.8 million t. The next two largest producers are Vietnam (0.53 million ha; 1.07 million t, mainly robusta) and Colombia (0.47 million ha; 0.89 million t, arabica) (FAO, 2011a). About 70% of the world crop is estimated to be grown on holdings less than 10 ha in area (DaMatta *et al.*, 2010).

The coffee plant is an evergreen, and leaves are produced throughout the year, at rates that depend on temperature and water availability, but are shed during periods of drought. According to Nunes *et al.* (1968), the optimum mean annual air temperature for *C. arabica* is in the range 18–21°C. Temperatures below 12°C for long periods inhibit growth and development, and above 24°C net photosynthesis begins to decrease and is negligible at 34°C. Prolonged exposure to high temperatures (*c.* 30°C) accelerates leaf loss and induces a general decline in tree health (Drinnan and Menzel, 1995). However, DaMatta and Ramalho (2006) believe that the arabica coffee plant possesses a higher temperature tolerance for photosynthesis (up to 30–35°C) than these figures suggest. As an example, the crop is now being grown successfully in so-called marginal regions (mean annual air temperatures 24–25°C) of north-east Brazil. In Brazil, Lima and Da Silva (2008) confirmed the higher temperature tolerance of *C. arabica*. They identified the base temperature for growth (from transplanting to flower formation) as 12.9°C and the optimum temperature as 32.4°C.

In equatorial areas (Kenya, northern Tanzania and Columbia), *C. arabica* is usually grown at altitudes above 1000 m. The crop is susceptible to frost damage, which limits the areas (latitude and altitude) suitable for commercial production. The ecological aspects of the physiology of the coffee crop have been reviewed by DaMatta and Ramalho (2006) and DaMatta *et al.* (2010), and follow reviews by Maestri and Barros (1977) and Carr (2001).

Crop development

In Ethiopia (the centre of origin of *C. arabica*), flower and fruit development occurs naturally in a sequence that maximises the likelihood that the fruit will expand during the rains and after a flush of new leaves. Hence floral initiation occurs during the cool, dry winter period; the flowers then remain dormant during the dry season, and blossom after the first showers, which invariably precede the main rains. The immature or 'pinhead' fruits then remain dormant until the main rains have begun and the new flush of leaves, triggered by the same 'blossom' showers, have expanded. Intense rainfall throughout the year (without dry seasons) can lead to scattered harvests and low yields (Cannell, 1985).

A similar cycle occurs in most non-equatorial coffee-growing areas (such as southern India, Hawaii, Central America, south-central Brazil, Malawi and Zimbabwe), but in Kenya and Columbia, countries close to the equator, fruit expansion can often coincide with drought. It is against this background of two contrasting ecological areas, where commercial *C. arabica* is predominantly grown, that the development, water relations, water requirements and water productivity of the coffee crop are considered.

Vegetative

Coffee shoots have two distinctive structural features, namely (a) the axil of each opposite and decussate leaf contains not one, but a series of buds, and (b) branching occurs in two distinct forms (i.e. it is dimorphic). The vertical (orthotropic) shoots produce horizontal (plagiotropic) branches from the topmost bud, whilst the lower buds remain dormant or produce more vertical shoots (or occasionally inflorescences). The inflorescences develop from the buds at each node on the horizontal branches. As these branches usually produce flowers only once, pruning has to be done to ensure a continued supply of flowering nodes (Cannell, 1985).

Water stress reduces rates of shoot extension, the number of nodes and the area of individual leaves (Boyer, 1969 for *C. canephora*; Fisher and Browning, 1979; Tesha and Kumar, 1979). Shoot growth rates can be restricted before there is any evidence of differential stomatal closure between irrigated and unirrigated trees (Wormer, 1965). In Kenya, new leaves emerge most rapidly during synchronous growth flushes that occur after rainfall, but usually only when the rain is accompanied by a rapid temperature drop (defined as 3°C in 40 minutes). Flushes are most marked during the hot dry season (January to March) and occur at this time after rainfall even on irrigated trees (Browning, 1975a), indicating the action of some stimulus associated with rainfall, but independent of soil water status. Trees irrigated after eight weeks of enforced drought flushed immediately, producing leaves faster than trees that had been irrigated regularly (Browning and Fisher, 1975). Trees irrigated after 12 weeks of drought did not respond in the same way: instead they produced 70% more lateral shoots than trees that experienced shorter periods of drought. Tesha and Kumar (1979) have observed similar compensatory vegetative growth following the relief of water stress, in Kenya, particularly at high levels of nitrogen fertiliser, as have Drinnan and Menzel (1994) in Australia. Browning and Fisher (1975) have postulated that this response might be due to a reduction in root resistance to water uptake following a build-up of abscisic acid under stress.

Flowering

There are four successive stages of coffee flower bud development: initiation, dormancy, growth and blossoming. After initiation, buds grow for several months, reaching an average length of 4–6 mm before becoming dormant, a state associated with high levels of endogenous abscisic acid. Flower bud dormancy is slowly broken by continuous water stress over a period of one to four months. This stress can be induced by: (a) dry soil, and/or (b) large evaporative demand associated with high temperatures and large saturation deficits of the air

(Browning, 1975b). After several weeks of water stress the flower buds, as with shoots, can be stimulated to grow again by the relief of water stress following rain (or irrigation) or by a sudden drop in temperature and/or increase in atmospheric humidity, changes associated with the onset of rain. During the three to four days after the stimulus has been received meiosis occurs and there is a large increase in the giberellic acid content that is thought to overcome the inhibiting effect of the abscisic acid (Browning, 1975b). During the next six to 12 days the water content of the flower buds increases rapidly and they grow in length three- to fourfold, developing to blossoming and anthesis at a rate that is temperature dependent. A period of water stress therefore appears to be mandatory for normal flower bud development (Alvim, 1960, 1973).

At Campinas in Brazil (22° 50′ S), Magalhaes and Angelocci (1976) attempted to quantify the level of water stress needed to allow flowering to occur by making simultaneous measurements of the water potential of dormant flower buds and the subtending pair of leaves under varying levels of water stress (cv. Mundo Novo). They found that buds had to experience critical water potentials below −1.2 MPa for flowering to be stimulated by irrigation. This threshold value was associated with a change in the direction of water movement, from a net flow into the buds from the subtending leaves to a net flow out of the buds. When irrigation was applied, there was a rapid influx of water into the buds and flowering occurred within seven to 10 days afterwards, but only if the buds had experienced water potentials below −1.2 MPa. In ecological terms, this synchronous blossoming mechanism may protect the sexual structures of the flower buds from water stress during dry periods, in the same way that winter dormancy in temperate areas protects plant tissues from frost damage (Browning, 1975b). In Brazil, Astegiano *et al.* (1988) subsequently confirmed that a period of water stress was an essential prerequisite for flowering to occur. But, using detached branches and tracers, they were unable to show that the subtending leaf, nor the water potential gradient between this leaf and the flower bud, affected this process. Instead they suggested that the signal for the break in dormancy occurred when buds become fully turgid.

Crisosto *et al.* (1992) reported the results of a similar study in Hawaii (21° 21′ N; alt. <100 m). Here the aim was to identify ways of improving the uniformity of flowering to facilitate mechanical harvesting. From a series of detailed field and glasshouse experiments they found that flower opening was stimulated by irrigation after one period of water stress when pre-dawn leaf water potentials declined below −0.8 MPa. Similar stimulation of flowering was observed when less severe but more prolonged water deficits were imposed ($\psi_l = -0.3$ to -0.5 MPa for two weeks). In both cases flowering only occurred in buds that were at the 'open white cluster' stage of development when water stress was imposed. At this time, secondary xylem tissue was in the process of being formed. In split-root experiments, where one half of the root system was droughted and the other well irrigated, flowering was still simulated in the same way as in plants in which both parts of the root system were kept dry. This indicated that it was a 'root signal', perhaps a cytokinin or a giberellin, from the dry part of the root system that stimulated flowering, independently of the leaf water potential. From these observations they concluded that, in leeward areas of Hawaii, frequent irrigation to prevent flowering followed by a controlled water deficit and then re-irrigation may represent a practical way of synchronising flowering and shortening the period of harvest. In a greenhouse study, Schuch *et al.* (1992) found, though, that gibberellic acid only partially compensated for insufficient water stress for flower initiation. Trees that experienced leaf water potentials below −2.65 MPa, and flower bud water potentials of about −4.0 MPa, flowered within about nine days after irrigation. These stress levels are much greater than those Magalhaes and Angelocci (1976) found to be necessary to stimulate flowering in Brazil.

Again with the aim of trying to synchronise flowering to allow non-selective mechanical harvesting, Drinnan and Menzel (1994) conducted experiments with potted plants (cvs. Catuai Rojop and Mundo Novo) in a heated greenhouse in Queensland, Australia (27°S). They found that water stress did not affect the timing of flowering, which only occurred when the photoperiod was less than 12 hours. Severe water stress (leaf water potential allowed to fall to −2.5 MPa before rewatering) reduced the number of inflorescences, indicating that irrigation was needed during floral initiation. Water stress accelerated floral development, but had no deleterious effects on floral differentiation. They concluded that stressing of trees in the late stages of floral development, after initiation is complete, provides a means of increasing the proportion of fully differentiated dormant flower buds (stage 4) on a tree. The effects of temperature on the initiation and development of flowers of several cultivars, grown in pots under glass, independent of water status have also been investigated by the same authors (Drinnan and Menzel, 1995). Floral initiation did not occur at temperatures above 28°C, nor when the photoperiod was longer than 13 h. There was little difference between cultivars in these responses. The difficulties of implementing controlled water deficit regimes under field conditions, in order to synchronise flowering and reduce the number of harvests needed, have been summarised by DaMatta *et al.* (2010).

Fruit

For the first six to eight weeks after fertilisation, the ovaries of coffee undergo cell division, but the fruits grow very little, remaining as so-called pinheads. From about six to 16 weeks after blossoming, the fruits increase rapidly in volume and weight mostly owing to pericarp growth. The two fruit locules (cavities) swell to full size and the endocarps, which line the locules, lignify so that during this swelling stage the maximum volume of the beans is determined (Wormer, 1966). The size to which the locules grow depends greatly on the water status of the trees at the time of lignification; fruits which expand in wet weather will have larger beans than those which develop during hot, dry conditions (Cannell, 1974). From 12–18 weeks after blossoming, the seeds (beans) are formed and begin to fill the locules, increasing rapidly in dry weight with little increase in fruit size. About 30–35 weeks after blossoming, the fruits ripen, losing chlorophyll, producing

Figure 5.3 Coffee. Fruits ripen, turning red, about 30–35 weeks after flowering – Malawi (MKVC). (See also colour plate.)

ethylene and turning red (Figure 5.3). During this time the pericarp increases in dry weight and volume. The growing fruits act as priority sinks for assimilates and minerals, and can draw carbohydrates from elsewhere in the tree. This may lead to dieback of branches when yields are excessive (see below).

In Kenya, the proportion of large, commercially valuable coffee beans (those retained on a 6.75-mm sieve) varies greatly within and between years. Cannell (1974) analysed the results of three long-term field trials at Ruiru (1°4′ S; alt. 1610 m) and found that about 50% of the between-year variation in large beans could be explained by the number of rainy days (>1 mm rainfall) between 10 to 17 weeks after flowering, the period when fruits were expanding most rapidly. Irrigation and mulching are the two most important field treatments that have a beneficial effect on bean size. In equatorial areas, such as Kenya, floral initiation can occur almost throughout the year (Wormer and Gituanja, 1970) and periods of fruit expansion will not always coincide with one of the two short annual wet seasons. By contrast, in areas away from the equator with a single, annual and reliable rainy season fruits are more likely to develop during periods of adequate rainfall (Clowes and Wilson, 1974; Clowes and Allison, 1982).

Components of yield

The seed yield (Y; kg ha^{-1}) of coffee can be represented by the following equation:

$$Y = P{\cdot}N{\cdot}F{\cdot}W_f{\cdot}S$$

where: P = tree density (ha^{-1}); N = fruiting nodes (tree^{-1}); F = fruits (node^{-1}); W_f = fruit weight (kg fruit^{-1}); and S = seed:fruit weight ratio.

Most of the world's coffee has been planted with fewer than 2000 trees ha^{-1} with either one, single stem, or two or three, multiple stems, per tree. These are usually pruned to 2 m height or less. However, the greatest yields per unit area are obtained with densities of 4000–10 000 trees ha^{-1} (Cannell, 1985). At the higher densities, there is a reduction in the number of fruits per tree (thought to be due to the effect of mutual shading on floral initiation), but the mean weight per seed remains fairly constant. The yield of individual trees at conventional spacing is highly dependent upon the number of potential flowering nodes produced the previous year. This number varies considerably from year to year and is the component affected most by treatments such as irrigation, mulching and fertilisation.

The number of fruits per node also varies from season to season, with fruit set in the range 20–80% of the flower buds formed. Poor fruit set can be caused by: (a) the development of atrophied flowers, attributed to prolonged drought or excessive rainfall during the initial stages of flower bud development, and/or (b) incomplete pollination or fertilisation owing to heavy rain, low temperatures or a shortage of pollinators at blossom time.

According to Cannell (1985), under favourable conditions 12–20 fruit can be set per node, each of which carries two 3000–4000 mm^2 leaves. Even when the leaves at non-fruiting nodes are included, coffee has the capacity to set more fruit than it can sustain. But, in contrast to other woody perennials, there is no mechanism of early fruit shedding in coffee to prevent excessive crop load, possibly because there was no evolutionary advantage in the natural forest habitat, where floral initiation is low due to heavy shade (Van der Vossen, 2005). Some fruit shedding does occur, principally during the period of rapid fruit swelling. This shedding is exacerbated by drought, as well as by other factors. Even after drought, numerous fruits may still remain and trees are said to 'overbear'. In extreme cases the vegetative shoots 'die back' and very few potential flowering nodes are produced for the next season. This problem is especially serious in young trees, which often need to be deblossomed. Culivars differ in their sensitivity to dieback with cvs. Caturra and Bourbon, for example, being more susceptible than cv. Mundo Novo or cv. Catuai (DaMatta et al., 2010).

Drought can therefore reduce: (a) the number of fruiting nodes per tree, (b) the number of fruits per node, and/or (c) the size of the seed, depending on its timing and severity. However, the effect on final yield will depend on whether or not the numbers of fruits that remain are more or less than the number that can be sustained by the tree, which is a function of leaf area. Overbearing, by contrast, will lead to a reduction in the potential number of flowering nodes, with implications for yields in the following year.

Drought symptoms

A summary of field observations by Clowes and Logan (1985) for growers in Zimbabwe (17–20° S; alt. <1500 m) is given below:

Figure 5.4 Coffee. Symptoms of water stress – northern Tanzania (MKVC).

- Drought reduces the rate of production of new leaves; leaves are smaller and internodes are reduced in length, especially on fruit-bearing branches (Figure 5.4).
- Old leaves turn yellow and are shed prematurely, particularly on sides of the tree exposed to the afternoon sun.
- Premature cessation of extension growth occurs, particularly on fruit-bearing branches, which restricts or delays flower bud initiation with adverse effects on the following year's crop.
- There may be insufficient leaf area to support the current crop. This not only reduces yield and quality but can also lead to dieback of shoots, with implications again for the next crop.
- During the early expansion phase (10–12 weeks after flowering) drought will cause fruits to become blue/green in colour before being shed; severe stress a little later causes fruits to go yellow, and the seeds turn black and wither.
- Stress at the end of this phase (up to 17 weeks after flowering) will reduce the final size of bean as the parchment skin is laid down. Severe wilting will cause fruit to feel spongy and beans may become desiccated.
- Stress when the seeds are developing results in misshapen (ragged) beans with a reduced mass.
- Stress also delays ripening and the fruits take longer to turn red. Severe water stress, leading to dieback of the branches, causes the fruit to go black.

Roots

Depth and distribution

Early studies on the root growth of coffee were made by Nutman (1933a, 1933b, 1934) and by Bull (1963) in northern Tanzania and Kenya (1–3° S; alt. 1250–1700 m), areas of bimodal rainfall (up to *c*. 1500 mm annual total). Nutman (1933b) described a 'normal' root system developed by transplanted seedlings as comprising one or more main vertical (or tap) roots, a superficial layer of horizontal lateral roots near the soil surface, and other deeper laterals arising from the tap roots but growing downwards at an acute angle to the vertical.

The depth and distribution of roots varies with many soil physical and chemical properties, as well as with cultural practices such as mulching, irrigation, nitrogen fertilisation and tree spacing. For example, Bull (1963) undertook a detailed study of the long-term (20 years) effect of mulch (banana trash) and irrigation on the root growth of two clones. The root systems of 20 trees were excavated (two per irrigation × mulch treatment combination for each clone), and the roots separated into the components, which were weighed and counted. Irrigation reduced the depth of penetration of the tap root (by 0.56 m or *c*. 20%) and the development of primary (defined as roots with axes >5 mm in diameter) and secondary roots (any root subtended by a primary root) in deeper layers of the soil profile, which was a deep volcanic clay loam. In the surface layers irrigation increased the *length* of lateral roots and the number of lateral secondary roots. Mulching, by contrast, increased both the size of lateral and 'sinker' roots (derived from lateral primary and secondary roots), and the depth of tap root penetration. There was a dense mass of fibrous roots in the surface layers beneath the mulch. The combination of mulch and irrigation gave the 'best' developed root system, but the maximum root depth was still reduced by about 0.5 m compared with the unirrigated treatment. However, in all the treatment combinations roots extended to depths below 2 m.

Wallis (1963), summarising earlier work in Kenya, stated that roots of unirrigated coffee could explore the top 3 m of soil (a deep red laterite clay loam), although the bulk (80%) of the roots were in the top 1.2 m. In studies of the water use of coffee, Pereira (1957), Wallis (1963) and afterwards Blore (1966) found that, in dry years, the soil at depths of 3 m would reach permanent wilting point.

In Veracrux, Mexico (19° 10′ N: alt. 1225 m), Garriz (1979) excavated roots of 24-year-old plants to depths of 1.8 m (cvs. Typica and Pluma Hidalgo) and 2.4 m (cv. Bourbon) in a loamy clay soil. About 50% of roots (dry mass), with diameters of 2 mm or less, were found in the top 0.6 m with all three cultivars.

At Chipinge in Zimbabwe (20° 13′ S: alt. 1130 m), Cassidy and Kumar (1984) found that, in the absence of any physical or chemical impediment, roots could extend to depths of 2.5 m (on deep, well-drained slightly acid loamy soils). On heavier soils with high silt contents, root development was severely restricted and axial roots rarely extended much below 1.0 m. Soil compaction, gravel strata and high water tables all restricted root growth. They too found that mulch enhanced root growth in the top 0.25 m, although mulching had less effect at high plant populations (>4000 ha^{-1}) due to mutual shading and self-mulching. They also reported that axial roots of six- to seven-year-old irrigated trees (cv. SL28) penetrated deeper (by about 0.5 m) at large plant populations (4000–6000 ha^{-1}) than at low densities. This has implications in terms of responses of traditional and high-density planted coffee to drought and to irrigation (Fisher and Browning, 1979). The lateral spread of surface roots was constrained by the influence of neighbouring trees, whether planted at high densities or as a cova (more than one tree at each station). Axial roots of trees planted in a cova also grow deeper, down to 3.5 m, than those of single trees at similar densities. Clowes and Logan (1985) subsequently summarised these results, and their practical implications, for growers in southern Africa.

At Ngapani in Malawi (14°S: alt. 1200 m), axial roots of young (unmulched, seven months from field planting), immature (third year after planting) and mature drip-irrigated coffee (more than four years after planting) extended to depths of 0.45, 1.2 and >1.5 m respectively (cv. Catimor 129, spaced 3.6×1.2 m). The majority of feeder roots were found in the top 0.5 m (young) and 1.25 m (immature) depth of the deep clay loam soil within 0.3 and 0.8 m from the main stem respectively (Sanders, 1997). Inter-row soil compaction restricted root pro-liferation in the topsoil. At this location where the annual effective rainfall is about 800 mm there was no evidence that drip irrigation during the extended dry season caused the roots to be concentrated in the wetted soil volume. Where soil depth restricted rooting (to 0.9 m), there was evidence (pre-dawn $\psi_l <-2.0$ MPa after 24 days without rain or irrigation) that these plants came under water stress earlier than plants ($\psi_l = -0.6$ MPa) growing where there were no depth restrictions (>1.2 m).

Extension

Huxley and Turk (1975) reported preliminary studies at Ruiru, in Kenya, on factors affecting the extension growth of the fine white roots. From observations made against glass windows in an underground root laboratory, they found that roots grew throughout the two-year study, fine roots remaining alive for at least this period of time. 'Clumps' of roots were active in producing fine roots at different times. Rapid rates of root extension generally preceded shoot extension. There was little effect of crop or irrigation on the growth of these roots. The authors acknow-ledged that these observations were made on a limited number of trees, and that there was a great deal of variability that could not be adequately explained. Cannell (1972) also found from a growth analysis study that the (<3 mm diameter) feeder roots continued to grow almost unchecked during periods of rapid shoot growth at the start of the main rains. Roots also grew rapidly during the cool dry weather, when shoot growth was slow. In general, roots grew more continuously than shoots. In another study at Ruiru, in Kenya, using a radioisotope of phosphorus placed at different depths (down to 1.8 m) and distances from the main stem (to the mid-row at 1.35 m), Huxley et al. (1974) found that the relative activity of roots changed markedly with season, except at depths of 1.8 m. After prolonged drought, relatively high root activity was found at mid-depth, near to the trunk, but after the soil was rewetted by rain most root activity occurred in the topsoil at quarter row distance. After the soil profile had been wet for some time, functional roots were more evenly distributed. The position of maximum root activity frequently did not coincide with the distribution of fine roots of this mature crop. The authors commented on the variability of the data obtained using this technique. The fruit-bearing capacity of the tree has also been shown to influence the development of the root system of irrigated *C. canephora* (cv. Conilon) in south-eastern Brazil, with large yields temporarily reducing the dry mass of the root system (DaMatta *et al.*, 2010, citing the work of Bragança, 2005).

Cuenca *et al.* (1983) stressed the importance of the superficial root system that extended into the leaf litter layer, including root hairs attached to the surfaces of decomposing leaves, in the absorption of mineralised nutrients. From a study in Miranda State in Venezuela (10° N; alt. 1400 m) on mature unfertilised coffee, grown in an acid soil under heavy shade, they found that about one third of the total dry mass of fine roots (less than one millimetre in diameter) to a depth of 0.5 m was located in the top 0.1 m of soil. Root production in the litter layer, though, was very variable and transient. According to Wrigley (1988) there is little evidence that mycorrrhizae are present on coffee roots.

Summary: crop development

1. Coffee has a dimorphic growth habit.
2. Water stress reduces rates of shoot extension, the number of nodes and the area of individual leaves.
3. A period of water stress, induced either by dry soil or dry air, is needed to prepare flower buds for blossoming, but the intensity and duration of the stress required have not been specified.
4. Blossoming is stimulated by rain (or irrigation), but buds must experience water potentials less than −1.2 MPa for this to occur.
5. Water must be freely available during the period of rapid fruit expansion to ensure large, high-quality seeds.
6. There is no mechanism for early fruit shedding to prevent excessive crop load, which can lead to the 'dieback' of vegetative shoots.
7. Roots can extend to depths of 2–3 m, but the greatest mass of roots is normally found in the top metre of soil.
8. Root growth tends to be continuous throughout the year.

Plant water relations

Physiological aspects of the water relations of coffee have been studied for more than 70 years, beginning perhaps with the classical work of Nutman (1937a, 1937b, 1941) in northern Tanganyika (now Tanzania). There were similar studies in Brazil, Costa Rica and Mexico by Alvim (1960), Butler (1977), Fanjul *et al.* (1985) and, in Kenya, by Wormer (1965), Browning and Fisher (1975), Fisher and Browning (1979) and Kumar and Tieszen (1980a, 1980b). Other names associated with this work include Nunes (1976) and Bierhuizen *et al.* (1969). More recently, the results of fundamental research in Hawaii by Meinzer and co-workers have been reported (Meinzer *et al.*, 1990; Crisosto *et al.*, 1992; Gutierrez *et al.*, 1994), and in Brazil by Barros *et al.* (1997) and DaMatta *et al.* (1997). Over the period covered, many new field techniques for measuring plant water status have been introduced, including the diffusion porometer, the pressure chamber (Angelocci and Magalhaes, 1977), the infrared gas analyser and the sap flow method for estimating transpiration. Much ingenuity has been shown by individual scientists, particularly by Nutman in his pioneering studies during the 1930s in East Africa. In this section, an attempt is made to reconcile the results of some of this work, carried out under contrasting conditions and with different facilities and techniques.

Stomata

Stomata are found only on the abaxial surface of *C. arabica* leaves at densities variously quoted in the range 150 to 330 mm^{-2} (Franco, 1939; Wormer, 1965; Josis *et al.*, 1983; Wrigley, 1988; DaMatta *et al.*, 1997). Stomata are also present in green fruits at densities of 30–60 mm^{-2} that may represent 20–30% of the photosynthetic surface on heavily bearing trees (Cannell, 1985).

Nutman (1937b) used a homemade recording resistance porometer to make measurements of diurnal changes in stomatal opening under shaded and sunny conditions in northern Tanzania (3° 30′ S; alt. 1370 m). He found that stomata opened early in the morning, but only remained fully open throughout the day when it was overcast, or when leaves were shaded from direct sunlight. If the incident radiation levels were large, there was a rapid reduction in stomatal conductance (even when all the other leaves on the tree were removed). On days when wilting of the youngest leaves was observed, stomata in the other leaves had closed by midday and stomatal conductance remained low for the rest of the day, even when leaves were shaded. The stomata were observed to respond to changing ambient conditions within three minutes. Subsequently, partial closure of the stomata was also observed on sunny days in Brazil and in Costa Rica (references cited by Maestri and Barros, 1977), and in a glasshouse experiment in the Netherlands (Bierhuizen *et al.*, 1969).

In Kenya, Wormer (1965) and later Browning and Fisher (1975) used the infiltration technique (liquid mixtures of isopropanol and distilled water) to measure the effects of soil water availability on stomatal opening. They too observed partial closure of the stomata during the day, even in irrigated trees. Wormer (1965) showed how increasing air temperatures (over the range 22 to 33°C), saturation deficits (0.2 to 2.4 kPa) and daily total solar radiation levels were each associated with a linear reduction in the degree of stomatal opening during the afternoon. For example:

$$IS = 18.5 - 0.365\,T$$

where IS represents the infiltration score (scale 1–14), a large number indicating that the stomata are wide open, and T the air temperature in the field (°C). This equation was found to be valid for observations made between 1130 and 1600 h (local mean time) and explained 93% of the variation in the measured infiltration score.

In pot experiments, Wormer showed that relative stomatal opening was closely related to the soil water content. However, in a field experiment, the relationship was only satisfactory during extended periods of dry weather that were not interrupted by rain. Like Nutman (1937b), Wormer too found that the degree of stomatal opening was always more in shaded leaves than in ones exposed to the sun. Interestingly, Wormer also observed that the application of nitrogen fertiliser (100 kg ha^{-1}) increased stomatal opening, particularly in irrigated trees.

Wormer found that the degree of stomatal opening in unirrigated trees did not return to the level observed in irrigated trees for several months after the rains began. By comparison, Bierhuizen et al. (1969), in a pot experiment, noted a lag of only four to five days after watering before there was full recovery. By contrast, Browning and Fisher (1975) reported that the degree of stomatal opening (as recorded at 1400–1500 h) increased, for up to four weeks after the relief of stress, to values 1.4 to 1.6 times greater than those recorded in trees previously irrigated at weekly intervals. This effect was most marked when drought had been imposed for about eight weeks, by which time the potential soil water deficit ($0.5E_{pan}$) had reached about 300 mm.

In Bahia, Brazil (15° S) Butler (1977) measured stomatal conductances on sunlit leaves of C. canephora (cv. Guarini) as low as 0.4–0.5 mm s^{-1} in the middle of the day, compared with 1.0–1.5 mm s^{-1} on shaded leaves at the same time. On cloudy days, values were typically 2.0 mm s^{-1}. Butler also found that temperatures of (horizontal) leaves were often 10–15°C warmer than the prevailing air temperature (29°C), but 1–2°C cooler when shaded. The temperature differential (δT,) was related to the net radiation absorbed by the leaf (R_n, W m^{-2}) as follows:

$$\delta T = 0.0264\,R_n - 1.07 \quad (r = 0.92)$$

Low stomatal conductances (0.4–0.5 mm s^{-1}) were associated with positive leaf/air temperature differences exceeding 10°C, with corresponding large saturation deficits of 4.5 to 6.0 kPa. When leaf and air temperatures were similar, the calculated stomatal conductance was about 1.8 mm s^{-1}, close to those measured on shaded leaves. These values were derived from an energy balance analysis, with assumed aerodynamic resistances of 35 s m^{-1} for individual leaves. In contrast to results reported for C. arabica, Butler found that after 'substantial' rain the stomata of sunlit leaves did not close, in spite of temperatures exceeding 30°C.

Insufficient information was given by Butler (1977) to assess the role of soil water availability, or atmospheric humidity, on the stomatal conductances.

Fanjul *et al.* (1985) were, though, able to demonstrate the sensitivity of stomatal responses of young seedlings (cv. Typica) to changes in the saturation deficit of air (at constant day/night, 25/15°C, temperatures). At low (shade) irradiance (200 μmol m^{-2} s^{-1}, photosynthetically active radiation, PAR, equivalent to about 44 W m^{-2}) raising the saturation deficit from 0.2 to 1.5 kPa, reduced stomatal conductances (measured with a continuous flow porometer) from about 3.0 to 0.7 mm s^{-1}.

Fanjul *et al.* (1985) reported similar measurements made in field experiments at two sites in Veracruz State, Mexico (19°27–31′ N; alt. 1225–1340 m). These confirmed that stomatal conductances were larger in shade-grown plants (cvs. Bourbon and Caturra) than in sun-grown ones (1100–1200 trees ha^{-1}). At dawn, values in sun-grown plants were large (12 mm s^{-1}), but they normally decreased during the day (to about 4 mm s^{-1}) as total irradiance (0 to 800 W m^{-2}), air temperatures (14 to 26°C) and saturation deficits (0 to 1.6 kPa) increased. At higher values of each these variables, though (e.g. 1000 W m^{-2}, 26–30°C, 1.6–2.8 kPa respectively), the stomata remained closed all day. By contrast, under shade, stomatal conductances continued to increase during the morning (reaching 20 mm s^{-1}) before declining during the afternoon.

More recently, Barros *et al.* (1997) reported the results of measurements (cv. Red Catuai) in south-east Brazil (20°45′ S; alt. 650 m). They found that stomatal conductances (measured with a diffusion porometer), during the main growing season, were relatively high in the early morning but declined throughout the day, as air temperatures increased from 20 to 30°C, and saturation deficits from low values to 2.0–3.0 kPa. In contrast, during the winter months, conductances were low throughout the day, observations that they attributed to low night temperatures.

In a detailed study in Hawaii (21°54′ N; alt. 98 m), Gutierrez *et al.* (1994), using a gas exchange system, also found that stomatal conductances (cv. Yellow Catuai) were typically high in the morning and declined after midday as the saturation deficit (leaf to air) and PAR levels increased. After adjusting for changes in PAR, the observed diurnal hysteresis in the relationship between conductance and the saturation deficit was removed, revealing a strong negative relationship with the dryness of the air. This was especially the case when canopy conductances, expressed on a unit leaf area basis, were plotted. They also showed that an increase in wind speed could reduce canopy conductance as a result of the transfer of dry air to the leaf surface. From these and other observations they concluded that stomatal control of water fluxes from the canopy of a well-watered coffee crop was strongly influenced by the interaction of wind and atmospheric humidity. Differences in the proportion of net radiation dissipated through transpiration from irrigated crops between two years at the same location were also attributed to the sensitivity of the stomata to the leaf to air saturation deficit (Gutierrez and Meinzer, 1994a).

In a comparative study of gas exchange in different species in Colombia (3°31′ N; alt. 1020 m), Hernandez *et al.* (1989) showed clearly how stomatal conductances in shaded plants (cv. Arabigo) declined rapidly, reaching about 90% of the initial value when the leaf to air vapour pressure difference reached 4.0 kPa. This was judged to be due to the inherent sensitivity of the stomata to dry air rather than to concurrent changes in the bulk leaf water potential, or to changes in intercellular carbon dioxide concentrations. In a controlled environment study in Japan, Kanechi *et al.* (1995) have also shown how stomatal conductances decline in well-watered plants as the saturation deficit of the air increases over the range 1.0 to 3.0 kPa, with almost complete closure of the stomata at the dry end of this range under natural daylight. There was a corresponding rise in leaf temperatures from about 25 to 35°C.

Photosynthesis

Nutman (1937a) also pioneered the measurements of photosynthesis in coffee. At Lyamungu, in northern Tanzania, he observed that apparent rates of assimilation of carbon dioxide were relatively constant during days when it was cloudy, but fell to low values during the middle of the day when the sun was shining. Rates of assimilation were maintained if leaves were shaded, naturally or artificially. He attributed the midday suppression of photosynthesis in sunlit leaves to the effects of large incident levels of solar radiation on stomatal opening. Later Bierhuizen *et al.* (1969) reported the results of a pot experiment carried out under controlled conditions (25°C day/20°C night; 70% relative humidity) in a glasshouse in the Netherlands. As the soil dried from field capacity to permanent wilting point, there was a reduction in the rate of photosynthesis. Unusually, there was a time lag before there was a corresponding reduction in transpiration (see below).

At Ruiru in Kenya, Kumar and Tieszen (1980a) studied the effects of light intensity and temperature on rates of photosynthesis of container grown (cv. SL28) seedlings. Leaf and air temperatures were controlled, by varying the temperature of the circulating water in the leaf chamber, but there was no corresponding control of the saturation deficit of the circulating air. The saturating photon flux density (PFD) for shade-grown plants was about 300 μmol m^{-2} s^{-1} (*c.* 66 W m^{-2}), half the value for sun-grown plants, and much less than the peak values of 2500 μmol m^{-2} s^{-1} (*c.* 550 W m^{-2}) recorded in the tropics. Rates of photosynthesis declined at temperatures above 25°C but, surprisingly in view of the results reported above, stomatal conductances remained constant over the temperature range 25–35°C. This was thought to be due to an increase in mesophyll resistance to carbon dioxide diffusion. In a similar field study, Kumar and Tieszen (1980b) found that rates of photosynthesis (at temperatures of 24 ± 2°C and irradiance of 300 μmol m^{-2} s^{-1}) closely matched ($r = 0.96$) concurrent changes in stomatal conductance (*g*, diffusion porometer) over a range (-0.6 to -3.5 MPa) of leaf water potentials (ψ_1). Rates changed from 1.6 *g* m^{-2} h^{-1} (when $\psi_1 = -0.6$ to -1.0 MPa and *g* =1.25 mm s^{-1}), through 1.2 *g* m^{-2} h^{-1} ($\psi_1 = -1.0$ to -2.0 MPa; *g* = 1.0 mm s^{-1}) to 0.4 *g* m^{-2} h^{-1} ($\psi_1 = -2.8$ to -3.5 MPa; *g* = 0.3 mm s^{-1}).

In Mexico, Fanjul *et al.* (1985) found that rates of photosynthesis of young seedlings (cv. Typica) grown in a controlled environment declined rapidly once stomatal conductances fell below about 2.0 mm s^{-1} at a low (200 μmol m^{-2} s^{-1}) irradiance (PAR) level. Changes in rates of photosynthesis tended to match changes in stomatal conductance (over the range 0.5 to 3.0 mm s^{-1}). Both conductance and photosynthesis declined as the saturation deficit of the air increased from 0.2 to 1.5 kPa. By contrast, at Palmira in Colombia (3°31′ N; alt. 1020 m), Hernandez *et al.* (1989) showed how the rate of uptake of carbon dioxide in shaded plants remained relatively unchanged as the leaf to air vapour pressure deficit increased from 0.5 to 1.5 kPa but then decreased almost linearly as it rose further to 3.5–4.0 kPa.

In Hawaii, water stress reduced the total leaf area of container-grown plants, but assimilation rates on a unit leaf area basis were always similar. Meinzer *et al.* (1992) suggested that this represented an important process by which coffee adjusts to reduced water availability. Measurements of carbon isotope discrimination (intrinsic) and gas exchange (instantaneous) failed to give consistent estimates of water-use efficiency. In Brazil, DaMatta *et al.* (1997) found that *C. canephora* (cv. Kouillou) was superior in photosynthetic performance to *C. arabica* (cv. Red Catuai) in stressed and unstressed plants. They attributed this in part to differences in stomatal density.

In Colombia (05°01′ N 75°36′ W; alt. 1425 m), diurnal changes in stomatal conductance, transpiration and photosynthesis were measured on container-grown plants (cv. Colombia) in relation to ambient weather conditions. Net photosynthesis declined from mid-morning onwards, matching changes in stomatal conductance. Transpiration remained virtually constant until late afternoon. Instantaneous water-use efficiencies were greatest in the morning. Midday air temperatures reached 29°C and saturation deficits 1.6 kPa (Gómez *et al.*, 2005).

According to Cannell (1985), there are four notable features concerning the photosynthesis rate of leaves of *C. arabica*, all of which seem to reflect its evolutionary history as a shade-adapted C3 species:

- The maximum net photosynthetic rates of *sun* (unshaded) leaves are low (around 7 μmol CO$_2$ m^{-2} s^{-1} at 20°C) but higher (up to 14 μmol CO$_2$ m^{-2} s^{-1}) for *shaded* leaves that contain more chlorophyll per unit area than sun leaves.
- The saturating irradiances for photosynthesis are low (500–600 μmol m^{-2} s^{-1} of PAR for sun leaves, and about half this for shade leaves). This compares with irradiance (PAR) levels of 2500 μmol m^{-2} s^{-1} at midday on sunny days in the tropics. These values approximate to about 110–130 and 550 W m^{-2} PAR.
- The net photosynthetic rates decrease markedly with increases in leaf temperatures above 20–25°C. This may be due to increased mesophyll resistance (Kumar and Tieszen, 1980a), but in the field high temperatures are often associated with low leaf water potentials, which cause midday stomatal closure and an increase in internal CO$_2$ concentrations. On sunny days in the tropics, leaves can reach temperatures of 35–40°C, as much as 10–15°C above ambient

(Cannell, 1971; Butler, 1977). Even with light-adapted sun leaves, their photosynthetic apparatus seems to be physically damaged by continuous exposure to high temperatures.

The validity of some of these conclusions has since been challenged by DaMatta (2004b) who suggested that under full sun photosynthesis is restricted by a decline in stomatal conductance (due to high leaf temperatures/large saturation deficits of the air) and not as a direct result of high irradiance levels. He also warned about the difficulties of scaling up from measurements made on a single leaf to the crop canopy where mutual shading occurs and leaves have time to acclimatise to changing irradiance levels.

In south-east Brazil, Ronquim *et al.* (2006) compared leaf gas exchange processes on clear and cloudy days in the wet season. Independent of irradiance levels, leaf water potentials were always higher than the minimum required (> -1.5 MPa) to affect net photosynthesis which, together with stomatal conductance and leaf transpiration, declined on a clear day in all three cultivars tested. When integrated over a day, the depression in leaf gas exchange and an index of photochemical efficiency (especially around midday) resulted in a large (*c.* 70%) reduction in daily carbon gain on a clear day compared with a cloudy day. Conductance and net photosynthesis were both negatively correlated with the saturation deficit of the air (range 1–3.5 kPa), but only on clear days. Irrespective of the irradiance level during the day, stomatal conductance played a central role in controlling gas exchange. Air temperatures during the day ranged from 24 to 35°C.

At the other end of the temperature range, Bauer *et al.* (1985) showed, in a detailed controlled environment study in Austria, how chilling of above-ground tissues at night (6°C and below for 12 hours) impaired the photosynthetic process recorded the next day (at 24°C). Depending on the chilling temperature, it took from two to six days for photosynthesis to return to the value for the control treatment. Chilling on successive nights at 4–6°C reduced photosynthesis progressively on each day following treatment. Conditioning the plants failed to reduce these effects. About 25% of the chilling effect was shown to be due to reduced stomatal conductance and 75% to impairment of the carboxylation process. These observations confirm the circumstantial evidence that chilling injury contributes to determining the limits to successful coffee production due to altitude and/or latitude. In a follow-up study, Bauer *et al.* (1990) found that cultivars differed in their responses to chilling stress. Using a derived index of susceptibility, relatively resistant standard cultivars were identified (K7, SL28 and K33), confirming field observations made in Zimbabwe during a cold spell.

The primary production rate of mature trees benefits from mutual shading, with up to 90% of the total radiation intercepted by the top 'layers' of leaves. Nevertheless, the radiation and heat load is sufficiently spread to enable four to five-year-old trees in Kenya to attain net assimilation rates similar to those of seedlings (Cannell, 1971). The potential annual primary production of closed canopies of unshaded coffee seems to be 20–30 t ha^{-1}, equivalent to those of tea, oil palm and many tropical forest plantations. Compared with other trees, coffee, of all ages, allocates a relatively large proportion (up to 40–45%) of net annual dry matter increment to leaves (Cannell, 1972). Leaf area indices of between seven and 11 have been recorded for closely spaced trees, with access paths.

Transpiration

In northern Tanzania, Nutman (1941) measured the transpiration rates of whole trees over five-minute intervals using an ingenious, sensitive, continuous weighing device. He found that at low total radiation levels (630 W m^{-2}) transpiration increased with radiation, but at higher levels (840 W m^{-2}) the relationship was less clear due to stomatal closure. Daily totals of transpiration correlated well with cumulative radiation, and with the mean saturation deficit of the air during daylight hours. Reducing radiation levels by 70% with a hessian cloth only cut transpiration by individual trees by about 10%, compared with unshaded plants. Using the stem heat balance technique in Hawaii, Gutierrez and Meinzer (1994a) recorded transpiration rates similar to those previously determined by Nutman (1941), over 50 years earlier.

In Colombia, Hernandez et al. (1989) showed how transpiration rates from individual (shaded) leaves rose as the leaf to air saturation deficit increased up to 1.0–1.5 kPa, but then declined rapidly as the dryness of the air increased to 4.0 kPa. By comparison, transpiration from sunflower leaves continued to increase, but at a declining rate, over the same range. In Japan, Kanechi et al. (1995) found that transpiration from one-year-old plants (cv. Typica) in a greenhouse, measured with a steady-state diffusion porometer, was always greater on a cloudy day compared with a sunny day, for both well-watered (especially) and droughted plants. These differences, which reached a factor of three, were attributed to the sensitivity of the stomata to the leaf-to-air saturation deficit, and not simply to radiation levels. Stomatal conductances declined logarithmically with increasing leaf temperatures and saturation deficits. When the values of these two variables exceeded about 30°C and 2.0 kPa respectively, the stomata were virtually closed even in well-watered plants.

Because of the capacity of stomata to control water loss, relative turgidity measurements do not appear to be a good measure of plant water status. Bierhuizen et al. (1969) recorded daytime values between 92%, when the soil was close to field capacity, down to 80% at permanent wilting point. Nunes and Correia (1983), in a controlled environment study, found that the Catuai group of cultivars could support smaller relative leaf water contents (76–81%) than the Caturra group (85–90%) without wilting.

Plant water status

In Kenya, Fisher and Browning (1979), Kumar and Tieszen (1980b) and Gathaara and Kiara (1984) have, however, successfully used a pressure chamber to measure leaf water potentials (ψ_l) during the day in the field. Kumar and Tieszen (1980b) found that when mid-afternoon temperatures were 25–30°C, and saturation deficits had reached 2.5–3.0 kPa, ψ_l fell to minima of about -1.5 MPa, even when the soil was close to field capacity, before increasing again in the late afternoon. When the soil profile was relatively dry (soil water content below 50% of field capacity in the top 1.2 m) the corresponding minimum value of ψ_l was -2.8 MPa, but recovery in the late afternoon was rapid. On cloudy or overcast days (saturation deficits 1.1 to 1.2 kPa) minimum values were only -1.0 to -1.2 MPa. By comparison, Fanjul et al. (1985) recorded minimum values of ψ_l in Mexico during the middle of sunny days of about -0.8 MPa, in sun-grown plants and sometimes down to -1.4 MPa even under natural shade. Both ψ_l and stomatal conductances were sensitive to changes in the saturation deficit of the air. However, on occasions when the stomata closed during the day, ψ_l could remain relatively high (-0.6 to -0.8 MPa) suggesting that stomatal closure limited the reduction in ψ_l, and prevented excessive transpiration.

Drought tolerance

Cultivars are known to differ in their responses to drought. For example, experience in Zimbabwe, reported by Clowes and Logan (1985), suggests that the dwarf cultivar Caturra (a mutant, first identified in Brazil) is relatively drought tolerant, despite a large amount of foliage. However, because it fruits late in the dry season it can suffer from drought at that time. Similarly K7, selected in Kenya and with a spreading habit of growth, appears to be able to withstand hot dry conditions well in Zimbabwe. By comparison, cultivars SL28 and SL34 (both also selected in Kenya) are thought to be more susceptible to drought than either Caturra or K7, but less than Geisha or Agaro, which are both of Ethiopian origin. Elsewhere, Wrigley (1988) stated that Laurina and San Ramon (both mutants) were considered to be relatively drought resistant, while Van der Vossen (2001) observed that SL28 wilted later than other Arabica cultivars, or accessions of Ethiopian origin, after a prolonged dry spell in Kenya.

Only a limited amount of work appears to have been done to identify the reasons for these differences between cultivars. Nunes (1976), in a review of the water relations of coffee, referred to some of her earlier work in which the responses to drought were compared in pot experiments. Those in the Caturra group appeared to have a faster rate of water use than others, and to be able to dry the soil to lower soil water potentials. Others differed in their capacity to retain leaves after the first sign of wilting, with Agaro and KP 423 (from Tanzania) responding well. Previously, in Uganda, Dancer (1963) had also identified apparent differences between seedlings in their responses to drought (measured in terms of vegetative growth).

In neighbouring Burindi (3° 12′ S; alt. 1570 m), Josis et al. (1983) compared 11 cultivars in terms of the relative changes in leaf water potential between the beginning and the end of the dry season. The trees had been planted in the field four years earlier at a density of 2500 ha^{-1}. They identified four cultivars which maintained higher water potentials than the others when droughted, namely Catuai Vermelho (compact growth habit, from Brazil), Mbirizi Temoin (Rwanda), Mysore (India) and ABK 5718 (Ethiopia). They recommended that these cultivars should be planted in marginal areas. For measuring leaf water potential they used a hydraulic press, which previous work had shown gave results very close ($r = 0.97$, $n = 58$) to those obtained with a pressure chamber (Renard and Ndayishimie, 1982).

From a comparative study with container-grown plants in Brazil, DaMatta *et al.* (1993) concluded that coffee is a 'water saving' rather than a 'dehydration tolerant' species. They recorded similar, but limited (−0.34 MPa, 22%), osmotic adjustment in five cultivars when they were subjected to water stress, but slightly more (−0.48 MPa) in the single *C. canephora* cultivar tested. The amino acid proline was an important component of this process (Maestri *et al.*, 1995). Effective stomatal control maintained high relative leaf water contents. Subsequently, Pinheiro *et al.* (2005), in a comparison of four *C. canephora* clones grown in large containers (0.8 m deep, 129 l volume) in south-eastern Brazil, identified differences in rooting depth, hydraulic conductance and stomatal control of water use as the key indicators of the drought tolerance of clones labelled 14 and 120 relative to clones 46 and 109A. Diurnal and seasonal measurements of leaf gas exchange and leaf water potential in south-east Brazil, recently described by Novaes *et al.* (2011), showed how four-year-old, field-planted composite plants (*C. arabica* scion (cv. Obata IAC 1669–20) grafted onto a *C. canephora* rootstock (cv. Apoata IAC 2258)) were less susceptible to dry air inhibition of photosynthesis than the non-grafted *C. arabica* genotype. The authors concluded that, by using *robusta* rootstocks in this way, the likely adverse effects of climate change on the productivity of *arabica* coffee in Latin America could be reduced.

Previously Meinzer *et al.* (1990), in Hawaii, had observed osmotic adjustment in field-grown plants when subjected to drought. This together with increases in tissue elasticity had reduced the leaf water potential at which turgor loss occured. There was, though, considerable variation between cultivars in the rate at which leaf water deficits developed, with Mokka showing the biggest decline in leaf water potential as the soil dried. In a controlled environment study, Renard and Karamaga (1984) compared the responses of two cultivars to drought. Caturra Amarelo, from Brazil, maintained its leaf water status, as measured by a number of techniques including osmotic potential, better than Harrar, from Ethiopia. Similarly Venkataramanan and Ramaiah (1987) showed how, in South India, cultivars differed in their responses to drought as a result largely of differences in the capacity of the young plants to adjust the osmotic potential through, mainly, the accumulation of proline. In general, cultivars of *C. arabica*, particularly San Ramon, could endure drought better than those of *C. canephora*.

Other attempts have been made to develop procedures for identifying potentially drought tolerant cultivars. Thus, Meguro and Magalhaes (1983) in Brazil compared the responses of five cultivars to water stress in terms of the activity of the enzyme nitrate reductase, leaf diffusion resistance and leaf xylem water potential (using a pressure chamber). The results, obtained with one-year-old seedlings that were transferred to a controlled environment immediately before the start of the experiment, showed a linear reduction in nitrate reductase activity as leaf water potentials declined. One cultivar (Catuai) was identified as possibly being more tolerant of drought than another (Nacional). There were no significant differences, though, between cultivars in the sensitivity of stomatal activity to changes in leaf water potential: stomatal conductances always decreased linearly (range 1.0–0.16 mm s^{-1}) as leaf water potentials fell (range -1.0 to -3.0 MPa) with correlation coefficients (r) ranging from 0.69 ($n = 23$) to 0.91 ($n = 26$).

In a detailed, short-term, field comparison of three morphologically (crown architecture) contrasting cultivars (Typica, San Ramon and Yellow Caturra) in Hawaii, Tausend *et al.* (2000) found that sap flow and conductances were greatest in Typica (3.7 m tall) and least in San Ramon (1.5 m), a cultivar that was relatively unresponsive to the withholding of irrigation. Differences between the three cultivars in the relative vulnerability of the xylem to cavitation when they were droughted were also recorded, leading to reductions in hydraulic conductivities. Typica was less susceptible to the loss of hydraulic conductivity than the other two cultivars, which was consistent with its low leaf water potential. In all cases, recovery was rapid on rewatering.

DaMatta (2004a) has briefly reviewed selected traits that may be explored in breeding programmes targeted at selection for tolerance to water stress. These include: (1) leaf area and gas exchange, with the warning to be cautious when screening genotypes by measuring instantaneous gas exchange under drought stress conditions, since a reduced leaf area (fewer, smaller leaves, leaf shedding) may allow net photosynthesis on a unit area basis to be maintained at a constant level, (2) crown architecture; cultivars with open canopies are more susceptible to drought (being 'coupled' to the bulk (dry) air) than those with closed canopies (uncoupled), (3) carbon isotope discrimination; considered to be a useful long-term indicator of water-use-efficiency, (4) water relations; including osmotic adjustment (insufficient evidence of its value in coffee; DaMatta and Ramalho, 2006), relative water content (not sensitive enough), stomatal conductance (an early indicator of water shortage), and hydraulic conductance (implies high rates of water use and should only to be considered for areas with brief periods of water deficit), and (5) root characteristics (despite benefits, too difficult to be used in a selection programme). As DaMatta (2004a) warns, much of the research on drought tolerance has been done on container-grown seedlings in a greenhouse and, as a result, has only limited relevance to field conditions where water stress develops more slowly.

Clearly differences exist between cultivars of both *C. arabica* and *C. canephora* (Montagnon and Leroy, 1993) in their responses to dry soil (and air) conditions; the mechanisms responsible though are less clear. As Jones *et al.* (1985) have pointed out, heterogeneous tree crops are much closer to their wild relatives than most short-term agricultural crops, because of the long-term nature of breeding programmes. For a grower it is the capacity of trees to survive and/or achieve a reasonable yield under adverse conditions that is important.

Summary: plant water relations

1. Stomata are present on the lower surface of coffee leaves (and on green fruits) at densities of 150–330 mm^{-2}.
2. Commercial cultivars have retained characteristics adapted to shady environments.
3. High leaf temperatures (>25°C), acting alone or together with dry air (saturation deficit >1.5 kPa) induce partial or complete stomatal closure during the day, even when the soil is close to field capacity.
4. On sunny days, leaf water potentials (ψ_l) can be as low as -1.5 MPa, even when the soil is wet, or -2.8 to -3.5 MPa if the soil is dry.
5. Photosynthesis and transpiration rates decline when leaf water potentials fall below about -1.0 MPa.
6. Fast rates of transpiration cannot be maintained when the evaporative demand is excessive.
7. Cultivars differ in their susceptibility to drought (Caturra and K7 are, for example, considered to be relatively drought tolerant). There is limited evidence for osmotic adjustment; drought-tolerant mechanisms are not fully understood.
8. Much of the work reported has been done on container-grown plants.

Crop water requirements

Evapotranspiration

Much of the early work on investigating and quantifying the water use of field-grown coffee was carried out in Kenya, close to the equator (1°4′ S; alt. 1610 m), at the Coffee Research Station, Ruiru (Pereira, 1957, 1967; Wallis, 1962, 1963; Blore, 1966). Comparisons were made between measured changes in the soil water content in the 3-m deep profile with evaporation from a tank of water (circular, 1.2 m diameter) either raised above the ground or sunken. Pereira (1957) initially used an empirical factor (f) to convert E_{pan} (sunken) to an estimate of potential coffee water use (ET_c). Its value ranged from 0.8 during the middle of the rains down to 0.5 during the dry season when the leaf area was least. For irrigated crops in the dry season, f was assumed to decline by 0.1 (from 0.8) for each 10-day period without irrigation (or when rainfall $>E_{pan}$) until it reached the value corresponding to that for unirrigated coffee for that month. It was assumed that, within the 3-m deep profile, all the water was available to the crop, and that the transpiration rate declined at the rate of 0.25 mm d^{-1} until the deficit reached 325 mm, when it ceased. If the soil was at field capacity (defined as the water content at 33 kPa suction), excess rain was assumed to drain immediately below the root zone.

Using this model, Wallis (1963) reported the results of 15 comparisons of actual changes in soil water content, made over the period 1950 to 1962. The general level of agreement between computed and measured soil water deficits (SWD) was good for unirrigated crops ($r^2 = 87\%$, range SWD 30–270 mm), but less good for irrigated crops ($r^2 = 65\%$, range SWD 0–115 mm). There was a tendency to overpredict deficits during periods of severe drought. This was thought by Wallis to be due to the false assumption that roots could extract easily all the available water within the 3-m deep profile.

Wallis (1963) showed that measurements of evaporation from a sunken 'Kenyan' pan, in comparison with a raised pan (with a grid), gave estimates of E_o closer to those obtained using the 1948 version of the Penman equation, particularly during periods of high evaporative demand. Blore (1966) developed this work with two further years of measurements of changes in soil water content, at a distance of about one metre from each tree, and derived the following relationship:

$$ET = (0.86 - 0.0033D)E_{pan}$$

$$\left(r^2 = 36\%; P<0.01\right)$$

where ET is the total water use over a 10-day period (mm), D is the mean soil water deficit at the end of this period (mm) and E_{pan} is the corresponding mean total evaporation (mm) from a sunken Kenyan pan.

When soil water was not limiting (at deficits taken to be less than 100 mm; Wormer, 1965) the relationship was simplified to:

$$ET = 0.86E_{pan}.$$

The first equation, when extrapolated, predicts that transpiration will cease when D reaches 270–295 mm. This work was done on clean-weeded (except for a flush of weeds at the start of the rains) coffee spaced at 2.7 × 2.7 m. Blore considered that his model predicted actual water use by an unirrigated crop more precisely than Pereira's method, but because of the smaller range of deficits and the difficulty of quantifying drainage, the estimates for an irrigated crop lacked the same precision. In this location, the rate of water use from irrigated crops ranged from about 2 mm d^{-1} in July and August to 3.8 mm d^{-1} in January to March. Annual water-use totals averaged about 850 mm for an unirrigated crop and 940 mm for an irrigated crop (range 825–1050 mm). This compared with E_{pan} (sunken) values of about 1500 mm, and annual rainfall totals of 1040 mm (Wallis, 1963).

Away from the equator, in Zimbabwe, the recommended method for calculating the water requirements of an irrigated crop has been to use an $ET_c:E_{pan}$ ratio for mature (more than three years old) trees in the range of 0.55 to 0.75. The exact value depended on the method of irrigation used, and whether or not the crop was mulched. E_{pan} in this case refers to evaporation from a standard United States Weather Bureau (USWB) Class A pan (Wilson and Pilditch, 1978). For younger trees, the $ET_c:E_{pan}$ ratio for a mulched crop was taken to be equal to the proportion of the soil shaded by the tree canopy, and could be as low as 0.2 in the first year after planting.

In Zimbabwe and Malawi (11–20° S), growers using drip irrigation were advised to include a 'canopy factor' to allow for the age/size/planting arrangement of trees when estimating irrigation water requirements. Defined as the 'ratio of the canopy area to the planted area' it can range in value, for example, from 0.03, for one-year-old, to 0.62 for five-year-old trees, spaced 3.05 × 2.4 m apart (1366 stations ha^{-1}) with two trees per station (Clowes and Logan, 1985). Thus:

$$ET_c = \left(0.6E_{pan}\right) \times \text{'the canopy factor'}.$$

Subsequently, Logan and Biscoe (1987) have suggested that, in Zimbabwe, the pan coefficient, when used with a USWB Class A evaporation pan, should be 0.8 and 0.7 for mature unmulched and mulched coffee respectively, rather than 0.6. This method was widely used for calculating the irrigation water requirements in southern Africa, although it fails to allow for the effects of, for example, climatic factors on stomatal opening. Hess *et al.* (1998) have also pointed out that this approach seriously underestimates the water use of young coffee since it only takes into account the canopy cover and ignores the effects of crop height and leaf area, and other microclimatic factors, which can influence the water use of widely spaced tree crops.

Detailed measurements of the water balance of a coffee crop have been made in Hawaii (21° 54′ N; alt. 98 m). In this island climate, Gutierrez and Meinzer (1994a) used the Bowen ratio and stem heat balance methods for estimating the actual water use of a commercial crop (cv. Yellow Catuai) at different stages (age related) of canopy development (leaf area index, $L = 1.4$–7.5). When the crop, grown in hedgerows, was well irrigated (drip), latent heat loss was the most important component of the energy balance at all stages of development (*c.* 60% of the net radiation at $L = 6.7$). Evaporation from the soil surface and from inter-row vegetation ranged from 40% of total evapotranspiration at $L = 1.4$ down to nearly zero at $L = 6.7$, when the crop almost completely shaded the soil surface. Interestingly the proportion of net radiation dissipated as latent heat was less in the year when the leaf to air saturation deficit was high (1.68 kPa) than when it was lower (1.35 kPa). This was judged to be the result of associated variations in stomatal conductances. Evaporation from the crop ($L = 6.7$) began to decline within four days of irrigation ceasing but returned to normal two days after it was resumed. This followed 25 days without irrigation. Wilting was observed when the proportion of the net radiation dissipated as latent heat fell from 60% to about 30% of the total.

Crop coefficients (K_c) were calculated using data derived from this study:

$$K_c = ET_c/ET_0$$

where ET_c is transpiration by the coffee trees plus evaporation from the inter-row and ET_0 is potential evapotranspiration calculated for a reference crop, grass or alfalfa, using the Doorenbos and Pruitt (1977) version of the Penman equation.

The weather data were collected from an automatic station. K_c averaged 0.55 ($L = 1.4$), in the second year after planting at spacings of 3.6×0.7 m, and reached 0.68 to 0.82 ($L = 5.4$–6.7) for plants more than two years in the field. In the first year ET_0 rates were in the range 4.4 to 6.6 mm d^{-1}. Interestingly, in the following year, for plants of a similar age ($L = 3.4$–4.2), K_c values were about 30% less (0.45). It was suggested that this was due to the higher ET_0 rates (7.4 mm d^{-1}) experienced in that year compared with the year before (Gutierrez and Meinzer, 1994b). When irrigation ceased, K_c values declined within four to five days, halving in value, from 0.8 to 0.4, after 18 days ($L = 6.7$). At this stage the leaves were visibly wilting ($\psi_l = -2.14$ MPa), but transpiration was continuing at about 30% of its original value, suggesting that substantial gas exchange was still taking place.

In the most recent FAO manual on crop evapotranspiration (Allen *et al.*, 1998) the tabulated K_c values presented for coffee are in the range 0.9–0.95 for a clean-weeded crop, and 1.05–1.10 for a crop with weeds, when using the FAO version of the Penman–Monteith equation to estimate ET_0. These values are for well-managed crops, 2–3 m tall, grown in a sub-humid climate (minimum relative humidity *c.* 45%). If, instead of a single coefficient, K_c is derived from its two constituent components, transpiration and bare soil evaporation, the corresponding values are given as 0.80–0.90 and 0.85–0.90 respectively, assuming a dry soil surface. The validity of these values remains to be seen.

Since then, Marin *et al.* (2005) have reported the results of a similar study to that undertaken by Gutierrez and Meinzer (1994a, 1994b) in Hawaii. They too used the Bowen ratio and sap flow measurements to estimate the water use (ET_c) and transpiration (T) over a five-month period by a five-year-old drip-irrigated crop planted in a hedgerow system (cv. Apoata) in the state of Sao Paulo, Brazil (22° 12′ S 47° 30′ W; alt. 546 m). ET_0 was estimated using the FAO version of the Penman–Monteith equation. The crop coefficient (K_c) had a value of 1.0, while the T/ET_0 ratio averaged 0.76, although there was a suggestion that, when ET_0 values exceeded 4 mm d^{-1}, T remained relatively constant.

When vegetation occupied the inter-row areas, T represented $0.68ET_c$, and when the soil was bare $0.87ET_c$. The linear relationship between ET_c and E_{pan} (USWB Class A) had a slope of 0.61 and for transpiration 0.45. There was again an indication that transpiration levelled out when ET_0 exceeded about 4 mm d^{-1} and E_{pan} was greater than 6 mm d^{-1}, implying that the stomata were controlling water loss when potential evaporation rates were high. DaMatta and Ramalho (2006) refer to the fact that stomatal control of transpiration decreases substantially as the scale of measurement increases from a single leaf to the crop canopy. This uncoupling of the crop from the effects of the air surrounding the canopy can occur as a result of humidification of the air within the canopy (transpiration), or an increase in the boundary layer resistance (low wind speed).

In a review of the ecology of the coffee crop, Barros *et al.* (1995) concluded that our knowledge of the actual water use of coffee crops grown in diverse ways, with contrasting energy balances and aerodynamic profiles, and depending on such factors as location, cultivar, spacing, pruning system, and in some places on the type and density of shade, was very limited. They cited values for the albedo, or reflection coefficient, (0.15 to 0.19) for several cultivars; the roughness lengths (Z_o) for Caturra at two spacings (2.0 × 2.0 m, tree height (Z) = 2.9 m, Z_o = 0.42Z; and 2.0 × 1.0 m, Z = 1.7 m, Z_o = 0.20Z). The corresponding zero planes of displacement (d) were 0.42Z and 0.55Z.

In Brazil, Filho et al. (2010) used the Allen et al. (1998) version of the Penman–Montieth equation with the appropriate crop coefficients to map the spatial and temporal water requirements of coffee in Minas Gerais State (14° 14′ S–22° 55′ S). Meteorological data were obtained from 42 weather stations for periods of up to 17 years. Annual ET_c totals ranged from less than 1000 mm in the south to more than 1400 mm in the north of the state.

Drought mitigation, planting density and shade trees

Before irrigation is considered every attempt should be made to minimise the adverse effects of drought through less expensive and appropriate water conservation practices. Clowes and Logan (1985) have made the following recommendations to growers in Zimbabwe.

- Planting two or more trees at each station (known as the *cova* system) is considered to be more appropriate in drought conditions than the hedgerow system (plants grown with close within-row spacing and wide intra-rows), partly because roots extend deeper in the soil (Figure 5.5).
- Ratooning is also seen as a way of reducing the effects of drought. This should be done on a five to eight year cycle; trees in the middle of this period, with large crops born mainly on primary branches, are considered to be susceptible to drought and to respond to irrigation more than trees at the beginning (new vegetative growth) or end (more higher order branches) of the growth cycle.
- Mulching reduces evaporation from the soil surface, and improves infiltration of water into the soil. However, the effects on yield seem to be additional to those of irrigation, probably through its influence on nutrient availability, aeration and perhaps soil temperature. Mulching is particularly important in young coffee. The importance of mulching has long been recognised in Kenya (Pereira and Jones, 1954; Njoroge and Mwakha, 1985), Tanzania (Robinson and Mitchell, 1964) and in China, where rice straw is recommended as a mulch (Cai et al., 2007).
- Weed control, effective and timely, is also important as a water conservation technique.

These are some of the measures open to growers who are either unable to irrigate or who have insufficient water to cover the whole area adequately. Many of the measures, though, are associated with good cultural practices whether the coffee is irrigated or unirrigated.

In Kenya, Fisher and Browning (1979) found that planting at high densities, over the range 5000 to 20 000 ha^{-1}, did not increase the susceptibility of individual

Figure 5.5 Coffee. Drought mitigation measures include tied-ridges, mulching and planting two plants at each station (cova system) – Malawi (MKVC). (See also colour plate.)

plants to drought, and may even have alleviated it, at least up to the highest practical density (8000 ha^{-1}). They based these conclusions on measurements of leaf water potential, stomatal opening (infiltration score), rates of extension of lateral branches and the number of nodes. Similarly, Gathaara and Kiara (1984) found no evidence, based on measurements of leaf water potentials, that increasing tree density from 1322 to 6610 ha^{-1} (cv. SL28) increased plant water stress in dry weather. These, perhaps unexpected, responses were explained on the basis that mutual shelter must have reduced the evaporative demand within the crop canopy and, as a result, water use per unit of land area did not increase with density. Factors other than concern about drought susceptibility will therefore contribute to decisions about the optimum planting density for any location.

Coffea arabica is a shade-adapted species (or according to DaMatta, 2004b, shade-tolerant), but the effects of shade trees on the physiology, water use and yield of this crop, and others, are complex (Willey, 1975). In terms of crop water relations the presence of shade trees can: (a) reduce the incident solar radiation; (b) reduce maximum air temperatures (e.g. by 5.5°C) and (c) increase minimum air temperatures (by 1.5°C); (d) reduce the saturation deficit of the air (by 0.2 kPa); (e) change the aerodynamic roughness of the cropped area; (*f*) reduce evaporation (by 40% from a Piche evaporimeter); and (*f*) modify the interception and throughfall of rainfall. The values cited are those measured by Barradas and Fanjul (1986) in a study comparing the microclimate in a heavily shaded (*Inga jinicuil*) crop with a similar unshaded area in central Veracruz state, Mexico ($19°31'$ N; alt.1225 m). The net effect on crop water use and water-use efficiency (taking into account water use by the shade trees themselves) will depend a great deal on the local situation and as yet there is no predictive model to enable this relationship to be quantified in a useful way. Shade trees do of course influence the ecological conditions, soil and microclimate, in which crops grow in many diverse ways (Huxley, 1999). For example, in Kenya, they are used to reduce weed growth in coffee where the costs of weeding are high, and to elevate night temperatures at high elevations in order to reduce chilling damage. The issue of shade trees in coffee has been the subject of a detailed review by DaMatta (2004b), who concluded by stating that, as a rule, the benefits of shading increase as the environment becomes less favourable for coffee cultivation. Van der Vossen (2005) has also usefully summarised the positive and negative effects of shade trees in coffee, particularly in the context of organic production.

In support of this practice, Baggio *et al.* (1997) and Caramori *et al.* (1996) have shown in Brazil how shade trees (*Grevillea robusta* and *Mimosa scabrella* Benth.) can reduce damage caused by radiation frosts, a major and unpredictable cause of yield loss when minimum air temperatures fall to -3 to $-4°C$. Frost protection in these areas ($20–24°$S) with sprinkler irrigation at night, as practised in some temperate fruit orchards, is not considered to be economic. Cultivars differ to a limited extent in their susceptibility to frost damage (Filho *et al.*, 1986).

Summary: crop water requirements

1. Our understanding of the actual water use of coffee crops grown in diverse ways is imperfect.
2. Present methods of estimating crop water requirements for the purposes of irrigation scheme design and management are imprecise, and probably subject to large errors depending on local circumstances and the system of irrigation used.
3. For a mature crop well supplied with water, K_c has a value of about 1.0 ET_0, or 0.7–0.8E_{pan} (Class A).
4. There are indications that if ET_0 is more than about 4 mm d^{-1} transpiration remains constant.
5. A range of drought mitigation practices has been identified, of which mulching is perhaps the best option.

Water productivity

In view of the importance of irrigation in commercial coffee production (in Brazil alone it is estimated that about 200 000 ha are irrigated, 10% of the total coffee area), it is perhaps surprising that there have been few experiments reported which quantify, with precision, the yield, quality and financial benefits that can be derived, and the parameters by which irrigation systems should be designed and managed. Assessing the irrigation need of coffee is not straightforward; in particular there are two principal ecological areas to consider (a) equatorial, with two short (sometimes unreliable) rainy seasons, and (b) sites away from the equator with single rainy seasons and seasonal climates. There are also the maritime climates of Hawaii, and how altitude affects air temperature and potential evaporation rates.

Bimodal rainfall areas

Wallis (1963) reported the results of experiments carried out at Ruiru, close to the equator, in Kenya between 1957 and 1961. Excluding the 1958 data, a total of 1900 mm of irrigation water was applied in the four-year period. This led to a total yield increase of 370 kg ha^{-1} (or 12%) of clean 'beans' (seeds), an average response of 0.77 kg ha^{-1} per mm of water applied. The biggest effect though was on the yield of large, grade 'A' beans which was increased by over 60% (see also Cannell, 1974). These experiments were conducted on conventionally spaced trees (1300 ha^{-1}). Yields from the unirrigated crop averaged 1.98 t ha^{-1}.

Previously, attempts had been made in northern Tanzania (3° 30′ S; alt.1370 m) to quantify the responses to irrigation from an experiment that began in 1940 and continued until 1955. Pereira (1963) highlighted the difficulties in interpreting the data when, for much of the period, there was no satisfactory way of estimating crop water requirements using sound scientific principles. For the first 10 years of the experiment there were cumulative yield benefits from irrigation, despite its arbitrary nature. For the next eight years there were no consistent benefits as a result, it is claimed, of unsound irrigation practices. Mulching (banana trash), though, was always beneficial (Robinson and Mitchell, 1964). Irrigation experiments are difficult to manage with any crop, but particularly with coffee when there is uncertainty about actual rates of water use and the role that water plays in controlling the time and duration of flowering.

In a later experiment in Kenya, the application of 400 mm of water increased red fruit (cherry) yield (cv. SL28) by 1.4 kg tree^{-1} in one year (Cannell, 1973). For a planting density of 2250 trees ha^{-1}, this is equivalent to 3150 kg ha^{-1}. Assuming a weight ratio of fresh fruit to sun-dried (c. 11% water content) clean green beans of 6:1 (a value that appears to remain relatively constant regardless of treatment), this represents a (bean) yield increase of 1.3 kg ha^{-1} for each millimetre of water applied, nearly double the figures derived from the data reported by Wallis (1963). The corresponding base yield (without irrigation) of clean beans was about 3.7 t ha^{-1}. These results apply to crops irrigated with over-tree sprinklers that wet the whole land area.

Subsequently, Gathaara and Kiara (1988) reported the results of a three-year (1984 to 1987) irrigation rate × frequency experiment, also at Ruiru in Kenya. Although the results were incomplete, irrigation during both dry seasons appeared to increase annual yields of clean beans (cv. French Mission, planted at a density of 1333 trees ha^{-1}), but by only about 9% (or 167 kg ha^{-1} year^{-1}) over the three years. There was a larger effect on the proportion of grade 'A' beans, which increased from 30% to 43% of the total seed weight. There were no significant effects of irrigation depth (38, 76 and 100 mm) or frequency (intervals of 14, 21 and 28 days) on bean yields that averaged about 2.1 t ha^{-1}.

Planting densities now range from 2600 to 5000 ha^{-1}, thereby influencing the responses to drought and hence to irrigation through effects on the size and shape of the crop canopy, and the depth and distribution of roots. Thus, Kiara and Stolzy (1986) reported that in the first year after tree establishment, at Ruiru in Kenya, drip irrigation increased yields (cv. SL28) from about 1200 to 1500 kg ha^{-1} at a low density (1322 ha^{-1}, but with two stems per tree). At a high density (5288 ha^{-1}) the absolute yield increase was larger than this (from 4100 to 5100 kg ha^{-1}), but there were no further yield benefits at the next highest density (6610 ha^{-1}). In the following year there was a smaller effect of density on clean bean yield, from 3100 to 3900 kg ha^{-1} for the irrigated crop, and from about 2200 to 2700 kg ha^{-1} for the rain-fed crop. Unfortunately insufficient information was given in the paper to allow the results to be interpreted with more precision than this. A previous paper, however, had reported how irrigation and tree density had influenced certain components of yield (Gathaara and Kiara, 1985). Irrigation (and high densities) increased vegetative growth but reduced the number of nodes and the number of fruits per node and per primary branch, especially in the lower parts of the canopy. This was thought to be the result of a reduction in the transmission of light within the canopy. The yield benefits resulting from irriga-tion in Kenya must therefore be associated with the production of large fruits and seeds (Cannell, 1974). Fisher and Browning (1979) had previously suggested that irrigation in Kenya should be withheld for a period during each dry season, in order to reduce vegetative growth and/or to enhance flowering so that the yield benefits could be maximised, but reducing stem extension could limit the potential crop in the following year. Their suggestion does not yet appear to have been evaluated in experiments.

On the basis of the results of the early experiments by Wallis (1963) growers in Kenya were advised to irrigate crops, rooting to depths of about 3 m in deep clay loam soils at normal spacing, when the soil water deficit reached 150 mm, and then to apply 100 mm (net) of water. A deficit of 150 mm represents a depletion of about 50% of the available water in the root zone. This means that the irrigation interval was about 40 days from June to August, 25 days from September to October and 20 days from December to March (assuming no rain). These were the recommendations for overhead sprinkler systems. Total net annual irrigation water requirements ranged from 140 to 800 mm (mean 400 mm) depending on the season. Later, following the results of the experiments described by Gathaara and Kiara (1988), it was concluded that 38 mm applied every 21 days was a realistic practical schedule to follow. This compared with the earlier recommenda-tions for growers to apply 76 mm at monthly intervals. Akunda and Kumar (1981) have suggested that the timing of irrigation of coffee could be based on the time taken for a dry cobalt chloride paper disc to change in colour from blue to pink, when attached to the abaxial surface of a leaf. If this exceeded four or five minutes during the middle of sunny days in Kenya, corresponding to leaf water potentials of about -2.0 MPa, then irrigation should be applied. Subsequently, Venkataramanan et al. (1998) have evaluated this technique in South India using seedlings and field-grown plants of C. canephora. They concluded that the critical leaf water potential was similar, about -1.9 MPa, which corresponded to 7.5 minutes for the colour change to occur.

Unimodal rainfall areas

South of the equator in Africa ($>8°$ S), by comparison, there is a single rainy season from November/December to March/April followed by a long dry season which is initially cool but afterwards, depending on the altitude, can be hot. The importance of these differences is illustrated by the following analysis by Wilson and Pilditch (1978). At Chipinge, in the Eastern Highlands of Zimbabwe ($20°13'$ S; alt. 1130 m) irrigation is deliberately withheld for up to two months towards the end of the dry season (peak $E_{pan} = 6–7$ mm d^{-1}) to break flower bud dormancy. Flowering is then stimulated by the onset of the rains. The exact duration of the dry period depends on the depth and type of soil. By contrast at Karoi, in north-west Mashonaland ($16°48'$ S; alt. 1280 m) peak evaporation rates can reach 9–10 mm d^{-1} in the same two months (September and October), and these hot dry conditions, before the onset of the rains, alone provide sufficient internal water stress to break flower bud dormancy. The increase in atmospheric humidity, or cooling, at the start of the rains then provides the stimulus to induce flowering. Figure 5.6 attempts to highlight these differences between locations close to and away from the equator in relation to stages of yield development and irrigation need.

On water-retentive soils (available water capacity >140 mm m^{-1}) deeper than one metre, the recommendation is to apply a net 50 mm of water at Chipinge, or 65 mm at Karoi, at the equivalent soil water deficits. Net average annual irrigation water requirements range from 380 mm in Chipinge to 740 mm in Karoi, for crops irrigated with over-tree sprinklers, with peak net monthly requirements of 75 mm and 135 mm respectively, and minimum irrigation intervals of 12–14 days (Wilson and Pilditch, 1978). On shallow or sandy soils, the recommendation is to apply 45 mm net at 10-day intervals. It is not clear on what experimental evidence these recommendations are based. There appear to be no published data describing the yield benefits to be derived from irrigation for these locations.

In the Australian states of Queensland ($17°$ S; alt. 15–700 m) and northern New South Wales ($28°$ S; alt. sea level), where coffee is largely machine harvested, the recommendations (Drinnan, 1995) are to irrigate trees regularly, at three to seven day intervals, during the periods of floral initiation and development (stages one to four), ensuring that there are no large fluctuations in plant water status during stage four as the flower buds approach maturity. Water stress is then applied by withholding irrigation until the first signs of wilting and the leaf water potential

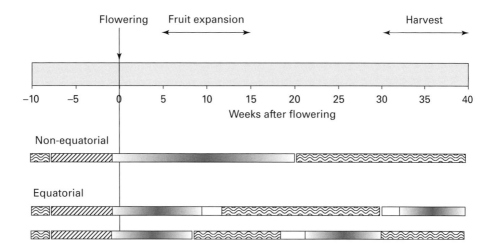

Figure 5.6 Coffee. A diagram illustrating the stages of yield development in coffee in Africa after flowering has been stimulated by the start of the rains, following a period of water stress, in non-equatorial and equatorial areas, either at the start of the 'long' rains (upper line) or the 'short' rains (lower line). The extended dry seasons when irrigation may be beneficial are indicated: it may also be necessary to supplement rainfall with irrigation during the periods of fruit expansion. (Redrawn from Carr, 2001.)

declines to −2.5 MPa (pressure bomb). This level of stress is then maintained for three to four weeks by daily (morning) applications of small quantities of water (1–2 l tree^{-1}, drip irrigation) in order to prevent tree damage and leaf fall. After eight to ten weeks of water stress the soil profile is then brought back to field capacity as quickly as possible, to encourage rapid uniform flowering. If rain interrupts this sequence:

- early in the cycle, that is during the first three weeks, the advice is to continue to impose stress even though some flowering will occur;
- in mid-cycle, four to six weeks after stress was initiated, the advice is to wet the top 0.15–0.20 m of soil if the flower buds are observed to begin to enlarge within two to three days of the rain; this will allow these flower buds to develop normally; water stress is then reimposed to trigger flowering of the remaining buds at the specified time; and
- late in the cycle, seven to ten weeks after stress has been imposed, the advice is to bring the soil back to field capacity immediately.

The aim in this region is to synchronise flowering so that it occurs over the shortest possible period of time in order to ensure that a large proportion of the total crop can be machine harvested, preferably in a single pass. This may require a series of drought cycles if rain falls during the period of stress. Supplementary irrigation is also applied, as needed, during the period of rapid fruit expansion.

Summary: water productivity

1. The need for irrigation, and its role in controlling the timing of flowering, varies depending on the rainfall distribution, and the severity of the dry season.
2. Two geographic areas, in particular, need to be distinguished: those close to the equator with a bimodal rainfall pattern and those at higher latitudes with a single rainy season and a single, extended dry season.
3. Yield responses to irrigation, and the associated financial benefits, have not yet been adequately described or quantified in ways that are useful to planners and others in either location.
4. Allowable soil water deficits have only been specified for a few situations, usually linked to conventional sprinkler systems of irrigation.

Irrigation systems

Because irrigation is largely supplementary, and the topography of coffee areas is uneven, conventional surface or flood irrigation is not commonly practised, but there is still a wide choice of irrigation methods suitable for coffee (Pilditch and Wilson, 1978). They can be divided into two groups:

- over-tree sprinkler systems which can be fully portable, semi-portable or fixed (a so-called solid set system), and
- under-tree systems including minisprinkler, microjet, drip and basin irrigation (Figure 5.7).

All these methods are currently used in coffee and each has its relative advantages and disadvantages in terms of: (a) capital cost, (b) labour requirements, (c) operating costs, (d) water application efficiency, (e) ease of operation and maintenance, and (f) flexibility to adjust for trees increasing in size. The choice of method will vary depending on the requirements and constraints (e.g. capital, labour, topography, water availability and quality, mechanisation, and technical support by manufacturers and suppliers) of individual producers. In this section the main criteria influencing the choice of system and its design, with particular reference to coffee, are considered.

Over-tree systems

These are based on conventional medium pressure (0.3–0.4 MPa) sprinklers spaced at, say, 9×12 m on tall risers. The main and lateral lines can either be portable or fixed. Using such a system, the whole land area will be irrigated,

Figure 5.7 Coffee. Young plants being established with 'water hose and basin' irrigation – Malawi (MKVC).

including the access paths and inter-row areas. It has been estimated in Columbia that about 5 mm of water are needed to saturate a coffee tree (3.5 years old, single stem, spaced 1 × 2 m) and to wet the lower surface of the leaves (Guzman and Gomaz, 1987). This depth of water is therefore needed, before water reaches the soil surface, each time the crop is irrigated. The crop is likely to remain wet for four to 17 hours after irrigation ceases, depending on the time of day when irrigation begins (at 0500–1700 h respectively). The operating efficiency of sprinkler systems of this type can be as high as 80–85%, depending on wind speed and direction, but with a crop of widely spaced rows like coffee, it is difficult to be precise. Over-tree systems allow chemicals to be washed off leaves, and associated cooling can induce flowering. Centre-pivot and linear-move irrigators have also been used successfully in coffee, for example in Zambia and Queensland respectively.

Under-tree systems

It is often more sensible to apply water directly to the soil below the tree, and to minimise the amount of water applied to the inter-rows, which only encourages weed growth. Small under-tree sprinklers mounted on skids can be attached to a long flexible small-diameter PVC pipe. The pipe is pulled to a new position

Figure 5.8 Coffee. Drip irrigation allows precise quantities of water to be applied at frequent intervals – Kawalazi Estate, Malawi (MKVC)

(usually at intervals of 6 m) from the end of a row at fixed times, and usually operates in alternate pathways at a 6 × 6 m spacing. Although low in labour requirements, this system still applies some water to the inter-row areas, and application rates can be high.

Microjets are an alternative system in which small, low-pressure sprinklers or microjets are attached directly to a small-diameter PVC pipe running down each row. Depending on tree spacing, these jets, mounted on their own plastic stand, are spaced at intervals of about 2 to 2.5 m. The wetted diameter is about 3 to 4 m, and application rates are about 14 mm h^{-1}. One sprinkler may serve more than one tree. With tree crops it is not necessarily important to wet all the rooting volume of soil, providing roots have access to sufficient water within part of the root zone. This will require careful checking in the field. Each row is irrigated in sequence. The lateral pipes can be moved from one row to another if necessary, but there is increased risk of damage to the irrigation system and perhaps to the trees.

Drip irrigation has become popular for coffee in recent years (Figure 5.8). This method allows precise quantities of water to be applied at frequent, but as yet unspecified, intervals. Providing the system is well designed, with good filtration, it is ideal for coffee and has the advantage that, if necessary, nutrients can be mixed with the water. The system can also be adjusted, by increasing the size or number of the emitters, to apply more water, as the trees get bigger. However, the critical issue with drip (or trickle) irrigation is the need to specify what proportion of the root zone needs to be wetted, and to ensure that the water spreads laterally in the soil. On sandy soils, there is a risk that the water will move vertically below the depth of roots with limited lateral spread: in such cases, microjets are likely to be more appropriate. The lateral line with emitters can either be placed on the soil surface or, if there is a risk of damage during weeding, for example, or to make it easier to see that all the emitters are working, the lateral pipe can be fixed about a metre above the ground for easy viewing and maintenance. Emitters are usually spaced between 0.7 to 1.0 m apart along the line, and apply water at rates of 2 or 4 $l\,h^{-1}$.

Although systems should be designed to match the crop water requirements, no detailed experiments have been reported to specify the optimum design or operating conditions for drip irrigation, including fertigation, of coffee. In Zimbabwe, the recommended minimum irrigation interval is about six days. Since coffee is still grown in areas with high rainfall, roots will extend throughout the profile during the rainy season, but it is possible that there will be a concentration of feeder roots within the volume of soil wetted by water from each emitter, especially if fertiliser is applied with the water. Gathaara et al. (1993), Azizuddin et al. (1994) and Ram et al. (1992) have reported the results of empirical studies of the use of drip irrigation in coffee in Kenya and in India. In the latter paper, the test crop was C. canephora and pitcher irrigation (buried earthenware pots adjacent to each tree that are filled with water), and microsprinklers were also evaluated.

Crops irrigated with drip systems still use the same amount of water as crops irrigated in other ways. The saving in water comes from more precise applications, providing the system is managed well enough to ensure, for example, that delivery from individual emitters meets the design specifications. This can be difficult in practice. If the soil surface is compact it may be necessary to build a small basin beneath each emitter to allow the water time to infiltrate and to prevent run-off. If cost is a major constraint, it is possible to use one lateral line to irrigate several rows by moving it from one row to the next, the so-called 'dragline' system. This method is especially useful in young coffee.

Basin irrigation is a traditional method. Water can be applied from a ditch, a hosepipe connected to a supply line, or from water containers carried into the field. Basins are formed around each tree or group of trees and a measured volume or depth of water applied to each basin. The system can be adapted to supply more than one tree at a time, using multiple outlets. Rates of application from hosepipes vary between 4 and 20 $l\,min^{-1}$. This is a low cost and efficient way of applying water, but again it is necessary to ensure that an adequate volume of soil is wetted within the root zone.

Other opportunities exist to conserve water. For example, in Ethiopia, Tesfaye et al. (2008) demonstrated how partial root-zone drying (in which two halves of a divided root system were alternately wetted and dried) increased the water-use efficiency of young seedling coffee plants, although nearly all aspects of vegetative growth were reduced relative to the well-watered control treatment. The water-use efficiency of this treatment was also better than the conventional deficit irrigation treatment, both of which received only half the quantity of water applied to the control treatment. The root:shoot ratio was highest in the partial root-zone drying treatment.

Summary: irrigation systems

1. As with nearly all crops, no single method of irrigation is necessarily better than another: each can be made to work well. The choice depends on the particular conditions, resources and the priorities of individual growers.
2. There is little published information, based on sound experimental work, on how to design and operate advanced drip or microjet irrigation systems to best advantage.

Conclusions

Water availability plays a dominant role in many aspects of the growth, development and yield of the coffee crop. Despite the increasing international importance of irrigation in commercial coffee production, many of the relationships have not been quantified in commercially useful ways. This is especially surprising in view of the detailed understanding of many aspects of the water relations of the coffee plant, particularly the mechanisms controlling the development of flower buds.

In order to interpret and to apply this scientific understanding of the role that water plays in the growth and development of the coffee plant to provide practical advice that can assist the grower to plan and to use water, whether rainfall or irrigation, effectively for the production of reliable, high-quality crops, there is a need for further research.

- Well-designed and managed field experiments should be conducted, over a range of typical sites, to quantify the yield responses of coffee to water. These are likely to vary with the cultivar, planting density and soil nutrient status.
- Adequate supporting measurements (crop, soil and prevailing weather conditions) must be taken to allow the results to be interpreted sensibly, and applied with confidence to other locations where the climate and soils may be different.
- Our understanding of the physiology of the coffee plant is such that the development of a process-based model should perhaps precede the experiments in order to set the parameters for the field research and to prioritise the measurements to be taken. For example, it appears that flowering can still be stimulated if only part of the root system is kept dry (Crisosto et al., 1992).
- Linked to this is the need to develop further our understanding of the factors influencing the actual rates of water use of coffee, building on the work of Gutierrez and Meinzer (1994a) and Marin et al. (2005) in Hawaii.
- The design and operating criteria for drip, and other appropriate, irrigation systems need to be specified with precision in order to optimise crop-yield/water-use efficiencies.
- Methods of drought mitigation need to be investigated further, including selecting genotypes for drought tolerance.

- By linking the outputs from this research to a geographic information system, a method for assessing the benefits of irrigation, in crop and financial terms could be developed, and used to justify investments in specific locations and farming systems.

The notorious lack of stability in world coffee prices is often caused by fluctuations, real or imaginary, in the predicted production levels. The profitability of coffee growing changes accordingly, and this impacts directly on the economies of individual countries, as well as on the livelihoods of the people involved in its production, including smallholders and the employees of large-scale commercial producers. Rainfall variability is one of the principal contributing factors to this instability, but we are still unable to quantify the effects of drought with precision, or to recommend with confidence where and when irrigation is worthwhile, and how it is best practised. For a commodity crop of such international commercial importance as coffee, and with increasing demands on all fresh water supplies, this is not good enough.

Summary

The results of research on the water relations and irrigation needs of coffee (*Coffea arabica* L.) are collated and summarised in an attempt to link fundamental studies on crop physiology to irrigation practices. Backgound information on the centres of production of coffee is followed by reviews of (1) crop development, including roots; (2) plant water relations; (3) crop water requirements; (4) water productivity; and (5) irrigation systems. Although there are differences in their responses to drought, commercial cultivars have retained many of the characteristics adapted to the shady environment of the forests in the Ethiopian highlands in which *C. arabica* is believed to have originated. These include partial closure of the stomata, even if the soil is at field capacity, when evaporation rates are high as a result of large leaf-to-air saturation deficits (>1.5 kPa). This is thought to be an adaptive mechanism that minimises transpiration at high irradiances when the leaves are light saturated.

A period of water stress, induced either by dry soil or dry air, is needed to prepare flower buds for blossoming, which is then stimulated by rain or irrigation. Although attempts have been made to quantify the intensity and duration of stress required, these have not yet been specified in ways that are commercially useful. Water must be freely available during the period of rapid fruit expansion to ensure large, high-quality seed yields. Depending on the time and uniformity of flowering this can occur at times when rainfall is unreliable, particularly in equatorial areas.

Our understanding of the actual water use of coffee crops grown in diverse ways is imperfect. For mature crops, well supplied with water, the crop coefficient (K_c) appears to have a value $1.0 \times ET_0$ or 0.7–0.8 times the evaporation from a US Weather Bureau Class A pan. There is some evidence that transpiration remains stable when evaporation rates are high ($ET_0 >4$ mm d^{-1}). For immature crops allowance has to be made for the proportion of the ground area shaded by the leaf canopy, but this alone may underestimate rates of water use. Present methods of calculating crop water requirements for the purposes of irrigation scheme design and management are imprecise, and probably subject to large errors depending on local circumstances.

The need for irrigation, and its role in controlling the timing of flowering, varies depending on the rainfall distribution, the severity of the dry season, and soil type and depth. Two geographic areas need to be distinguished in particular: those close to the equator with a bimodal rainfall pattern and those at higher latitudes with a single rainy season and an extended dry season. Despite the international importance of irrigation in the production of the coffee crop, the benefits to be derived from irrigation in yield and financial terms, in either location, have not been adequately quantified. Allowable soil water deficits have been specified for deep rooting crops (2–3 m) on water-retentive soils, usually linked to conventional over-tree sprinkler irrigation systems. Other, potentially more efficient, methods of irrigation are now available for coffee growers to use, in particular microjets and drip. However, there appears to be little advice, based on sound experimental work, on how to design and operate a drip-irrigation system, for example, to best advantage.

There is a need to interpret and apply the scientific understanding of the role that water plays in the growth and development of the coffee plant to provide practical advice that can assist the grower to plan and to use water efficiently, whether rainfall or irrigation, for the production of reliable, high-quality crops. Future research opportunities are identified.

Figure 1.1 Irrigated coffee estate – Kilimanjaro Region, northern Tanzania (MKVC).

Figure 1.3 A well-managed irrigated tea estate surrounded by indigenous forest – Mufindi, Tanzania (MKVC).

Figure 2.1 Banana. In the tropics, bananas are grown for subsistence and as a cash crop by smallholders, here alongside maize and tea – southern Tanzania (MKVC).

Figure 2.3 Banana. The floral phase is considered to extend from floral initiation to inflorescence (bunch) emergence from the throat of the pseudostem (shown here), the start of the fruiting phase – cv. Prata Anã (AAB Prata subgroup) – Espirito Santo state, Brazil (LM).

Figure 2.4 Banana. A fruit bunch contains a cluster of fruits at a node (known as a hand), which in turn is composed of individual fruits (or fingers) – cv. Prata Anã (AAB) – Espirito Santo state, Brazil (LM).

Figure 3.3 Cocoa. Flowers form in meristematic tissues located above leaf scars on the main stem and woody branches – Bahia, Brazil (AJD).

Figure 3.4 Cocoa. Fruits remain attached to the tree until harvested – Bahia, Brazil (AJD).

Figure 4.1 Coconut. The 'tree of life' is the source of many different products – Kerala, India (MKVC).

Figure 4.2 Coconut. Irrigated Green Dwarfs – Brazil (LM).

Figure 4.5 Coconut. An inflorescence is initiated up to 44 months before the fruit is harvested – Tanzania (MKVC).

Figure 4.8 Coconut. Drought symptoms; petiole breakage of the lower fronds of a mother palm in a seed garden (cv. Malayan Yellow Dwarf) – Gunung Batin, Lampung, Indonesia (XB).

Figure 5.1 Coffee. New estate planted on the contour; note the run-off control measures – Roscommon Estate, Zimbabwe (MKVC).

Figure 5.3 Coffee. Fruits ripen, turning red, about 30–35 weeks after flowering – Malawi (MKVC).

Figure 5.5 Coffee. Drought mitigation measures include tied-ridges, mulching and planting two plants at each station (cova system) – Malawi (MKVC).

Figure 6.1 Oil palm. Aerial view of 'undercropped' oil palm – Univanich, Thailand.

Figure 6.2 Oil palm. Each palm has a single stem with one apical vegetative growing point. A mature tree (>8 years) can produce 20–25 leaves (fronds) each year – New Britain Palm Oil Ltd, Papua New Guinea (RHVC).

Figure 6.4 Oil palm. Individual fruits are tightly packed in large ovoid bunches – Univanich, Thailand (RHVC).

Figure 6.5 Oil palm. The two principal components of yield are bunch number and bunch weight – Univanich, Thailand (RHVC).

Figure 7.2 Rubber. It is now predominantly a smallholder crop – Java, Indonesia (MKVC).

Figure 7.3 Rubber. When the tree is five to seven years old the latex can be harvested (or 'tapped') at regular intervals by cutting a spiral groove in the bark and draining the latex into a collecting cup until it begins to coagulate and the flow ceases – Kerala, South India (MKVC).

Figure 8.1 Sisal. A source of coarse leaf fibres used by industry in the manufacture of twines and ropes, carpet-backing, bags and matting – Tanzania (MKVC).

Figure 8.2 Sisal. It is still predominantly a large-scale estate crop in Tanzania (MKVC).

Figure 9.1 Sugar cane. Irrigated estates – Swaziland (MKVC).

Figure 9.4 Sugar cane. Prior to harvest and after 'drying off' the cane may be burnt to reduce the amount of leaf 'trash' – Swaziland (MKVC).

Figure 9.5 Sugar cane. Priming syphons for furrow irrigation – Swaziland (MKVC).

Figure 10.1 Tea. In Africa away from the equator (e.g. in southern Tanzania, Malawi and countries to the south) there is a single dry season that can last from four to six months. In these areas, tea is often irrigated during the dry season – Malawi (MKVC).

Figure 10.12 Tea. In general, the sensitivity of tea to drought declines six or seven years after planting – southern Malawi (MKVC).

Figure 10.13 Tea. The shape of the annual yield/water-use or irrigation response functions changes with time from planting, and varies between clones (experiment N10) – Ngwazi Tea Research Station, Tanzania (MKVC).

Figure 10.19 Tea. Locally designed and built mechanical harvesters operating in Mufindi, southern Tanzania. The circular brown patches are caused by uneven water applications, the result of operating sprinklers at low pressures (MKVC).

Figure 10.20 Tea. A planned new tea estate will be irrigated with water applied through centre-pivot irrigators – southern Tanzania (MKVC).

Figure 10.21 Tea. Smallholder tea farms – Mufindi, Tanzania (MKVC).

Figure 11.1 A selection of weather stations of variable quality from around the world. They provide the data on which so much depend (MKVC).

6   Oil palm

Introduction

The centre of origin of the oil palm (*Elaeis guineensis* Jacq.) is the tropical rainforest of West Africa where its natural habitat is believed to be in swamps and along river banks. The main economic product is palm oil obtained from the mesocarp in the fruit. Palm kernel oil, which has a different fatty acid composition, is produced in smaller quantities. Kernel cake, produced after the kernel oil has been extracted, is used in animal feeds. Oil palm is the highest yielding oil crop (t ha^{-1}) in the world. Traditionally, palm oil was used in soap, margarine and cooking fat, but is now largely used in food products. It has also become the source of more diverse materials (Corley and Tinker, 2003). When grown as a smallholder crop, the oil palm contributes much more than oil to the communities living in the areas where it is cultivated. For, example, the sap can provide raw material for sugar and alcoholic beverages. The palms are also a source of building materials, and some of the tissues are important sources of fibre.

The oil palm has long been important to village economies throughout its area of distribution in the forests of West Africa, where semi-wild palm groves became established around homesteads (Nouy *et al.*, 1999). The export of palm oil and kernels from Africa grew rapidly in the late nineteenth century, while the first large-scale plantations in Malaysia and Indonesia were established in the early years of the twentieth century. These were followed in the 1920s by plantations in the Congo and later in other parts of West Africa (Corley, 1976a). It is now grown between latitudes 19° N (Dominican Republic) and 15° S (Brazil). In 2009, the two largest palm oil producers by far were Indonesia (85 million t fresh fruit bunches from 5 million ha) and Malaysia (83 million t from 3.9 million ha). This represents a substantial proportion of the estimated total global area of 14.7 million ha. In terms of fruit production these two countries produce 81% of the world total with fruit bunch yields averaging 17–21 t ha^{-1} (containing 3.4–4.2 t ha^{-1} of mesocarp oil) (FAO, 2011a).

The oil palm has therefore been taken from its natural forest habitat in West Africa to become a large-scale commercial tree crop now centred in South-east Asia (Figure 6.1). This is an example of a transfer process that is continuing to this day as the crop moves away from the humid tropics into drier regions including those found in Thailand, and parts of West Africa and South America, where

Figure 6.1 Oil palm. Aerial view of 'undercropped' oil palm – Univanich, Thailand. (See also colour plate.)

there are regular dry seasons and water stress becomes an important yield-limiting factor (Gerritsma and Wessel, 1997).

The climatic conditions prevailing in the highest yielding regions (Indonesia and Malaysia) can be summarised as follows (Hartley, 1988; Lim *et al.*, 2008):

- Annual rainfall of at least 2000 mm spread evenly during the year.
- Mean maximum and minimum air temperatures of 29–33 °C and 22–24 °C respectively.
- Relative humidity >85% (equivalent to a saturation deficit <0.6 kPa).
- Bright sunshine averaging 5 h d^{-1} throughout the year rising to 7 h d^{-1} in some months (or solar radiation of 16–17 MJ m^{-2} d^{-1}).

The books by Hartley (1988) and Corley and Tinker (2003) include tables summarising temperature, sunshine and rainfall data for some of the main oil palm growing centres, together with typical yields.

The international importance of the oil palm as a source of vegetable oil is a relatively recent phenomenon, mainly the result of Malaysian enterprise and initiative over the last 30–40 years, not only in establishing the crop but also through progressive research support to the developing industry by both the state and, particularly, the private sector. Oil palm production is dominated globally by the large-scale plantation sector, with an estimated 80% of oil production coming from estates.

Figure 6.2 Oil palm. Each palm has a single stem with one apical vegetative growing point. A mature tree (>8 years) can produce 20–25 leaves (fronds) each year – New Britain Palm Oil Ltd, Papua New Guinea (RHVC). (See also colour plate.)

A great deal of research has been reported over the same time period on the water relations of oil palm, and more recently on the irrigation need. Various aspects of this topic have previously been reviewed by Corley (1996, irrigation); Lim *et al.* (2008, climate); Henson (2009a, ecophysiology); and Corley and Tinker (2003, all aspects).

Crop development

The oil palm has a single stem with a single apical vegetative growing point (meristem) (Figure 6.2). It can eventually grow to a height of 30 m or more, although old palms become difficult to harvest and are usually replaced in commercial plantations once they start to exceed 12 m (Figure 6.3). In a mature palm, the meristem produces a new leaf primordium about every two weeks. This leaf then takes about two years to develop and for the leaflets to unfold in the centre of the palm crown. In each leaf axil there is an inflorescence primordium. These develop into separate male and female inflorescences. Following pollination, the female inflorescence develops into a fruit bunch from which mesocarp and kernel oil can be extracted (Corley and Gray, 1976a).

Figure 6.3 Oil palm. Although palms can grow to a height of 30 m or more, they are usually replaced in commercial plantations once they start to exceed 12 m, when they become difficult to harvest – New Britain Palm Oil Ltd., Papua New Guinea (RHVC).

Canopy

A mature oil palm (>eight years) can produce 20–25 leaves (fronds) each year (more when younger). These may live for two years after emergence, remaining photosynthetically active for at least 21 months (Corley, 1985). The crown consists of 40–50 open fronds, each up to 5 m in length, with another 50–60 in various stages of development. During the dry season, leaflet opening is delayed and the rate of spear leaf extension is reduced (see e.g. Henson, 1991b; Henson *et al.*, 2005). Although new leaves continue to be initiated and to elongate slowly, they accumulate unopened as 'spear' leaves. When the rains start, there is a flush of leaf opening followed by a return to the normal pattern (Corley and Gray, 1976a). As the fronds age, they collapse downwards and are then usually removed. Drought does not have much effect on leaf area (and light interception) of mature palms, but may slow development in young palms (Corley and Tinker, 2003). According to Corley (1976b), the optimum leaf area index for yield varies with location from, for example, about 4 in west Cameroon to 6–7.5 in Malaysia, the precise figure depending on the leaf pruning regime (removal of unproductive leaves) and planting density.

Flowering

The early development of the inflorescence, when it is completely enclosed by leaves, takes on average two and a half to three years. Shortly before anthesis the inflorescence

Figure 6.4 Oil palm. Individual fruits are tightly packed in large ovoid bunches – Univanich, Thailand (RHVC). (See also colour plate.)

emerges from the leaf axil (Corley and Gray, 1976a). Pollination is mainly by insects, in particular by a weevil (*Elaeidobius kamerunicus*) that was introduced into South-east Asia from Africa in 1981. Previously hand pollination was practised.

Fruit

The fruit (botanically a drupe) consists of an orange-coloured mesocarp which contains 'palm oil', a hard lignified shell (endocarp) and a white kernel containing 'kernel oil'. There are three fruit forms: *dura*, *pisifera* and *tenera*. The *dura*, originally introduced into South-east Asia from Africa, has a thick endocarp with correspondingly less oil-bearing mesocarp tissue. By contrast, the *pisifera* has little or no endocarp and is commonly female sterile. The discovery of single gene inheritance of shell thickness allowed *duras* (female) to be successfully crossed with *pisferas* (male) to produce high oil yielding *teneras*. Today, all commercially produced seed and all estate-scale plantings are Dura × Pisifera crosses (Hardon, 1995; Corley and Tinker, 2003). The individual fruits are tightly packed in large ovoid bunches (Figure 6.4).

Yield

The first fruits normally ripen during the third year after planting. Yields per palm rise for the next few years and then slowly decline as interplant competition

Figure 6.5 Oil palm. The two principal components of yield are bunch number and bunch weight – Univanich, Thailand (RHVC). (See also colour plate.)

occurs. The two principal components of yield are bunch number and bunch weight; bunch weight increases with plant age, while bunch number declines (Figure 6.5). Bunch number per palm depends on the leaf production rate (each leaf subtends one inflorescence), the sex ratio (the ratio of female inflorescences to the total, which is reduced by drought), the abortion rate (also associated with dry conditions) and the bunch failure rate (the sudden failure of bunches to develop 2–4 months after anthesis). Each of these components is determined sequentially from about 25 months before harvest (sex differentiation) to 1–3 months before harvest (bunch failure). The components of bunch weight are probably determined in the following sequence: spikelet number (16–18 months before harvest), flowers per spikelet (12–15 months before), inflorescence abortion (10–11 months before), fruit set (at anthesis, 5–6 months before), mean fruit weight (2–5 months before). Young palms develop fruit faster than older ones (Turner, 1977). The yield of oil depends on the oil weight to bunch weight ratio (Corley and Gray, 1976b; Corley and Tinker, 2003), which normally ranges between 19% and 26%, depending on genotype and harvesting standards. In addition, palm kernels constitute about 5% of bunch weight and contain about 50% oil (2.5% of bunch weight).

The effect of water stress on fruit yield varies depending on its timing and severity relative to each of these development stages, a process further complicated by the fact that the potential for fruiting is continuous. Water stress also influences the distribution of yield during the year. This too is of commercial importance and is largely governed by bunch number. In countries like Malaysia, with less marked seasonal changes, the yields are relatively uniform, with only up to 12–15% of the annual crop being harvested in the peak months. By contrast, in regions where there is a marked dry season, like Benin, this figure can increase to 35–40% with as little as 1% of the crop being harvested in the dry months. This has cost implications in terms of the harvesting and processing facilities needed to cope with the yield peaks (Nouy et al., 1999; Corley and Tinker, 2003). In a recent paper, Legros et al. (2009a) suggested that day-length (photoperiod) may play a role in controlling seasonal peaks in flowering, even in regions near the equator. They based this suggestion on observations made at two sites, one very close to the equator $0°$ $55'$ $0''$N (Sumatra), where water is not a limiting factor, and the other located at $3°$ $12'$ $15''$S (South Kalimantan), where droughts occur. The photoperiod-sensitive phase was estimated to occur nine months before bunch maturity (or a function of this time interval), while the sensitive phase for drought was estimated to be 29 months before bunch maturity. It is difficult to believe that the palm is sensitive to differences in day-length of as little as one minute.

Roots

The depth and distribution of roots affect the amount and availability of water in the soil. The oil palm has an adventitious root system categorised by Tinker (1976) as follows: primary roots (6–8 mm in diameter) that develop from the base of the trunk (or bole) and either spread horizontally or descend into the soil; secondary roots (2–4 mm) that develop on the primary roots; tertiary roots (1.7–1.2 mm) arising on the secondaries; and unlignified quaternaries (0.1–0.3 mm) on the tertiaries. By making a number of assumptions, Tinker (1976) estimated the total effective nutrient-absorbing root length (taken to be restricted to the tertiary and quaternary roots) in the top few centimetres of soil to be about 60 km palm^{-1}. This analysis ignored roots deeper in the profile, which are important for water uptake in dry weather. Roots have been traced to depths of 5 m where there were no physical restrictions or high water tables. The greatest mass of fine roots is generally found in the surface layers, down to 0.60 m (variable). Root distribution is influenced by irrigation. For example in the Côte d'Ivoire, irrigation virtually doubled the density of fine roots, below 0.20 m down to at least 1.0 m, the limit of measurement, compared with the unirrigated control (Prioux et al., 1992).

Also in the Côte d'Ivoire, Dufrêne et al. (1992) traced primary, secondary, tertiary and quaternary roots of 14-year-old palms, growing in a deep ferralitic sandy soil, to depths of 4.8 m. The fine roots (<1 mm diameter) were mainly found in the top layers (49% within 0.40 m of the surface). The total root biomass (dry) equated to 30 t ha^{-1}.

Subsequently, in a remarkably detailed study, also in the Côte d'Ivoire, Jourdan and Rey (1997a) identified eight morphologically different types of roots on mature oil palms growing in what is commonly known there as a 'tertiary sand'. The relative positions and growth characteristics of all these roots were described. The horizontal (plagiotropic) primary roots grew at an average rate of 30 mm d^{-1}, reaching a maximum length of 25 m, while the primary vertical (orthotropic) roots, growing at a similar rate, reached depths of at least 6 m (the limit of observation). At the same site, Rey *et al.* (1998) traced roots (secondary, tertiary and quaternary), gradually declining in numbers, to depths of at least 4 m, although water extraction still occurred at depths below 5.0 m. Primary (horizontal?) roots were found in the surface horizons to depths of 0.60 m, 50% being in the top 0.20 m.

Based in part on these field observations and measurements, Jourdan and Rey (1997b) developed a 3-D model of the complete root system of oil palm with which they were able to: (1) simulate the spatial distribution of roots under plantation conditions, (2) estimate root biomass and (3) locate and quantify root-absorbent surfaces. The results, some of which were presented in elegant illustrations, were compared with actual root distribution data (to depths of 1.0 m). Examples of the outputs from the model included: (1) root competition (horizontal primary roots) between neighbouring plants (triangular 9 m spacing,) occurred in the topsoil as early as five years after planting; (2) total root biomass was estimated to be 3 t ha^{-1} after four years and 55 t ha^{-1} after 16 years (this is an exceptionally high figure) at a planting density of 143 palms ha^{-1}; and (3) quaternary and tertiary roots provided 84% of the total absorbent root surface (total = 1480 m^2 ha^{-1}).

In Papua New Guinea, Nelson *et al.* (2006) used soil water depletion as a measure of root activity beneath palms on two soil types. Water uptake was greatest under the weeded zone close to the stem. Water was extracted to a depth of 1.6 m, the limit of measurement. The total dry mass of roots within this depth equated to 20 t ha^{-1}. In the weeded zone, 75% of the total root mass occurred in the top 0.39 m. This, and other similar observations, probably explains why oil palm is often described (mistakenly?) as shallow rooted, although, as noted above, roots can reach depths of 5–6 m.

Summary: crop development

1. The oil palm has a single stem with one growing point which, in 'mature' palms, initiates one leaf primordium every two weeks. It takes about two years for a leaf to emerge and expand fully.
2. During dry weather, leaflet opening is delayed and 'spears' accumulate in the crown. When the rains start there is a flush of leaf opening. The rate of leaf *initiation* is thought to be relatively unaffected by drought.
3. The optimum leaf area index for yield varies with location (range 4–6.5).
4. There is one inflorescence in each leaf axil. The flowers are borne on separate male and female inflorescences. Sex differentiation occurs about 25 months before harvest.

5. The proportion of female inflorescences (the sex ratio) is an important determinant of yield, and is sensitive to water stress. Maleness is favoured by stress conditions.
6. Abortion of inflorescences is associated with dry conditions.
7. Fruiting is a continuous process; the timing of water stress relative to stages of development is important for predicting its impact on yield. Day-length, as well as rainfall, may play a role in determining seasonality of yield.
8. The oil palm has an adventitious root system.
9. Plagiotropic roots can spread laterally >25 m from the trunk. Orthotropic roots can extend to depths of at least 6 m. The greatest concentration of roots is in the 0–0.6 m layer.
10. Primary roots can grow at rates of about 30 mm d^{-1}.

Plant water relations

Gas exchange

In order to understand better how the oil palm responds to water stress, factors influencing stomatal conductance and rates of photosynthesis have been studied using a range of techniques. Stomatal conductance, in particular, has been found to be a sensitive indicator of plant water status.

The juvenile palm has stomata on both leaf surfaces but, as the plant develops, there is a progressive loss of adaxial stomata on the later-formed leaves, and an increase in abaxial stomata number and size. For an older palm (>2 years after planting), stomatal densities are in the range 130–200 mm^{-2} (lower, or abaxial, surface only; mean 175 mm^{-2}) depending in part on the position of the leaflet within the frond (Henson, 1991a).

Factors influencing the degree of stomatal opening in oil palm were first studied in Benin by Wormer and Ochs (1959). In a series of measurements using the infiltration technique (which gives a measure of the relative stomatal opening), summarised by Ochs and Daniel (1976), they found that:

- Changes in stomatal opening as the soil dried were related to changes in transpiration.
- The critical soil water deficit at which the stomata first began to close varied with soil type.
- The rate of unfolding of new leaves declined two to three months after the stomata first began to close (at midday).
- Relative stomatal opening could be used to quantify the number of 'dry days' (see below).
- Relative stomatal opening could be used to schedule when to irrigate (see below).

Some of the related research in the Côte d'Ivoire, undertaken by the French research organisation IRHO, now known as CIRAD, was summarised by Caliman (1992), citing Dufrêne (1989). In particular, the relationship between photosynthesis rates and stomatal conductance, and the close match between stomatal opening (infiltration score) and the relative depletion of available soil water, were described.

The infiltration technique was also used by Rees (1961) to monitor diurnal and seasonal changes in stomatal opening in oil palm in Nigeria. In the wet season, the stomata opened early in the morning and remained wide open throughout the day before closing in the early evening. In the dry season, partial closure occurred during the middle of the day. A detailed analysis showed that stomata began to close rapidly when the air temperature (as measured in the shade of the leaf canopy) exceeded about 32°C, with complete closure at 35–36°C. Closure was also observed during periods of very low humidity, when it was cooler. This was associated with the *Harmattan* (desiccating, dust-laden winds from the Sahara desert). Subsequently, Corley (1973), using a diffusion porometer, observed a similar pattern of midday stomatal closure in drought-affected palms in Malaysia.

Later, a combination of field studies (in Colombia) and controlled environment measurements (in the UK) by Smith (1989) showed how stomatal conductance (measured with an infrared gas analyser) declined when the saturation deficit of the ambient air exceeded 1.7 to 2.0 kPa even when the soil was well watered. These observations were, however, unable to explain the diurnal changes in stomatal opening observed in irrigated palms in the field. In contrast to the results reported above, in the irrigated palms partial closure during the morning was interrupted by partial reopening in mid-morning followed by partial closure again in mid-afternoon. In the unirrigated treatment, stomatal conductance reached a minimum at midday before increasing in the late afternoon. It is not clear whether the atmospheric conditions were similar for irrigated and droughted plants when the comparisons were made.

In Malaysia, Henson and Chang (1990) and Henson (1991a) found that stomatal conductance (measured with a diffusion porometer or by infrared gas analysis) rose to a peak in mid-morning and then progressively declined to low values in the afternoon. The same diurnal pattern, obtained under clear sky conditions, was followed regardless of the age of the palm, although actual conductances increased with age. Photosynthesis rates followed a similar trend.

Similarly, Henson (1991b) found that on clear days stomatal conductance in young palms declined progressively from early morning to 1300 h (the limit of reported measurements) under both wet and particularly under dry soil conditions. In contrast to previous measurements, there was no mid-morning peak. The decline was linearly related to increases in the saturation deficit of the air (range 0 to 2.5 kPa). After allowing for the effect of the saturation deficit at the time of measurement, there was a linear decline in conductance as the potential soil water deficit increased (up to 150 mm) during the dry season. Photosynthesis rates also declined in a similar way.

Previously, Henson and Chang (1990) had observed a recovery in conductance following a midday minimum, on a day when humidity increased (i.e. saturation deficit declined) in mid-afternoon, thus replicating the midday partial closure observed by Rees (1961), Corley (1973) and Smith (1989).

Care has to be taken when making direct comparisons of stomatal behaviour when different measurement techniques are used. For example, the relationship between infiltration score and conductance (depending on the porometry method used) is complex and inconsistent, as Burgess (1992a) demonstrated with the tea crop.

However, Dufrêne and Saugier (1993), during detailed measurements in the Côte d'Ivoire, confirmed the sensitivity of stomatal conductance to changes in the saturation deficit of the air. They found an exponential decline over a range of 1.0 to 4.5 kPa, with a corresponding, but linear, reduction in transpiration rate. By contrast, photosynthesis of individual light-saturated leaflets did not decline until the saturation deficit exceeded 1.8 kPa. This decrease in transpiration with no corresponding reduction in net assimilation implies an increase in water-use efficiency at moderate saturation deficits (<1.8 kPa), a climatic condition commonly observed in the riparian forests of West Africa where the oil palm originated. Subsequently, Henson (1995) extended this observation on leaflets to the whole canopy. Canopy photosynthesis was halved when the saturation deficit of the air increased from 0.8 to 2.0 kPa.

In Sumatra (Indonesia), Lamade and Setiyo (1996) compared photosynthetic rates and stomatal conductances of three contrasting clones (MK04, MK10 and MK22). All three were sensitive to the saturation deficit of the air, with net photosynthesis declining linearly, and stomatal conductance exponentially, over the range 1–5 kPa. Clone MK22 was the most sensitive to the saturation deficit, and also to high temperatures. For all three clones, maximum rates of both photosynthesis and stomatal conductance occurred at 33°C. Although fully exposed leaves can reach temperatures 10°C above ambient (Hong and Corley, 1976), the importance of this in relation to whole canopy photosynthesis is not known. However, as Corley and Tinker (2003) stated, 'the inhibitory effects of temperatures between 33°C and 40°C on photosynthesis may be largely due to saturation deficit induced stomatal closure'. There is evidence that short-term imbalances in source (assimilation rates restricted by drought) to sink (demands of the developing fruit) relations can be compensated for by mobilisation of starch and glucose reserves in the stem (Legros et al., 2009b).

There is some evidence obtained from plants grown in containers that abscisic acid, generated in roots growing in a drying soil, may play a role in mediating stomatal responses (and photosynthesis) to water stress (Henson et al., 1992).

Drought impact assessment

Finding ways of detecting and quantifying the level of water stress in oil palms early in the extended fruit development process could assist in yield forecasting and irrigation decision making. Various approaches have been proposed.

Although oil palm exhibits some definite symptoms of water stress during the dry season, there is no visible wilting because of the nature of the leaves (fibrous, thick hypodermis and well-developed cuticle) (Rees, 1961). An index for determining the effects of extreme drought on oil palm was developed by Maillard *et al.* (1974), based on a weighted assessment of the visible effects of drought on the foliage of individual palms within a population (N). Three stages of water stress were identified (S_1–S_3), with a fourth comprising any palms that had died (D). In order of severity these stages were used to label palms in which:

- five or six spear leaves had accumulated in the centre of the crown – S_1
- four to six green leaves had collapsed or broken, accompanied by drying of fruit bunches – S_2
- all the leaves at the base of the crown had dried, and foliage in the centre of the crown had collapsed – S_3.

From this labelling procedure, a Stress Index was calculated in the following way:

$$\text{Stress Index} = \frac{10D + 5S_1 + 3S_2 + 2S_1}{N}$$

In this way, it was possible to compare numerically seasons, sites and cultivars. It is not known how widely this index has been used, or how it relates to productivity. Stages S_2 and S_3 are rarely seen outside the driest parts of West Africa.

Although leaf water status is used for other tree crops (e.g. cocoa and coconut) as a measure of water stress, there are few published values for oil palm. In Costa Rica, Villalobos *et al.* (1992) recorded leaf water potentials with a pressure chamber in adult (17-year-old) irrigated palms as low as −1.7 MPa. Surprisingly, the corresponding values for unirrigated palms were higher (−1.0 MPa) due to stomatal closure. This stomatal control of leaf water status was considered to be the reason why oil palm can survive long dry periods. By contrast, young palms (10 months after field planting) were unable to maintain a high leaf water potential under severe drought conditions (midday values were down to −1.95 MPa). Subsequently, Kallarackal *et al.* (2004) monitored the diurnal changes in leaf water potential at three sites in India, also using a pressure chamber. Under well-watered conditions values declined from −0.1 MPa early in the morning to minima of −1.4 to −1.5 MPa by the middle of the day before recovering in the late afternoon. Similarly, Henson and Chang (1990) in Malaysia reported values typically ranging between −0.4 MPa early in the morning to −1.4 MPa at midday. The seasonal changes in stomatal conductance due to drought were much greater than the corresponding changes in leaf water potential (only 0.3 MPa). A larger range of leaf water potentials is obtained when container-grown plants are subjected to water stress (Henson *et al.*, 1992).

Other methods of assessing crop water status have been tried. For example, Henson (1991c) demonstrated the potential value of measuring the leaf:air temperature difference for detecting and partly quantifying crop water stress provided that there was a well-watered crop nearby to act as a control. Henson (1998), following Dufrêne (1989), also found that sap flux probes provided a sensitive means of assessing relative, but not absolute, transpiration rates in relation to soil water availability and potential evapotranspiration.

With the objective of identifying a suitable assessment method, Henson *et al.* (2005) listed a selection of possibilities and evaluated some of them over one dry season in Malaysia. In terms of timescale (minutes to years) and sensitivity of measurement (high to low), the selection included: leaf or canopy temperature, leaf or canopy gas exchange, leaf water potential, sap flux, evapotranspiration, soil water depletion, spear leaf extension, spear leaf accumulation, frond production rate, inflorescence abortion, reduced sex ratio, bunch number reduction, reduced fruit yield, and death of the palm. Based on the sensitivity of the response, ease of detection and simplicity of measurement spear leaf extension rate (relative rate over a 24-h period, simple but with accessibility issues) and the canopy-air temperature difference (ΔT, which requires instrumentation, but which can be automated) were selected as being the best practical options. The next stage is to relate yield to these measurements, which is not an easy task.

Summary: plant water relations

1. Stomata occur on the lower surface of leaves of mature palms at densities of 130–200 mm^{-2}.
2. Relative stomatal opening is a good indicator of soil water availability provided allowance is made for levels of solar radiation and saturation deficit.
3. Stomata begin to close rapidly if the air temperature exceeds 32–33°C and/or when the saturation deficit of the air exceeds a critical value (there is uncertainty about the exact value, which ranges from zero to 1.7 kPa).
4. In a wet soil, stomata open early in the morning and remain fully open throughout the day before closing in the evening *or* partially close mid-morning, then reopen, then partially close mid-afternoon or open in the morning and then progressively close during the day.
5. In a drying soil, stomata partially close during the middle of the day *or* open in the morning and then progressively close during the day.
6. Conductance declines linearly (or exponentially) with increases in the saturation deficit above the critical value.
7. Some of the differences in stomatal responses reported are probably in part linked to the measurement technique used.
8. Rates of photosynthesis decline in concert with declines in stomatal conductance once a critical value (*c*. 1.8 kPa) has been exceeded.
9. There is some (limited) evidence that maximum instantaneous water-use efficiency occurs at saturation deficits of <1.8 kPa.
10. In a wet soil, the leaf water potential of field-grown palms can fall to −1.5 MPa in the middle of the day. Differences in leaf water potential between

watered and droughted palms are often small in comparison with differences in conductance or photosynthesis.

11. Leaf to air temperature differences and spear leaf extension rates have been identified as practical options for rapidly assessing crop water status in the field.

Crop water requirements

In this section, attempts to measure the actual water use of palms in the field are reviewed, together with estimates of the value of the crop coefficient (K_c) that relates actual water use (ET_c, transpiration plus evaporation from the soil surface) to potential reference crop evapotranspiration (ET_0). Problems with interpretation arise due in part to the different terminology and abbreviations, which are not always well defined, used by researchers. Similarly, the ways of calculating and defining evaporation are not always clearly specified (for example, which version of the Penman equation is used). These may be some of the reasons why Corley (1996) could not identify a consensus view on the value of the crop coefficient. A further complication is that a proportion of rainfall is intercepted by the foliage and evaporated directly from the crop surface (measured as the equivalent of 1 mm d^{-1} by Nelson et al., 2006), thereby substituting for transpiration. For mature oil palms, this can represent up to 22% of potential evaporation (Henson and Chang, 2000, citing others), while 15% of the rainfall volume can appear as stem flow (Nelson et al., 2006). For comparison, measurements in the Côte d'Ivoire on mature palms, made over two and a half years, suggested a much smaller value for stem flow, namely 4% of incident rainfall, with throughfall at 82%, and, by difference, intercepted rainfall at 14% (Dufrêne et al., 1992). In Malaysia, Squire (1984) found that intercepted rainfall represented 11% of the total. Clearly there is a range of values for interception and stem flow depending on the size of the palm, and the size and intensity of the rainfall event.

Radersma and Ridder (1996) allowed for canopy interception of rainfall when computing the water use of oil palms in La Mé, Côte d'Ivoire (c. 5° 20′ N 4° 02′ W; alt. 35 m). The ET estimates were derived from the Penman–Monteith equation (Allen et al., 1998), using published values for the key parameters (including crop, aerodynamic and surface resistances) controlling transpiration and evaporation from crop and soil surfaces. With an annual rainfall of 1500 mm, of which 13% was intercepted by the palms, daily transpiration rates (T) were estimated to be between 3.3 and 6.5 mm d^{-1} during the rains, depending on net radiation levels and the saturation deficit of the air, and from 1.3 to 2.5 mm d^{-1} during the dry season. The corresponding seasonal and annual ET totals were 623 mm (wet season), 395 mm (dry season) and 1118 mm (total).

Previously, at the same site, Dufrêne et al. (1992) had also used the Penman–Monteith equation to assess the transpiration rate (T) of mature palms. When soil water was freely available, the ratio (K_c) between actual evapotranspiration (ET, derived from a water balance analysis) and potential evaporation

(ET_0 = 3.2 mm d^{-1}, Penman–van Bavel) averaged 0.81 but, during the dry season (ET_0 = 3.5 mm d^{-1}) this declined to 0.56. The corresponding T/ET_0 ratios were about 0.70 and 0.35 respectively. ET first fell below ET_0 when 40% of the available water in the top 0.80 m of soil had been depleted (equivalent to a soil water deficit of c. 30 mm), with the ET/ET_0 ratio then falling, on a daily basis, to 0.10–0.20. The maximum depth of water extraction was about 5 m. This decline in ET/ET_0 was matched by corresponding reductions in the stomatal conductance (measured with a diffusion porometer) from about 6 to 1.5 mm s^{-1}.

In Kedah, Malaysia, Henson *et al.* (2005) monitored the water use of young (4–5 years) palms over one dry season (2002/03) with a Delta-T Profile moisture probe. The ratio of actual ET (estimated using the water balance method) to potential evapotranspiration rates (ET_0, Penman) declined from about 1.0 to 0.1 as the dry season progressed before returning to 1.0 following the start of the rains. ET_0 varied between 2.7 and 4.5 mm d^{-1}. The crop coefficient fell below unity when about 15% (c. 15–20 mm) of the available water in the top metre of soil had been depleted. On the deep sandy clay loam soil, water was extracted from depths below 0.60 m. In a follow-up study, Henson and Harun (2007) monitored gas exchange and water use at the same site during a later dry season (2005/06), which was interrupted by appreciable rain (up to 40 mm) at approximately monthly intervals. The impact of the dry conditions on gas exchange and growth was mitigated by the rain, with actual evapotranspiration rates for different drought periods varying between 3.9 and 2.7 mm d^{-1}, and the corresponding ET/ET_0 ratios between 0.85 and 0.50.

At the same site, Henson and Harun (2005) had earlier used a micrometeorological approach (the eddy correlation or covariance method) to monitor fluxes of carbon dioxide and water vapour above oil palms (60% ground cover). As might be expected, canopy conductance, photosynthesis and evapotranspiration were all lower during the dry season than during the rains. Thus, for example, evapotranspiration rates averaged 1.3 mm d^{-1} (0.3ET_0, Penman) compared with 3.3–3.6 mm d^{-1} (0.8–0.9ET_0) in the rains. On a daily basis, the carbon dioxide flux peaked early in the morning and then generally declined, quite sharply during dry periods. This response was again associated with changes in the saturation deficit of the air, with negative correlations (linear) being found between carbon dioxide flux and saturation deficit over the approximate range 1 to 4 kPa. Using a similar approach, Henson (1999) had previously distinguished evapotranspiration and carbon dioxide flux from the soil and ground flora (below canopy) from that originating from the oil palm canopy. Over a 10-d period in December when the soil was wet, below-canopy evapotranspiration averaged 0.47 mm d^{-1} and above-canopy ET_c 3.84 mm d^{-1}, with an average ratio of 0.13:0.87. For comparison, the ET_0 over the same period averaged 4.10 mm d^{-1}.

A similar approach was used by Kallarackal *et al.* (2004) to assess the water use of irrigated oil palms (4–5 years old) in three relatively dry areas of India. The microclimate above the crop was monitored and the data used to calculate potential evapotranpiration (ET_0, Penman, as modified by Van Bavel), while the Penman–Monteith equation was used to estimate actual transpiration (T). Photosynthesis and stomatal conductance were measured with an infrared gas analysis system. Diurnal changes in stomatal conductance followed a pattern similar to that described above, with the stomata opening in the early morning but then progressively closing from about 0800 h onwards. This closure was again associated (negative exponential) with increases in the saturation deficit of the air over the range <1 to 4 kPa. Transpiration rates varied between 2.0 and 5.5 mm d^{-1} (equivalent to 140–385 l palm^{-1} d^{-1}), with lower values being obtained in the dry season (associated with dry air) despite irrigation. The corresponding actual transpiration (T) to potential evapotranspiration (ET_0) ratio varied between 0.70 and 0.90.

A large-scale catchment water balance approach was used by Yusop *et al.* (2008) to estimate *ET* for oil palm in Johor, Malaysia. Rainfall and run-off were monitored over eight months in three catchments, each of which was planted with oil palms of different ages – two, five and nine years old, together with cover crops. Although there were inconsistencies between catchments in the estimates of *ET*, when averaged across catchments and expressed on an annual basis, estimated *ET* was a realistic 1200–1300 mm. Earlier, a similar catchment-scale study was undertaken in Pahang, Malaysia. In a comparison of a natural forested catchment with an adjacent one converted to oil palm, the annual *ET* (1525 mm) for 7–9-year-old palms represented 71% of the annual rainfall (2150 mm), equivalent to 4.2 mm d^{-1} (Henson, 2009b, citing others).

On a completely different scale, Foong (1993) used a drainage lysimeter (8.8 m diameter, 1.5 m deep) containing one irrigated palm to measure evapotranspiration rates over a 15-year period (1976–90) in Peninsular Malaysia. On an annual basis, ET_c during the first seven years averaged 4.5–5.0 mm d^{-1}, and for later years, 5.0–5.5 mm d^{-1}. During the monsoon season, monthly mean values fell to 3.0–3.5 mm d^{-1} and reached 6.5–7.5 mm d^{-1} in the dry season. The corresponding values for ET_0 or K_c were not reported.

In the Côte d'Ivoire, 15-year-old palms extracted water from depths below 5 m during the dry season when actual *ET* rates averaged 2.5 mm d^{-1} (range 4.6–0.6 mm d^{-1}) (Rey *et al.*, 1998). The total depth of available water in the 5-m deep profile was estimated to be only 250 mm. Based on stomatal conductance measurements (diffusion porometer), about 70% (175 mm) of this was considered to be easily available.

For ease of reference, the crop water use data are summarised in Table 6.1.

Summary: crop water requirements

1. Evapotranspiration rates averaged 4.1 mm d^{-1} (range 3.5–5.5 mm d^{-1}) in the rains (equivalent to an annual total of 1500 mm) and 1.9 mm d^{-1} (range 0.6–2.9 mm d^{-1}) in the dry season at various locations.
2. Under well-watered conditions for mature palms the crop coefficient (K_c) averaged 0.9 (range 0.8–1.0). During the dry season K_c values for young palms can fall to 0.1 (range 0.1–0.7).
3. A significant proportion of evapotranspiration (13%) can come from below the palm canopy (soil and ground flora).
4. Water can be extracted from depths >5 m in the dry season.

Table 6.1 Oil palm. Crop water requirements, summary table. Further details are in the text

Location	Method	Season	Age (year)	ET_0 (mm d^{-1})	ET (mm d^{-1})	T (mm d^{-1})	ET/ET_0	T/ET_0
Côte d'Ivoire[1]	Modelled	Rains	Mature		1.3–2.5			
		Dry			3.3–6.5			
Côte d'Ivoire[2]	Water balance	Rains	14–17	3.2			0.81	0.70
		Dry		3.5			0.56	0.35
Malaysia[3]	Water balance	Rains	2–3	2.7–4.5			1.0	
		Dry					Down to 0.1	
Malaysia[4]	Micromet	Rains	3		3.3–3.6		0.8–0.9	
		Dry			1.3		0.3	
Malaysia[5]	Micromet	Rains	11	4.1	3.8 (+0.5)		0.9 (1.1)	
India[6]	Micromet/ modelling	Rains Dry/ irrigated	4–5			2.0–5.5		0.7–0.9
Malaysia[7]	Catchment	Annual	2, 5, 9		1200–1300 mm (3.4)			
Malaysia[8]	Catchment	Annual	7–9		1525 mm (4.2)			
Malaysia[9]	Lysimeter	Rains Dry/ irrigated	0–15		3.0–3.5 6.5–7.5			
Côte d'Ivoire[10]	Water balance	Dry	15		4.6–0.6			

[1] Radersma and Ridder (1996)
[2] Dufrêne *et al.* (1992)
[3] Henson *et al.* (2005)
[4] Henson and Harun (2005)
[5] Henson (1999)
[6] Kallarackal *et al.* (2004)
[7] Yusop *et al.* (2008)
[8] Henson (2009b)
[9] Foong (1993)
[10] Rey *et al.* (1998)

5. Both single leaf and canopy level measurements confirmed that stomatal conductance, photosynthesis and transpiration are all reduced when the air is dry (saturation deficits from <1.0 to 4.0 kPa), even if the soil is wet.
6. The limiting soil water deficits above which transpiration is reduced, on sandy clay loam/sandy soils, are in the range of 15–30 mm.

Water productivity

Yield forecasting

As Corley and Tinker (2003) made clear in their review of the role of water in the productivity of oil palm, there is no simple relationship between rainfall totals and yield. The commonly accepted way of comparing sites and seasons is to use the concept of a soil water deficit, a measure of the relative dryness of the soil. The critical value of the deficit above which yield is lost varies with the soil type and the depth and density of rooting, and also with the stage of growth. Citing the work of others, including a review by Turner (1977), Corley and Tinker (2003) suggest that there is broadly a 10% reduction in fresh fruit bunch yield for each 100 mm increase in the soil water deficit. The responses may differ between locations depending on whether dry seasons occur annually (e.g. West Africa), or at irregular intervals (e.g. Indonesia). As fruiting is continuous, there are always delayed effects of drought on yield. The yield loss for an individual harvest depends on when it occurs in relation to the stage of inflorescence development, the most critical stages being floral initiation, sex differentiation and the abortion-sensitive period (Turner, 1977). Using data from Indonesia and West Africa, Caliman and Southworth (1998) developed statistical relationships between water deficits at certain growth stages and subsequent yields with the intention of being able to forecast the effects of drought on monthly yields. This followed a similar approach to yield forecasting, also based on water deficits, developed by Dufour et al. (1988) in West Africa. How transferable these outputs are is not known.

For north Kedah, Malaysia (6.27° N 100.29° E), a region with a distinct and sometimes prolonged dry season, Henson et al. (2007) used a mechanistic computer model (OPRODSIM) to simulate, with mixed success, the growth and fruit yield of oil palm for the first six years after planting and to provide predictions for the next four years. The model provided a useful first approximation of the effects of climate on dry matter production, yield and water use in a seasonally dry environment. It was also used to predict the likely yield increases from irrigation, which averaged about 20–25 kg fresh fruit ha^{-1} mm^{-1} when 200 mm or more (effective) irrigation was applied during the dry season (I. E. Henson, personal communication). The same model was also used to compare different approaches to scheduling irrigation, namely only applying varying proportions of the amount of water needed to return the soil profile to field capacity, or applying the full amount at different threshold soil water deficits (Henson, 2006). The effect on yield of these two approaches was similar. The yield response per unit of irrigation was greater when irrigation was only applied during the dry season rather than over the whole year. To quote the author: 'the model results do not seem too improbable'.

Yield responses to irrigation

Corley (1996) reviewed the results of 13 oil palm irrigation experiments conducted in several countries since 1967, four of which are described below as examples. Although the quality of the results reported is very variable, and generally of limited generic value, making direct comparisons difficult, Corley attempted to develop relationships between fruit yield responses to irrigation (YR, t ha^{-1}) and soil water deficit (WD, mm, as calculated using the IRHO method[1]) in the unirrigated control treatments. Assuming that a realistic target yield of fresh fruit with irrigation was 30 t ha^{-1}, the following equation was proposed as a basis for planning:

$$YR = 0.0288 \times WD.$$

In an early oil palm irrigation experiment at Grand Darwin in the Côte d'Ivoire, two irrigation treatments were compared with an unirrigated control over the two-year period 1966/67 and 1967/68. Applying 150 mm each month (total 1130 mm) increased annual bunch yields from 10.5 t ha^{-1} to 23.5 t ha^{-1}, while a similar yield increase was realised (but with much less water, total 650 mm) by scheduling irrigation using the degree of stomatal opening (infiltration score maintained >10) to decide when to irrigate. On the basis of these data, Ochs and Daniel (1976) concluded that a yield response of 25–30 kg bunches ha^{-1} mm^{-1} was possible in areas where the potential soil water deficit was 200–600 mm (IRHO method). The yield benefit from irrigation was largely the result of an increase in leaf production, fewer abortions and an improved sex ratio, all leading to increased bunch number.

In Benin, Taffin and Daniel (1976) reported the preliminary results of an evaluation of drip irrigation of oil palm. Although no direct comparisons were possible, drip irrigation providing virtually a full alleviation of the soil water deficit was judged to be an effective way of applying water to palms (despite problems with clogging of the emitters). Apart from days when temperatures exceeded 33°C and the air was dry, the stomata were fully open during the dry season, and good yields were obtained (30 t ha^{-1} fresh fruit in the third year of treatment). When irrigation was delayed, the stomata took several weeks to respond to water application and then failed to reopen fully.

Following this experience, a 900-ha estate was established in a marginal area (for oil palm) of Benin (potential soil water deficit = 800 mm, IRHO method) (Chaillard et al., 1983). The first 400-ha section was irrigated in 1977, and the remainder in 1979. Water was applied directly to the sandy clay soil through a low-level pipe network into rills (small trenches) at least 6 m in length alongside each palm. The scheme was designed to apply 5 mm d^{-1} (later restricted to 2.5 mm d^{-1} because of water shortages). Maintenance of the pipelines and rills was a constant problem. Because of the long time delay between sex differentiation of the flowers within the inflorescences and harvesting the ripe fruit, it is only possible to begin to judge the effectiveness of irrigation from about 28 months after it begins, providing it is sustained over the whole time period. The target fresh fruit yields, believed to be achievable, were only 18 t ha^{-1}. The economics of the project were not considered in their paper.

In the Côte d'Ivoire, Prioux et al. (1992) summarised the results of several trials conducted over a 12-year period on a commercial estate. On average irrigation (with two microsprinklers per tree) increased fresh fruit bunch yields from about 18 to 22 t ha^{-1}, and mesocarp oil from 3.99 to 4.84 t ha^{-1}. Irrigation from planting onwards also advanced the first harvest by about one year, and evened out crop distribution.

Following a five-year trial (1974–79) in Central Johore, Malaysia, Corley and Hong (1982) concluded that irrigation was unlikely to be economic in such areas where dry seasons only occur infrequently. A small increase in oil yield (average over three years = 0.52 t ha^{-1} y^{-1}; or $+8\%$) as a result of an increase in bunch number, which was attributed to a lower abortion rate, and in the mesocarp oil to bunch ratio, was not enough to justify the costs of irrigation.

In southern Thailand, where there is a three to four month long annual dry season, four different methods of irrigation (furrow, sprinklers, microsprinklers and drip) were compared in a commercial context (Palat et al., 2000, 2009). Of these, drip (Figure 6.6) was considered to be the 'best' from a practical perspective, that is in terms of operating cost and ease of management, although fresh fruit yields were similar for each method. With drip irrigation, different application rates were compared, from 150 to 450 l palm^{-1} d^{-1} at two levels of fertiliser. Irrigation was applied daily during the dry season whenever the soil water deficit exceeded 30 mm. Fruit yield increases of up to 10 t ha^{-1} (from 18 to 28 t ha^{-1}) were obtained, the highest yield coming from the treatment receiving 450 l palm^{-1} d^{-1} (equivalent to 6.4 mm d^{-1}) at the high fertiliser level (twice the commercial rate). Over the 10-year period (1995–2005) the cumulative soil water deficit during the dry season averaged 235 mm (based on pan evaporation, 4.2 mm d^{-1}, less rainfall). There was a linear relation between yield and daily application rates. Unfortunately, the total quantities of water applied were not reported. The yield response (following an increase in bunch number) in any one year was statistically related to the soil water deficit in the first quarter of that year (when abortion might have occurred), and to the deficit two years earlier (the time of sex differentiation).

When installing the drip irrigation, adverse effects on palms were observed when the tubing was buried within one metre of the trunk (causing damage to roots) but not when the distance was increased to 2 m. Emitters were spaced at one-metre intervals along the pipe, with the expectation that pipes between alternate rows of palms would be sufficient. The economics of irrigation were sensitive to the price of palm oil.

Irrigation in such areas has other advantages, for example allowing young palms to come into production early. Thus, irrigated palms in progeny trials in Thailand (Rao et al., 2009) started yielding fruit six months earlier than unirrigated palms (within two years of planting compared with 30 months). Production of fruit in the first harvest year, recorded from 30 months after planting, averaged 18 t ha^{-1}. increasing to 32 t ha^{-1} for the years four to seven, with the best progeny yielding over 40 t ha^{-1} in some years. For comparison, unirrigated palms averaged 22 t ha^{-1} over the same four to seven year period.

Excess water

In some poorly drained locations, such as coastal areas and valley bottoms, there may be excess water associated with a high water table. Little is known about its

Figure 6.6 Oil palm. Drip irrigation of a young palm – Univanich, Thailand (RHVC).

impact on the productivity of oil palm, although prolonged flooding is known to reduce stomatal conductance and gas exchange processes (photosynthesis and transpiration) and to kill young palms (Lamade *et al.*, 1998). The optimum depth of a water table for different soils has yet to be defined, as has the maximum allowable duration of exposure to flooding without yield loss (Henson *et al.*, 2008).

Drought mitigation

For reasons unknown (but presumably due to better root growth), removal of young inflorescences appeared to increase the drought resistance of young oil palms in Benin (Ochs and Daniel, 1976, citing Bénard and Daniel, 1971). Disbudding increased vegetative growth including roots and, in the first 24 months of harvesting, disbudded palms produced as much fruit as the controls yielded in 42 months. The duration of 'physiological drought', defined as when the relative stomatal opening was below five (recorded using the infiltration technique at midday on a scale from zero, closed, to 12, wide open) was reduced from 50 to 30 days. Keeping the soil surface weed free also reduced the number of 'dry days' experienced. Planting *Stylosanthes* as ground cover resulted in 10–15% more 'dry days' during the dry season than using *Brachiaria*. At the end of five years, cumulative fruit yield from trees growing on bare soil was double that from trees in soil planted with a legume cover crop.

Based on experience in the Côte d'Ivoire (Caliman, 1992) and Benin (Nouy *et al.*, 1999) the following drought-mitigation measures have been identified:

- Selecting a good site with deep water-retentive soils.
- On sloping land, planting on the contour together with adopting appropriate soil and water conservation measures (e.g. bunds, terraces) can lead to improved growth in young palms.
- Where the soil is compacted subsoiling can improve root growth and distribution, both vertically and laterally, and water availability, and fruit yield increases averaging 8% have been obtained.
- Leaving strips of bare soil along the planted row, killing off the cover crop between the rows at the start of the dry season, and leaving small patches to regrow when it rains. *Pueraria phaseoloides* regrows quickly, while *Calopogonium mucunoides* is drought resistant and develops well under the shade of oil palms. Yield benefits of >30% have been achieved from bare soil treatments.
- Mulching with organic materials (including empty fruit bunches at 30 t ha^{-1}) reduces evaporation from the soil surface (Lim *et al.*, 2008, citing others). (However, as 30 t ha^{-1} is about five times the annual production of empty bunches, this cannot be widely applied).
- In very dry locations (e.g. Benin) reducing the plant density to 100 palms ha^{-1} can reduce damage by drought.
- Removing some or all the inflorescences on young oil palms stimulates vegetative growth, particularly root development, and this may improve drought tolerance (Bénard and Daniel, 1971).

Summary: water productivity

1. Because of the long period between the initiation of the inflorescence and fruit maturity, the full response to irrigation will only be seen in the third year.
2. The allowable soil water deficits at different crop development stages are not known.
3. There are no absolute figures for quantifying the yield responses to irrigation but they are in the region of 20–25 kg fresh fruit ha^{-1} mm^{-1}, equivalent to a yield loss of about 10% for every 100-mm increase in the potential soil water deficit.
4. The main effect of irrigation is to increase the number of bunches by increasing the sex ratio of the inflorescences and reducing abortion losses.
5. Irrigation can advance the time when fruiting begins in young palms by at least six months and increase yields in the early years (e.g. years 4 to 7) after planting.
6. Irrigation offers an opportunity to improve crop distribution during the year (still to be confirmed).
7. Drip irrigation and microsprinklers are considered to be suitable methods for irrigating oil palm.

8. Recommended drought-mitigation measures include: maintaining bare soil along the rows, mulching, reducing plant populations, and removing young inflorescences.

Conclusions

The oil palm originated in West Africa where there are annual dry seasons of variable duration. The principal centres of production are now Indonesia and Malaysia where dry seasons are less regular and intense, although the crop is moving into drier regions of Malaysia, and north into Thailand and elsewhere in the world. Much of the research reported on the physiology and water relations of oil palm has been done in Malaysia and previously in Francophone West Africa. Central to understanding how water stress impacts on yield is a detailed know-ledge of the development stages of the inflorescence, which are spread over a long time. While these have been well defined, their timing can be variable. Quantifying the relation between water availability at each stage and yield has yet to be achieved. There is convincing evidence at both the single leaf and canopy levels that dry air reduces stomatal conductance, even when the soil is wet, with similar reductions in the rates of photosynthesis and transpiration. Less is known about the actual water use of oil palm at the field level, and the minimum amount of water needed to obtain good yields.

With respect to irrigation, the following view expressed by Henson in 2006 still applies:

Despite many trials, what constitutes an optimum or adequate water supply to maintain yield is still poorly defined. The issue has been complicated and deductions hindered by the variability of conditions under which the trials have been conducted, by poor experimental design and inadequate controls, by a lack of comprehensive monitoring of soil and atmospheric conditions, and by differences in methodology relating to definitions of soil water status.

All irrigation experiments are notoriously difficult to perform, including interpretation of the data in ways that are of practical value beyond the place and time that they were undertaken. For a tree crop like oil palm, it is particularly difficult, and compromise is always necessary between what is desirable and what is practical. Nevertheless, there is scope to develop functional relations between yield and crop water-use to aid rational irrigation planning and water management in the dry areas where oil palm is now being increasingly grown.

Summary

The results of research on the water relations and irrigation need of oil palm are collated and summarised in an attempt to link fundamental studies on crop physiology to drought-mitigation and irrigation practices. Background information is given on the centres of origin (West Africa) and of production of oil palm (Malaysia and Indonesia), but the crop is now moving into drier regions. The effects of water stress on the development processes of the crop are summarised followed by reviews of its water relations, water requirements and water productivity. The majority of the recent research published in the international literature has been conducted in Malaysia and in Francophone West Africa. The unique vegetative structure of the palm (stem and leaves) together with the long interval between flower initiation and the harvesting of the mature fruit (c. three years) means that causal links between environmental factors (especially water) and yield are difficult to establish. The majority of roots are found in the 0–0.6 m soil horizons, but roots can reach depths greater than 5 m and spread laterally up to 25 m from the trunk. The stomata are a sensitive indicator of plant water status and play an important role in controlling water loss. Stomatal conductance and photosynthesis are negatively correlated with the saturation deficit of the air. It is not easy to measure the actual water use of oil palm, the best estimates for mature palms suggesting crop evapotranspiration (ET_c) rates of 4–5 mm d^{-1} in the monsoon months (equivalent to 280–350 l palm^{-1} d^{-1}). For well-watered mature palms, crop coefficient (K_c) values are in the range 0.8–1.0. Although the susceptibility of oil palm to drought is well recognised, there is a limited amount of reliable data on actual yield responses to irrigation. The best estimates are 20–25 kg fresh fruit bunches ha^{-1} mm^{-1} (or a yield loss of about 10% for every 100-mm increase in the soil water deficit). These increases are only realised in the third and subsequent years after the introduction of irrigation and follow an increase in the number of fruit bunches as a result of an improvement in the sex ratio (female/total inflorescence production) and a reduction in the abortion of immature inflorescences. There is no agreement on the allowable depletion of the available soil water or on the associated optimum irrigation interval. Drip irrigation has been used successfully on oil palm.

Endnote

1 This method assumes (crudely) that in months with 10 or more rain days evaporation = 120 mm. In months with <10 rain days evaporation = 150 mm. It also assumes that there are 200 mm of available soil water. Deficits only accumulate when this water has been used. See Corley and Tinker (2003) for a critical analysis of this approach.

7   Rubber

Introduction

The rubber tree of commerce (*Hevea brasiliensis* Muell. Arg.) is indigenous to the
Amazon rainforest, within 5°N and 5°S latitude of the equator. Its properties
were well known to the Indians of South and Central America long before the
arrival of the Europeans in the sixteenth century. It is cultivated for its latex,
which is used in the production of natural rubber,[1] 60% of which is utilised in the
manufacture of tyres (Figure 7.1). Latex is a cellular fluid consisting of a suspen-
sion of rubber hydrocarbon particles, represented by the formula $(C_5H_8)_n$, in an
aqueous medium. The nineteenth century saw the vulcanisation of rubber (heating
with sulphur allows rubber to retain its physical properties unchanged over the
temperature range 0–100°C), the development of specialist machinery and techniques for manufacturing rubber goods, the rise of commercial trade in rubber,
and the first efforts to cultivate rubber when the demand for raw rubber began to
exceed the supply from wild trees in Brazil (Varghese and Abraham, 2005). In
1876 seeds were gathered from the rainforest and taken to Kew Gardens in
London. Subsequently, seedlings were sent from London to Sri Lanka and afterwards onward to Singapore where they formed the basis of the rubber-producing
industry that developed throughout the twentieth century, particularly in South-
east Asia (Purseglove, 1968).

The total annual production of natural rubber in 2008 was about 10.6 million t
from 8.9 million ha. The principal producers are Thailand (3.0 million t, 1.8
million ha), Indonesia (2.8 million t, 2.9 million ha), Malaysia (1.2 million t,
1.25 million ha), India (0.82 million t, 0.45 million ha) and Vietnam (0.61 million t,
0.63 million ha). The largest producer in West Africa is the Côte d'Ivoire (0.18
million t), and in South America it is Brazil (0.11 million t). In 2008, South-east
Asia produced 94% of the world's crop (FAO, 2010b; IRSG, 2010).

The rubber industry in Malaysia was initially dominated by large estates, but now
smallholders prevail there (>500 000), as well as in Indonesia and Thailand and other
countries in Asia (Manivong, 2007). Large plantations have given up rubber partly
because of its high labour requirements. In terms of world production, smallholders
account for over 70% of the total area under rubber cultivation (Figure 7.2).
The long immature phase can be a major constraint, particularly to smallholders,
when costs accumulate without any returns (Gunasekara *et al.*, 2007b).

Figure 7.1 Rubber. A sheet of natural rubber prepared from latex produced by a smallholder – Kerala, South India (MKVC).

Figure 7.2 Rubber. It is now predominantly a smallholder crop – Java, Indonesia (MKVC). (See also colour plate.)

Although principally grown in the humid tropics (between latitudes 15° N and 10° S, with annual rainfall totals in the range 1500–4000 mm) rubber cultivation, in response to increasing demand, is being extended into drought-prone areas. This is partly the result of land scarcity and competition from other crops (mainly oil palm) in the traditional areas. New areas include the Central Highlands of Vietnam (12° N), north-central Vietnam, northern India (29° N), south-west China (22° N), the southern plateau of Brazil (23° S) and north-eastern Thailand (19° N) where there is a long dry season (Silpi *et al.*, 2006).

The key environmental constraints to rubber production in different regions of the world have been described by Priyadarshan (2003). These include low temperatures associated with high latitudes and altitude, extended dry seasons and wind damage.

Various aspects of the ecophysiology and productivity of rubber have been reviewed by Rodrigo (2007) who focused on genotype selection, planting density and intercropping.

Crop development

Hevea brasiliensis is an erect tree with a straight trunk. The latex is formed and stored in rings of latex vessels (laticifers). These occur between the inner cambium tissue and the outer hard bark layers. When the tree is five to seven years old the latex can be harvested (or 'tapped') at regular intervals by cutting a spiral groove in the bark and draining the latex into a collecting cup until it begins to coagulate and the flow ceases (Figure 7.3). In the wild, the tree can grow to a height of 40 m, but under cultivation it seldom exceeds 25 m, because wood growth is restricted by 'tapping'. Trees are usually replanted after 25–35 years when latex yields become uneconomic. At the end of their life rubber trees provide a valuable end product as a medium-density tropical hardwood. In Malaysia timber characteristics are now among the selection criteria for clones.

Rubber trees are grown mainly as clones grafted on to seedling rootstocks, or as seedlings. The latter are derived from seeds produced by natural crossing between selected clones in isolated seed gardens. Clonal rootstocks as well as clonal scions are now being recommended, for example in Brazil (Cardinal *et al.*, 2007). The rootstock can have a positive effect on the scion, and the scion can also have a positive effect on the rootstock. The yield potential and adaptability of a selection of clones have been evaluated in north-east India, the highlands and coastal areas of Vietnam, southern China and the southern plateau of Brazil (Priyadarshan *et al.*, 2005). One clone (RRIM[2] 600) produced consistent moderate yields across all suboptimal sites, while others were adapted to specific regions.

Increases in the stem height of rubber trees are discontinuous, being characterised by a period of elongation, towards the end of which a cluster of leaves is formed. This is followed by a 'rest period' during which scale leaves develop around the terminal bud. By repetition of this sequence (four or five times in a

Figure 7.3 Rubber. When the tree is five to seven years old the latex can be harvested (or 'tapped') at regular intervals by cutting a spiral groove in the bark and draining the latex into a collecting cup until it begins to coagulate and the flow ceases – Kerala, South India (MKVC). (See also colour plate.)

year), leaves are produced in tiers or whorls separated by lengths of bare stem. Although the elongation of stems is intermittent, stem girth increases continuously (Webster and Paardekooper, 1989). The base temperature for the initiation of a leaf flush has been estimated in Brazil to be 16°C (for clones RRIM 600 and GT 1) with 420 degree days (summed above a mean air temperature of 16°C on a daily basis) needing to be accumulated between successive leaf flushes. The corresponding base temperature for stem extension is 19°C (Filho *et al.*, 1993).

Canopy

Trees older than three to four years shed all their leaves annually, a process known as 'wintering'. This renders the tree leafless for a short while (up to four weeks) before new leaves emerge from the terminal bud. 'Wintering' is believed to be induced by dry, or less wet, weather. Where there is a marked dry season, the period of defoliation is short and refoliation occurs before the rains begin, triggered it is suggested by an increase in day-length, following the equinox (Guardiola-Claramonte *et al.*, 2008). By contrast, if the dry season is less pronounced leaf fall occurs gradually, new leaves develop slowly and, although the trees are never completely leafless, latex yields are reduced more than in situations

Figure 7.4 Rubber. The optimum plant density for rubber is in the range 500–700 trees ha^{-1} – Kerala, South India (MKVC).

where complete defoliation occurs. Leaf disease can be a problem if the old and new leaves are present simultaneously (Rao, 1971). Clones differ considerably in their 'wintering' behaviour.

The optimum plant density for rubber is in the range 500–700 trees ha^{-1}, based on comparisons mainly done under wet conditions (Figure 7.4) (Rodrigo, 2007). Competition for light between trees begins in about the fourth year after planting, which leaves opportunities for short-term intercropping in the early years. For convenience in crop management rubber trees are usually grown in alleys or clusters. Under good conditions, a leaf area index of five or six is reached within five or six years from planting with a recorded peak of 14 at 10 years. Trees grown from seed reach the end of the juvenile phase of growth when branches begin to form on the main stem (Figure 7.5). The scion (providing it is taken from mature wood) on budded rootstocks does not pass through a juvenile phase (Webster and Paardekooper, 1989).

The architecture of the rubber tree can be considerably modified by irrigation. For example, in the Konkan region of western India (20°04' N 72°04' E; alt. 48 m), where there is an extended dry season, eight years of irrigation increased all aspects of growth by 20–30% compared with rain-fed trees (clone RRIM 600; 10 years old), namely: girth and height, width and depth of the canopy, and number and diameter of primary branches. In addition, the angle of the branches to the main stem was greater in the irrigated trees (58°) than in the rain-fed ones (44°). The amount of water applied was not specified (Devakumar *et al.*, 1999).

Figure 7.5 Rubber. Trees grown from seed reach the end of the juvenile phase of growth when branches begin to form on the main stem – Sri Lanka (MKVC).

Roots

The results of studies in Malaysia of root systems of rubber trees grown as seedlings or as clones grafted on to seedling rootstocks were summarised by Webster and Paardekooper (1989). Within three years of planting, tap roots had reached depths of 1.5 m, and by seven or eight years 2.4 m. In the same time intervals, roots had spread laterally 6–9 m from the trunk and >9 m respectively, well beyond the spread of the branches. The majority of these laterals were within 0.3 m of the soil surface. Feeder roots (c. 1 mm diameter) with hairs grew from the lateral roots. In a detailed study in Malaysia, Soong (1976) found that feeder root development varied considerably between different scion clones (e.g. clone RRIM 605 had 80% more roots by dry mass than clone RRIM 513 in the 0–0.45-m soil depth) while soil texture also had a marked effect (sandy soils contained two and half times more roots than clayey soils). Feeder root growth in the surface layers also varied seasonally, being at its maximum when the trees were refoliating after leaf shedding in the winter (February–March), and at a minimum prior to leaf fall (August–December). At greater depths root growth was at its peak three months later than in the upper soil layers. Root density declined with soil depth such that the dry mass of roots in the 0.30–0.45-m layer represented only 10% of the total in the top 0.45 m.

The growth pattern of the root system of young seedling rubber trees was described by Thaler and Pagès (1996) in relation to shoot development. Having taken detailed measurements of roots in root observation boxes over a three-month period, they found that whereas shoot growth was typically rhythmic, root development was periodic. Thus, root elongation was depressed during leaf growth, while branching was enhanced. During leaf expansion, the tap root continued to grow but at a reduced rate, and the emergence and elongation rate of secondary roots also declined. Tertiary root elongation ceased altogether at this time. Root types were considered to differ in their capacity to compete for assimilates, while root branching was promoted by leaf development. This study followed detailed observations of the architecture of the root system of rubber plants (Pagès *et al.*, 1995). The positions where root axes develop and their morphogenic characteristics were described under different conditions as a prelude to the construction of a mathematical model (Thaler and Pagès, 1998). This model successfully simulated periodicity in root development as related to shoot growth, and reproduced differences in sensitivity to assimilate availability in relation to root type. The apical diameter of a root was considered to be a good indicator of root growth potential (i.e. the larger the diameter the greater the potential for growth).

In south-east Brazil, Mendes *et al.* (1992) traced roots of five-year-old trees to depths in excess of 2.5 m and laterally up to 3.4 m from the trunk (trees spaced 7 m × 3 m). By far the most roots (50%) were within 0–0.30 m from the soil surface.

In Kerala, George *et al.* (2009) used the ^{32}P soil injection technique to determine active root distribution of mature (18 years old) rubber trees (clone RRII 105). They based their observations on a radio assay of the latex serum after 45 days. Of the four depths of application compared, 55% of ^{32}P uptake was from a depth of 0.10 m, 25% from 0.30 m, 13% from 0.60 m and 6% from 0.90 m. These values were assumed to represent similar differences in relative root activity. Uptake was comparatively uniform with distance from the trunk (up to 2.5 m).

After eight years of differential irrigation in the Konkan region of western India, where there is an extended dry season, the total root biomass was similar in both irrigated and rain-fed treatments within the volume of soil sampled (to a depth of 0.45 m and up to 1.5 m from the trunk). The rain-fed trees had a greater concentration of roots in the top 0–0.15-m layer than the irrigated trees. Most roots were also within 0.5 m from the trunk (Devakumar *et al.*, 1999).

In north-east Thailand, Gonkhamdee *et al.* (2009) traced fine roots of rubber trees (clone RRIM 600) in a sandy loam soil to depths of 4.5 m. Using a permanent root-observation access well, active root growth was monitored at the onset of the dry season in November at depths of 1.0 to 4.0 m. This was followed by a 'rest' period before roots resumed growth at 3.0–4.0 m depth at the same time as the leaves were flushing in March. With the onset of the rains in May active root growth occurred in the topsoil above 1.0 m. The greatest root length density was in the top 0.5 m. Fine roots growing at all depths had a life expectancy measured in months rather than weeks, while the decay rate of dead roots was slow, particularly at depth (also cited by Isarangkool Na Ayutthaya, 2010).

Yield

The annual yield of rubber per tree (dry mass, Y) is the product of the yield of rubber per tree per tapping (y) and the number of tappings per year (N):

$$Y = y \times N.$$

where y is the product of the volume of latex per tapping (L) and its dry rubber content $(r, \%)$:

$$y = L \times r.$$

Where L is a function (f) of the initial flow rate (F), the length of the cut (C) and the plugging index (P), which is an indirect measure of the duration of latex flow:

$$L = f(FCP).$$

The supply of latex is dependent on the pressure potential (turgor pressure) within the latex vessels, and this varies with the time of day and the rate of transpiration. Tapping is best started very early in the morning (subject to interference by rain) when the pressure potential is high. The flow of latex then declines as transpiration increases and the pressure potential falls. On a diurnal basis, there is a close negative relationship between the rate of latex flow and the saturation deficit of the air (Paardekooper and Sookmaark, 1969; Paardekooper, 1989). By contrast, the dry rubber content of the latex follows a reverse trend, being higher at midday than at night. The plugging index is not affected.

The complexity of the interactions between the commencement of tapping, the frequency of tapping and genotype on dry rubber yield and its components was highlighted in Sri Lanka ($6°32'$ N $80°09'$ E) by Gunasekara *et al.* (2007b). For example, commencing tapping early, at a stem girth of 400 mm (1.20 m above the bud union) instead of the normal 500 mm, increased yields (over 3½ years total; density 500 trees ha^{-1}) of one clone (RRIC 121, from 6.90 to 10.44 kg tree^{-1}, averaged across all other treatments), had no effect on another (RRISL 211), and reduced yields on a third (RRIC 102, from 8.16 to 6.39 kg tree^{-1}). Clones also differed in their responses in terms of tapping treatment effects on tree girth increment. The yield benefits reported for clone RRIC 121 resulted from increases in the dry rubber content that exceeded the reductions in latex volume.

In order to determine seasonal changes in both latex and wood production, displacement sensors (dendrometers) were successfully used by Silpi *et al.* (2006) to monitor the effects of water shortage on radial growth of rubber trees (clone RRIM 600) in an area of Thailand with a marked dry season ($13.4°$N $101.4°$E). In untapped trees, radial growth began with the onset of the rains and ceased completely during the dry season. When refoliation began in the middle of the dry season, there was a net shrinkage of the trunk. In tapped trees, radial growth slowed considerably within two weeks of the start of tapping so that by the end of the season cumulative growth was about half that of untapped trees. In the second year, the yield of latex increased but wood production was reduced.

In a paper comparing the productivity of several tropical perennial crops, Corley (1983) considered the annual dry matter production of rubber (above ground) for a well-managed crop to be 26 t ha^{-1} with a rubber yield of 2 t ha^{-1} (harvest index = 0.08). The corresponding best yields recorded were 36 t ha^{-1} and 5 t ha^{-1} respectively (harvest index = 0.14), while the highest harvest index reported was 0.37 for dry latex or 0.34 for rubber (rubber represents 90% of the dry matter in latex). The potential yield was estimated to be 46 t ha^{-1} (total dry matter) and 15 t ha^{-1} (rubber). When assessing the potential productivity, Corley (1983) assumed that: (1) leaves remain on the tree for nearly 12 months, and then are shed; (2) trees may remain leafless for as long as one month and (3) newly emerged leaves may not become photosynthetically active for at least one week after emergence (Samsuddin and Impens, 1979a, 1979b).

Quoting Templeton (1969), Corley (1983) noted that clones with the highest harvest indices were all susceptible to wind damage (trunk breakage) because insufficient proportions of assimilates were allocated to trunk growth. A realistic target yield for breeders was considered to be 15 t rubber ha^{-1} (from clones selected with short fat trunks). Because the laticifers delivering the latex become blocked in response to being severed, the yield from a rubber tree is considered to be 'sink' limited rather than 'source' limited (an inadequate supply of assimilate) (Squire, 1990).

Summary: crop development

1. The rubber tree has a single straight trunk, the growth of which is restricted by 'tapping' for latex.
2. Increases in stem height are discontinuous, a period of elongation being followed by a 'rest' period during which leaves emerge.
3. Leaves are produced in tiers separated by lengths of bare stem.
4. Trees older than three or four years shed all their leaves annually (a process known as 'wintering').
5. 'Wintering' is believed to be induced by dry or less wet weather; trees may remain (nearly) leafless for up to four weeks. The more pronounced the dry season the shorter (usually) is the period of defoliation.
6. Refoliation begins before the rains start, perhaps triggered by an increase in day-length.
7. Roots can extend in depth to >4 m and laterally >9 m from the trunk.
8. The majority of roots are found within 0.3 m of the soil surface.
9. Root elongation is depressed during leaf growth whereas root branching is enhanced.
10. The supply of latex is dependent on the pressure potential in the latex vessels: the rate of flow is negatively correlated with the saturation deficit of the air.
11. In tapped trees radial growth of the stem declines, relative to untapped trees, within two weeks of the start of tapping.

Plant water relations

Stomata

Stomata are only present on the lower epidermis of leaves at densities ranging, in a sample of 12 clones, from 278 (clone RRIM 605) to 369 mm^{-2} (IRCI 10) (Gomez and Hamzah, 1980). Expressed in another way, these densities are equivalent to

2.2 and 3.5 million stomata per leaflet respectively. Senanayake and Samara-nayake (1970) also reported large (+70%) differences between clones (25) in the density of the stomata, but unfortunately recorded their data on a per unit leaf area (unspecified) basis. In neither piece of research were any obvious links found between yield, or other attributes of growth, of an individual clone (or groups of clones) and stomatal density.

In a detailed comparison of the leaf anatomy of two clones on the same rootstock, Martins and Zieri (2003) recorded 296 stomata (clone RRIM 600) and 364 mm^{-2} (clone GT 1). For the same two clones, Samsuddin (1980) reported densities of 465 and 372 mm^{-2} respectively. Stomatal size and frequency, as well as the structure and distribution of leaf waxes, of three seedling *Hevea* families were described by Gomes and Kozlowski (1988). In their greenhouse study in Wisconsin, USA, the average density of stomata was about 700 mm^{-2}. Both leaf surfaces were covered with heavy deposits of amorphous wax, except near the stomatal pores. In Côte d'Ivoire, Monteny and Barigah (1986) recorded stomatal densities of three clones (RRIM 600, GT 1 and PB 235) in the range 389–568 mm^{-2} (units assumed). There is clearly variability in density with a range from 280 to 700 mm^{-2} depending on such factors as leaf age, size and exposure to the sun.

Leaf water status

In Kerala, South India (9° 22′ N 76° 50′ E), Gururaja Rao *et al.* (1990) compared the responses of two 10-year-old clones to water stress in terms of (among others) leaf water status, stomatal conductance and yield. Clone RRII 105 maintained higher mid-afternoon (and also pre-dawn) leaf water potentials (measured with a psychrometer) during the dry season (*c.* −1.3 MPa) than clone RRII 118 (−2.4 MPa) as a result of a reduction in stomatal conductance, confirming its relative drought tolerance (ability to conserve water) compared with RRII 118. Clone RRII 105 was also able to maintain a faster latex flow in the dry weather than RRII 118 as a result of higher pressure potential in the latex vessels, and the osmotic potentials were less.

Xylem cavitation

In a container-based study in Thailand, Sangsing *et al.* (2004a) found that the xylem vessels in immature rubber trees under drought stress were relatively vulnerable to cavitation (particularly in the leaf petiole), that clones differed in their degree of vulnerability, and that stomata, by closing, played an essential role in the control of cavitation. On the basis of these observations (based on two clones only), the authors predicted that cavitation-resistant clones would exhibit less xylem disfunction after a drought than susceptible clones, and that this could be an important attribute for drought survival in dry areas of northern Thailand. In a related paper, Sangsing *et al.* (2004b), on the basis of a comparison of the same two clones (RRIM 600 and RRIT 251), postulated that variations in xylem hydraulic efficiency between clones may explain differences in stomatal conductances and xylem water potentials and, hypothetically, growth performance. In a similar study in China, Chen et al. (2010) also found that stomatal closure reduced the risk of cavitation induced by water stress, and that the leaf petiole acts as a safety valve to protect the hydraulic pathway in the stem. Previously, Ranasinghe and Milburn (1995) had shown in a glasshouse-based study in Australia how cavitation occurred in *Hevea* clones when the relative leaf water content fell to 85% (corresponding to a xylem water potential of -1.8 to -2.0 MPa). This resulted in a reduction in the hydraulic conductivity in the petiole to about 40% of the original value, since gas bubbles blocked the flow of water inside many of the conduits. When specimens were rehydrated the conductivity increased again. They also concluded that xylem transport in *Hevea* is disrupted relatively easily under water stress.

Gas exchange

Using single-leaf net photosynthesis light intensity response curves, Samsuddin and Impens (1978) first showed that it was possible to differentiate between rubber genotypes in photosynthetic rates.

In the Côte d'Ivoire, Monteny and Barigah (1986) monitored the changes in rates of photosynthesis over the lifespan of individual leaves of three clones. Clone GT 1 maintained a steady rate for longer (up to 180 d) than the others (RRIM 600 and PB 225). When water stress was imposed on the container-grown plants, photosynthesis rates dropped sharply and did not return to their original values after rewatering.

In Kerala, *Hevea* clones were shown to differ in single-leaf net photosynthetic rates, particularly at low light intensities. Of the 12 immature container-grown clones compared, two in particular stood out (RRII 203 and RRIC 100) as having higher instantaneous water-use efficiencies than the others, and as being less dependent on stomatal conductance than on the capacity of the mesophyl to regulate photosynthesis (Nataraja and Jacob, 1999).

The attempts made by Nugawela et al. (1995) to develop a method for screening genotypes at an early stage in the selection process, based on photosynthetic parameters, have been summarised by Rodrigo (2007). A complication in this approach is that the canopy architecture of juvenile plants is very different from that of a mature rubber tree because of the high level of light attenuation within the mature tree. This makes it difficult to predict the yield potential of mature trees. Similar problems arise in attempts to select for water-use efficiency on the basis of instantaneous measurements of photosynthesis and transpiration.

During the early stages of growth the leaves of young rubber plants are often fully exposed to incident light at levels above the light saturation for photosynthesis (>1000 μmol PAR m^{-2} s^{-1}; PAR levels can reach 2000–2500 μmol m^{-2} s^{-1} in the tropics). This results in light-induced inhibition of photosynthesis and, taken together with shade adaptation by the exposed leaves, explains why early growth of rubber is enhanced by shade/intercropping (Senevirathna *et al.*, 2003; Rodrigo, 2007).

In a comparison of two mature eight-year-old clones in Sri Lanka, canopy photosynthesis of one (RRISL 211) was 20% greater than the other (RRIC 121). This was due primarily to greater light-saturated photosynthetic rates and a larger leaf area index in the top layer of the canopy of RRISL 211 (18–22 m above ground level). Tapping increased canopy photosynthesis in one clone (RRISL 211), but this was not reflected in the yields of dry rubber obtained as these were similar for both clones. RRISL 211 partitioned more of its assimilates to the *volume of latex* produced whereas RRIC 121 partitioned more to the *rubber content*. There was a gradual increase in stomatal conductance (and transpiration) with increasing light intensity (0 to 1200 μmol PAR m^{-2} s^{-1}) particularly with RRISL 211 when tapped (Gunasekara *et al.*, 2007a).

Drought tolerance

In the Konkan region of western India (20° 04′ N 72° 04′ E; alt. 48 m) where there is an extended dry season, Chandrashekar *et al.* (1998) monitored the monthly, seasonal and annual changes in girth of 15 immature clones over a six-year period. They identified five that they considered to be more drought tolerant than the remainder, namely RRII 208 (an Indian hybrid), RRIC 52 (a primary clone from Sri Lanka), RRII 6 (a primary clone from India), RRIC 100 and RRIC 102 (both hybrids from Sri Lanka). Clone PR 261 (a hybrid from Indonesia) was particularly drought susceptible. In this location, only RRII 208 was considered to have reached maturity (defined as 50–70% of the trees having reached a girth of 500 mm at a height of 1.25 m) within nine years from planting. A primary clone is one chosen from a polycross population, while a hybrid clone is from a single cross of known parentage.

Summary: plant water relations

1. Stomata are only found on the lower surface of the leaf, at densities from 280 to 700 mm^{-2}.
2. The xylem vessels of rubber trees under drought stress are vulnerable to cavitation, particularly in the leaf petiole.
3. By closing, the stomata play an essential role in limiting cavitation.
4. Clones differ in their susceptibility to cavitation.
5. Cavitation occurs in the xylem of the leaf petiole at water potentials in the range −1.8 to −2.0 MPa.
6. As a result of stomatal closure, clone RRII 105 has the capacity to maintain higher leaf water potentials than other clones, supporting its reputation for drought tolerance (through water conservation).

7. Clones differ in their photosynthetic rates. Light inhibition of photosynthesis can occur, particularly in young plants, which can therefore benefit from shade.
8. Girth measurements can be used to identify drought-tolerant clones.

Crop water requirements

In a catchment level study, Guardiola-Claramonte *et al.* (2010) proposed a modified method for estimating actual water use by rubber trees during the dry season when there is leaf shedding followed by a leaf flush. Measurements were made at two sites in South-east Asia (northern Thailand, 19° 03′ N 98° 39′ E, and China, 22° N 101′ E), where there is concern about the impact that any expansion of rubber planting will have on the water balance of the catchments. To allow for the changes in the phenology of the rubber tree during the dry season, the energy-based Penman–Monteith estimate of reference crop evapotranspiration (ET_0, Allen *et al.*, 1998) was combined with a new empirical crop coefficient (K_{rubber}). These changes in phenology were believed to be influenced by three variables, namely: the saturation deficit of the air, temperature and photoperiod. After incorporating this revised estimate of crop evapotranspiration for rubber ($ET_c = K_{rubber} \times ET_0$) into a hydrological model, the belief that water use during the dry season, after rubber trees were planted, was greater than that from indigenous vegetation was upheld. This was believed to be a result of (day-length induced) refoliation by rubber trees, ahead of the onset of the rains, in contrast to tea, secondary forest and grassland (Guardiola-Claramonte *et al.*, 2008). The predicted mean annual ET_c for the northern Thailand site was 1050 mm, about 20% more than the estimate based on crop cover (leaf area index, L). Replacing natural vegetation with rubber trees in these catchments would, by increasing ET_c, deplete water storage in the subsoil and reduce discharge from the basin. It is difficult to reconcile the validity of the assumptions made in this analysis, which are open to debate.

Similarly, it is also difficult to judge the validity of the assumptions made by Rodrigo *et al.* (2005) in their estimates of the wateruse of immature rubber, using the Penman–Monteith equation, grown as a sole crop or intercropped with banana in Sri Lanka. Estimates of transpiration by the sole rubber crop (with a cover crop) were exceptionally low (5 mm week^{-1}) even at 122 weeks after planting, when the leaf area index had reached 0.41.

In north-east Thailand (15° 16° N 103° 05′ E), Isarangkool Na Ayutthaya *et al.* (2009) monitored transpiration (T) of mature rubber trees (clone RRIM 600; the average trunk girth at 1.5 m above soil = 550 mm; maximum leaf area index = 3.9) during the dry season using a modified, and successfully calibrated, sap flow technique (Do and Rocheteau, 2002). They compared these estimates with others based on changes in soil water content. The sap flow measurements indicated that transpiration declined from about 1.6 mm d^{-1} to 0.4 mm d^{-1} as the dry season progressed. The corresponding estimates based on soil water depletion were of similar orders of magnitude but numerically different at 2.5 and 0.1 mm d^{-1}

respectively. The errors associated with each technique were discussed (e.g. calibration, soil evaporation, depth of rooting and water extraction).

As part of the same PhD study, Isarangkool Na Ayutthaya *et al.* (2011) monitored the effects of intermittent dry periods (up to 20 days) during the rainy season on transpiration rates (T). When the reference crop evapotranspiration rate (ET_0, Penman–Monteith) was less than about 2.2 mm d^{-1}, T matched ET_0. But when this ET_0 value was exceeded (maximum 4.2 mm d^{-1}), T fell below ET_0 even in a wet soil (less than 50% depletion of the available soil water, corresponding to a pre-dawn leaf water potential, $\psi_{\text{l predawn}} = -0.45$ MPa). At 70% depletion, T was reduced by 40% and at 90% depletion by 80%. Since, regardless of the soil water status, the daytime minimum leaf water potentials on sunny days were relatively stable at *c.* -1.95 MPa ($= \psi_{\text{l critical}}$), the decline in transpiration rates could be explained, using a simple model, based on the hydraulic limitation hypothesis, by a reduction in the hydraulic conductance of the whole tree (K_{tree}) and this critical minimum leaf water potential:

$$T = \left(\psi_{\text{l predawn}} - \psi_{\text{l critical}}\right) \times K_{\text{tree}} \times a$$

where *a* is a coefficient to transform midday sap flow densities to total transpiration per day per unit soil area.

This model was tested further during the 'wintering period', from the end of the rains through the dry season, when the green leaf area was changing rapidly (defoliation followed by leaf flushing). The validity of this approach to understanding how the rubber tree responds to drought (atmospheric or soil induced) was confirmed (Isarangkool Na Ayutthaya, 2010). This whole-tree, hydraulic response approach to estimating transpiration hides the complex short-term (e.g. stomatal closure, xylem cavitation) and long-term (e.g. defoliation, root growth at depth) adjustments that plants make in response to drought.

In south-east Brazil, Mendes *et al.* (1992) found that rainfall interception of five-year old trees did not exceed 5% of the total rainfall.

Summary: crop water requirements

1. Very little research on the water requirements of rubber has been reported.
2. It is difficult to judge the validity of the assumptions made in some of the methodologies reported.
3. The actual maximum evapotranspiration rates reported are generally lower than might be expected for a tree crop growing in the tropics (< 3 mm d^{-1}).

Water productivity

Attempts are being made to extend the cultivation of rubber into the North Konkan region (20° N) on the west coast of India. Although the average annual rainfall is 2175 mm, this is concentrated into the June to September period, and there is a long dry season. Potential evapotranspiration over the year is 2250 mm. It was in this location that Vijayakumar *et al.* (1998) attempted to quantify the responses to irrigation of immature clone RRII 105. In a complicated experiment, three levels of basin irrigation (0.5, 0.75 and $1.0ET_c$) were compared with three levels of drip irrigation (0.25, 0.5 and $0.75ET_c$) and a rain-only control treatment over a three/four year period (the exact duration of the experiment is not made clear). The Penman equation (modified) was used to estimate potential evapotranspiration, together with a crop coefficient (K_c) for rubber of 1.25 (it is not clear why this value was chosen), to give ET_c. Allowance was made for increases in crop cover as the trees matured. (How exactly the quantity of water to apply to each treatment was determined is not explained clearly.) Tree growth (biomass production) was estimated from measurements of tree girth at a height of 1.5 m above the bud union. With basin irrigation, growth rates were similar at all three water application levels, while with drip irrigation both the 0.5 and $0.75ET_c$ treatments outperformed $0.25ET_c$. Total biomass production from the best irrigated treatments was 2.8 times greater than from the rain-only control. Overall, on this oxisol soil, basin irrigation was more effective than drip. Supporting physiological measurements indicated that osmotic adjustment occurred in the laticifers of trees in the drier treatments. The authors concluded that with irrigation the immaturity period in this region could be reduced from more than 10 years to six. They also stated that the total water requirement in the dry season, once canopy cover was complete, was around 1340 mm (or 33 500 l tree^{-1} at 400 trees ha^{-1}) but applying only half this total could be just as effective. Perhaps this interesting finding was a result of using a high value for the crop coefficient (1.25). Irrigation increased the leaf area index and light interception by the canopy (Devakumar *et al.*, 1998). A similar, but less conclusive, study in the same region of India had been previously reported by Krishna *et al.* (1991).

Prior to this experiment, Jessy *et al.* (1994) reported the results of a similar experiment conducted in central Kerala (9° 32' N 76° 86' E) where there is a similarly extended dry season from December to April. Basin and drip irrigation (fabricated with locally available materials) were compared at two water application rates (30% and 50% replacement of water lost by evapotranspiration) over a five-year period from the year after planting (with clone RRII 105) in 1986 until 1992. ET_c was estimated using a modified version of the Penman equation (not specified), in which the crop factor was assumed to have a value of 1.0 with an allowance made each year for changes in the crop cover. The soil was a well-drained laterite with a water-holding capacity of 77 mm m^{-1}. Drip irrigation was applied daily during the summer and basin irrigation was applied once a week for the first three years and afterwards every four days. In year five, the application rates were equivalent to 1.5 and 2.1 mm d^{-1} for the two deficit irrigation treatments. For comparison, the ET_0 estimate averaged over the dry season was 5.0 mm d^{-1}. After five years the girth measurements (at a height of 1.5 m above the bud union) were similar in the drip and basin treatments at both watering levels (drip 364 mm;

basin 347 mm) and both were significantly greater than the control, unirrigated, treatment (300 mm). The implications of these results were not discussed, other than to recommend drip irrigation because of its greater conveyance and application efficiences. Treatment effects on root distribution to depths of 0.60 m after four years were described at 0.10 m, 0.50 m and 1.5 m from the trunk in two directions.

Summary: water productivity

1. Virtually no research on the yield responses of rubber to water has been reported.
2. With the crop now being planted in drier regions this lack of an evidence base is surprising.
3. In these drier areas, irrigation can reduce the immaturity period from >10 years to six years.
4. Other methods of drought mitigation need to be researched, particularly the selection of drought-resistant composite clones (scions and rootstocks).

Conclusions

The structure of the industry plays a role in determining research priorities. Viswanathan (2008) reported an interesting detailed analysis of smallholder rubber farming systems in north-east India and southern Thailand. Smallholders dominate rubber production in South-east Asia, 90% of production in Thailand, 89% in India and Malaysia and 83% in Indonesia. These four countries together represent 77% by area and 79% by production of the world's rubber industry. Assuming these figures are correct, smallholders in these countries produce about 70% of the world's natural rubber. The average size of a holding (area of rubber) is about 1 ha in north-east India and 2 ha in southern Thailand and the corresponding yields of rubber are in the range 950 to 1240 kg ha^{-1} (compared with the best commercial yields of 5000 kg ha^{-1}). The number of trees available for tapping is similar across these two regions, averaging about 380 ha^{-1}. The rubber farming systems vary within the South-east Asia region with a predominance of monoculture in Malaysia and southern India, coexistence of rubber and agroforestry in Indonesia, and integrated farm livelihood systems (consisting of rice, other crops and livestock with rubber) in Thailand and north-east India. Smallholder rubber monoculture is viable, so long as the price is right, while in the integrated systems rubber provides the dominant input to the household income. In so doing it contributes resilience during financial and other crises (i.e. it contributes to sustainable livelihoods). Other issues that affect the viability of smallholder rubber production include land tenure, shared cropping, and the marketing of rubber.

An interesting and useful framework for analysing and explaining structural changes in the production of plantation tree crops is that proposed by Barlow

(1997). Using rubber as the case study, he identified five stages of development, beginning with when a plantation crop is first introduced into a subsistence economy (e.g. Malaysia in 1870) and ending with its demise when, as the economy develops a manufacturing base, it is no longer profitable to grow a plantation tree crop in the traditional way except in remote settings, although existing trees may still be exploited (e.g. Malaysia since 1985). Rubber-producing countries are at different stages on this continuum with Malaysia and Thailand probably the most 'advanced'.

This therefore is the context within which research on the water relations of rubber has to be considered. Income from rubber is central to the livelihoods of several million people. Compared with most of the plantation crops reviewed in this book very little research has been reported in the literature on the water relations and irrigation requirements of rubber. When the crop was confined to the humid tropics this may not have been surprising, but with its expansion into regions with extended dry seasons one might have expected more emphasis to have been placed on this aspect of the agronomy of the crop, especially the selection of drought-tolerant clones. It is not known, for example, what the yield losses due to drought are in the different areas where rubber is grown (or conversely the likely benefits from irrigation). This is essential information for rational planning purposes.

Summary

The results of research on the water relations of rubber are collated and summarised in an attempt to link fundamental studies on crop physiology to crop management practices. Background information is given on the centres of origin (Amazon Basin) and of production of rubber (humid tropics, South-east Asia), but the crop is now being grown in drier regions. The effects of water stress on the development processes of the crop are summarised, followed by reviews of its water relations, water requirements, and water productivity. The majority of the recent research published in the international literature has been conducted in South-east Asia. The rubber tree has a single straight trunk, the growth of which is restricted by 'tapping' for latex. Increases in stem height are discontinuous, a period of elongation being followed by a 'rest' period during which leaves emerge. Leaves are produced in tiers separated by lengths of bare stem. Trees older than three to four years shed senescent leaves (a process known as 'wintering'). 'Wintering' is induced by dry, or less wet, weather; trees may remain (nearly) leafless for up to four weeks. The more pronounced the dry season the shorter the period of defoliation. Refoliation begins before the rains start. The supply of latex is dependent on the pressure potential in the latex vessels while the rate of flow is negatively correlated with the saturation deficit of the air. In tapped trees radial growth of the stem declines, relative to untapped trees, within two weeks of the start of tapping. Roots can extend in depth to >4 m and laterally >9 m from the trunk. The majority of roots are found within 0.3 m of the soil surface. Root elongation is depressed during leaf growth whereas root branching is enhanced. Stomata are only found on the lower surface of the leaf, at densities from 280 to 700 mm^{-2}. The xylem vessels of rubber trees under drought stress are vulnerable to cavitation, particularly in the leaf petiole. By closing, the stomata play an essential role in limiting cavitation. Clones differ in their susceptibility to cavitation, which occurs at xylem water potentials in the range -1.8 to -2.0 MPa. Clone RRII 105 is capable of maintaining higher leaf water potentials than other clones as a result of stomatal closure, supporting its reputation for drought tolerance. Clones differ in their photosynthetic rates. Light inhibition of photosynthesis can occur, particularly in young plants, when shade can be beneficial. Girth measurements have been used to identify drought-tolerant clones. Very little research on the water requirements of rubber has been reported, and it is difficult to judge the validity of the assumptions made in some of the methodologies described. The actual evapotranspiration rates reported are generally lower than might be expected for a tree crop growing in the tropics (<3 mm d^{-1}). Virtually no research on the yield responses to water has been reported and, with the crop now being grown in drier regions, this is surprising. In these areas, irrigation can reduce the immaturity period from >10 years to six years. The important role that rubber plays in the livelihoods of smallholders, and in the integrated farming systems practised in South-east Asia, is summarised.

Endnotes

1 cf. 'synthetic rubber' derived from chemicals sourced from petroleum refining.
2 RRIM represents the origin of a clone, namely the Rubber Research Institute of Malaya/Malaysia; similarly RRII corresponds to India, and RRIC/SL to Ceylon/Sri Lanka.

8 Sisal

Introduction

The sisal plant (*Agave sisalana* Perrine) is a source of coarse leaf fibres. These are used by industry in the manufacture of twines and ropes, carpet-backing, bags and matting (Figure 8.1). The principal areas of production are Brazil, where it is predominantly a smallholder crop, and East Africa, where it is a large-scale plantation crop (Figure 8.2). The importance of the crop, which is adapted to dry areas, has declined since the mid 1960s due mainly to competition from synthetics made from oil-based polypropylenes. As a result, little has changed in production methods in the last 50 years, and resources for research have been limited (Shamte, 2001).

Sisal belongs to a botanically complex group of American plants, the *Agavaceae*. It differs from many other plantation crops (except pineapple) in that it has a photosynthetic adaptation (crassulacean acid metabolism, CAM) that facilitates the uptake of carbon dioxide at night. This dramatically improves its water-use efficiency when it is grown under dry conditions.

Sisal is believed to have originated in Central America and Mexico where it is still found growing wild. As a hardy crop of the tropics and subtropics, sisal was introduced into Tanzania in 1893 from Florida, and the first plantations were established on the coastal plains in 1900 (Purseglove, 1972). In the mid 1960s, Tanzania (2°S to 10°S) was the leading producer of sisal in the world, with more than one quarter (peak production 234 000 t) of the total world output (*c*. 850 000 t). Since then the industry in Tanzania has been in more or less continuous decline for a number of reasons, not only because of competition from synthetics, but also because of political interference in the industry and poor crop management (Kimaro *et al.*, 1994; Hartemink and Wienk, 1995). However, in Tanzania and Kenya there are now attempts to encourage greater participation by smallholders (Machin, 2009) and to develop new products (CFC, 2005).

From 1970 onwards Brazil has been the principal sisal producing country, with an annual production (in 2008) of around 106 000 t out of a world total that has declined to 248 000 t. For comparison, Tanzania produced 33 000 t in the same year. Other major producers are China (60 000 t), Kenya (22 000 t) and Venezuela (10 500 t) (FAO, 2009b). In East Africa, sisal is grown at altitudes from sea level to 2000 m, although most is found below 900 m. In Bahia state, north-east Brazil

Figure 8.1 Sisal. A source of coarse leaf fibres used by industry in the manufacture of twines and ropes, carpet-backing, bags and matting – Tanzania (MKVC). (See also colour plate.)

Figure 8.2 Sisal. It is still predominantly a large-scale estate crop in Tanzania (MKVC). (See also colour plate.)

(10–12° S), where most (90%) of the sisal is produced at an altitude of about 600 m, the average annual rainfall is 500–600 mm, and the mean air temperatures are in the range 20–27°C. It is claimed that sisal grows best with a rainfall total of 1000–1250 mm, but it is often grown in drier areas than this. Excessive rainfall is harmful, sisal being very sensitive to waterlogging. It can be grown on a wide range of soils (Purseglove, 1972; Kimaro *et al.*, 1994).

The area planted to sisal peaked at more than one million hectares worldwide in the 1960s but in 2009 it was estimated to be 434 000 ha, of which 273 000 ha were in Brazil (FAO, 2009b).

In Tanzania, most plantations grow an *Agave* hybrid (11648) and the true sisal *Agave sisalana*. The hybrid was bred in the 1940s by cross-pollinating *A. amaniensis* with *A. angustifolia* followed by back-crossing and self-pollination (Wienk, 1970; Kimaro *et al.*, 1994). It is higher yielding than sisal (faster leaf emergence, more leaves) but in the first years after planting produces more small leaves and is also susceptible to diseases.

Crop development

The stem (known as a bole) is short (maximum height about 1.2 m) and thick (maximum diameter about 200 mm) with an apical meristem, and has a close rosette of about 100 leaves if they are not harvested. It serves as a storage organ as well as a main axis. Immature leaves are packed tightly around the meristem. In its lifetime, a plant produces 200–250 leaves before flowering (Figure 8.3). These are usually about 1.2 m long. Each leaf contains 1000–1200 fibres of two types: mechanical fibres (75%), which are located beneath the epidermis and keep the leaf rigid, and ribbon fibres (25%), which are associated with the conducting tissues (phloem and xylem). Fibres are bundles of thick-walled fusiform cells, each having a narrow lumen, which are strongly bonded together (Lock, 1962).

The complicated spiral leaf arrangement (phyllotaxy = 5/34) around the meristem exposes the upper leaves to full sunlight, while the rosette of unfurled leaves can function like a funnel by channelling rain falling on the water-repellent, wax-covered leaf surfaces to the base of the plant – an adaptation to desert conditions. The axil of each leaf protects a bud which can develop into either a rhizome or roots (Lock, 1962).

Rhizomes

Sisal produces *rhizomes* (underground stems) from buds at the base of the stem, which are situated below ground level. Each rhizome (and there can be 5–10 per plant) has numerous scale leaves each protecting a bud. Apart from the terminal bud, which develops into a sucker (a young plant), these buds usually remain dormant until the sucker is removed. Frequently, a rhizome is 2 m long before the rhizome turns upwards and a sucker is produced. Normally this first occurs after the

Figure 8.3 In its lifetime, a sisal plant produces 200–250 leaves before flowering – Rio Grande do Norte state, Brazil (AHLS).

parent plant is one year old. During cultivation the suckers are removed, and can be used as planting material. They form roots upon being replanted (Lock, 1962).

Roots

Sisal, being a monocotyledon, has an adventitous root system with no tap root. The roots arise from the base of the leaf scars at the bottom of the stem (bole). Two types of roots are distinguished: 'bearers' and 'feeders'. Bearers (diameter 2–4 mm) are long, well-branched main roots, which extend radially 1.5 to 3.0 m, and even up to 5 m, from the plant. They are concentrated in the top 0.3–0.4 m of soil. They rarely go deeper than 1.5 m. Feeder roots (1–2 mm diameter) arise on the bearer roots. They have root hairs. Feeder roots deteriorate when the flower stem emerges (Glover, 1939; Lock, 1962).

Inflorescence (pole)

Sisal only flowers once during a life of several (5–12) years, then it dies (botanically it is a perennial monocarp). The emergence of the flower stem (or pole) is preceded by a shortening of the uppermost leaves. The pole increases in height rapidly (100–120 mm d^{-1}) before reaching a maximum height of 5–6 m. Just before this height is reached, flowering branches emerge from buds low on the stem. The whole inflorescence is called a panicle. The flowers rarely set seed. Plantlets (known as bulbils) develop on each flower stalk. Consisting of rudimentary leaves and a root system, bulbils are the principal source of planting material (Lock, 1962; Wienk, 1970).

Yield

Harvesting begins when the lowest leaves that start withering have reached a length of 600 mm or more. At the first cut (usually 2–3 years after planting), 25 leaves are left on the plant and this number is reduced to 20 in subsequent cuts. Cutting is normally carried out at 12-month intervals, except for the second cut which is earlier, and continues until 'poling' occurs. Normally 3–4 leaves are unfurled each month, except during a drought. Until flowering starts, each successive leaf is 6–8 mm longer than its predecessor. The total number of leaves produced is about 210–230 of which about 170–190 (those > 600 mm in length) are harvested. The fibre content of a leaf of an uncut plant is about 3% (fresh weight), but cutting increases the fibre content of newly unfurled leaves, due to an increase in the coarseness of the fibres, so that over the entire cycle it averages about 4%. Long leaves yield proportionately more fibre than short leaves, the weight of a leaf being proportional to the square of its length. The usual planting density is 5000–6000 ha^{-1}. The average yield on the best yielding estates is 2000–2500 kg ha^{-1} (dry fibre) per year over the productive life of a plant (Wienk, 1970; Hartemink and Wienk, 1995). Since the extracted fibre only represents 2% of the plant dry mass, 98% of the biomass and short fibres (which have no value) are called 'waste'. This is a resource that can be used, for example, as a source of energy, for animal feed, and as a soil improver (Shamte, 2001) or as a mulch in the nursery (Lock, 1969). The recommended plant density is 4000–5000 ha^{-1} under less favourable conditions (annual rainfall <1000 mm) increasing to 6000 ha^{-1} under the best soil conditions (Lock, 1962, 1969).

Water relations

Gas exchange

Stomata occur on both surfaces of the leaf (in pits) in equal numbers. The average density of stomata for CAM succulents is 27 mm^{-2}, and 30–50 mm^{-2} for agaves, while *Agave sisalana* has 10 mm^{-2} fewer on the upper surface than the lower one (Nobel, 1988). These are much lower values those reported for the other plantation crops described in this book. The stomata are open throughout the night, with peak opening at dawn, followed by closure throughout most of the day. Compared with mesophytic plants, stomatal conductances are also much lower in CAM succulents. When droughted, the stomata close, although closure may be delayed by several weeks because of the large quantities of water stored in the tissues that act as a buffer (Ting, 1987). Since the stomata are closed when evaporation rates are high, the ratio of carbon gained to water lost is greatly increased. Typically, the water-use efficiency of CAM plants, expressed as CO_2 fixed per unit of water lost, may be three times higher than that of C_4 plants (e.g. sugar cane) and at least six times higher than that of C_3 species (Borland et al., 2009).

In the context of promoting the potential of plants with crassulacean acid metabolism on marginal lands, Borland et al. (2009) described in detail the biochemistry and regulation processes involved. In the dark, CAM plants open their stomata and perform PEPC (phospho*enol*pyruvate carboxylase)-mediated atmospheric and respiratory CO_2 uptake to form malic acid. The malic acid is subsequently broken down to release CO_2, which is fixed by the enzyme rubisco during the following day behind closed stomata. Rubisco mediates the fixation of carbon in C_3 plants, whilst PEPC mediates carbon assimilation in C_4 plants, both processes occurring in the daylight. This temporal separation of C_3 and C_4 carboxylation underpins CAM. The closure of the stomata in the light and the concomitant, almost complete, cessation of transpiration from the shoot surface explain the high water-use efficiency of CAM plants. Annual above-ground biomass production is comparable to that in C_3 and C_4 crops, but with only 20% of the water required for the cultivation of those crops (Nobel, 1991; Borland et al., 2009). In a review paper on water-use efficiency, Stanhill (1986) cited values for the transpiration ratio (transpiration/mass of above-ground dry matter) for 14 C_4 plants (320 ± 43 g g^{-1}), 51 C_3 plants (640 ± 165 g g^{-1}) and for five CAM species (103 ± 41 g g^{-1}). The average value for pineapple (the only CAM crop plant) was 69 g g^{-1}.

Despite the low gas exchange conductances associated with the succulent tissues, high productivities are therefore achievable by CAM plants in habitats where rainfall is seasonal or intermittent. This is partly because of their capacity to store large quantities of water as a result of having (a) relatively dilute cell sap with a corresponding high osmotic potential (close to -1 MPa), and (b) thin (elastic) cell walls. Furthermore, CAM plants can lose 80–90% of their water content and still survive long periods without rain. They also have the capacity to prevent the reverse flux of water from the storage tissues into the soil. This is achieved (a) by isolating their roots from the soil (by shrinkage of the root cortex and, in older roots, due to the presence of a sclerified epidermis), together with (b) cavitation in the xylem vessels in the root, while (c) aquaporins in the cortex and endodermis also regulate the flow of water (Nobel, 1988; Borland et al., 2009).

Water productivity

To date there appears to be only one published paper on the results of a sisal irrigation experiment (Shalhevet et al., 1979). This experiment was conducted in the northern Negev in Israel ($31° 19'$ N $34° 39'$ E) on a loessial sierozem soil over a three-year period. It consisted of 16 treatments, including winter irrigation only, summer irrigation only and irrigation throughout the year. Only the results of these three treatments, together with a no-irrigation control, were reported. Water was applied in furrows, with enough water at each irrigation event to wet the top 0.9 m of soil to field capacity. Leaves were harvested annually in July/August. Only leaves that were at an angle of 45° with the central axis were cut.

The highest yielding treatment, averaged over the three years, was the one that received 250 mm of irrigation (two applications, June and August, summer) in addition to 200 mm of mainly winter rain. Applying more irrigation water (580 mm, eight irrigations, all year round) reduced yields by 20%, whereas applying less (160 mm, one application in July) reduced yields by 30% compared with the best treatment. The control treatment (no irrigation, rain only, 200 mm) yielded only 60% of the highest yield.

The peak rate of actual water loss (ET), after a July water application, was only 1.4 mm d^{-1}, whereas in the winter it fell to 0.4 mm d^{-1}. Annual ET totals increased from 180 mm (rain only) to 380 mm (wettest treatment). Evaporation from a USWB Class A pan (E_{pan}) totalled about 1600 mm over a year. Estimates of ET were obtained from gravimetric soil sampling and/or with a neutron probe. The $ET:E_{pan}$ ratio averaged just 0.22.

The annual fibre yields ranged from 0.27 kg plant^{-1} (rain only) to 0.44 kg plant^{-1} (best treatment). At a plant density of 2222 ha^{-1}, these equate to yields of 600 kg ha^{-1} and 980 kg ha^{-1} respectively. The highest water-use efficiency obtained was 1.60 g fibre plant^{-1} mm^{-1} (increase in ET), between rainfall only and the single irrigation treatments, or 0.03 g fibre plant^{-1} mm^{-1} (irrigation applied). There was no effect of treatment on fibre quality, length, thickness or strength. Unfortunately, no statistics are included in the report of this experiment (Shalhevet *et al.*, 1979).

Summary

1. Sisal originated in the region of Central America and Mexico.
2. Brazil is the main producing country (previously it was Tanzania).
3. Leaves unfurl at a rate of 3–4 per month, producing 200–250 leaves (the source of a coarse fibre) before flowering (and dying).
4. Roots can reach depths of 1.5 m.
5. Sisal is a succulent with a crassulacean acid metabolism (CAM) photosynthesis pathway, in which CO_2 uptake occurs at night.
6. Stomata occur on both surfaces of the leaf in roughly equal numbers (30 mm^{-2}).
7. Stomata are closed during the day and open at night.
8. Stomatal conductances are much less in CAM succulents than in mesophytic plants.
9. CAM succulents can store considerable quantities of water in their tissues.

10. CAM succulents also have the capacity to prevent the reverse flux of water from storage tissues into the soil.
11. Above-ground dry biomass production is similar to that from C_3 and C_4 species, but with only 20% of the water (= high water-use efficiency).
12. Only one irrigation experiment has been reported (in Israel). Peak rate of water use (in July) was only 1.4 mm d^{-1}. Highest water-use (increase in *ET*) efficiency obtained was 1.6 g (fibre) plant^{-1} mm^{-1}.
13. Very little research appears to have been undertaken on water relations, which is specific to sisal.

9   Sugar cane[1]

Introduction

Sugar cane (*Saccharum officinarum* L., the so-called Noble Cane because of its fine thick stem) is believed to have originated in the islands of the South Pacific, probably New Guinea (2–10°S) having evolved through human selection from strains of two wild species *S. robustum* and *S. spontaneum* and hybridisation with *S. sinense* (Purseglove, 1972; Bull and Glasziou, 1976; Julien *et al.*, 1989; Jones *et al.*, 1990; Simmonds, 1998). Because of its natural sweetness, it has been grown for chewing since ancient times in the Pacific and South-east Asia. The production of sugar from sugar cane began in India, followed by China, Persia (Iran), Egypt and Spain and elsewhere around the Mediterranean. In the seventeenth century the first plantations were established in the West Indies, and the resultant need for labour, particularly for harvesting, led to sugar cane's links with the slave trade.

Most of the commerce between Europe and the sugar regions of the west that followed was subsequently based on the outward shipment of slaves and the homeward carriage of sugar, molasses and rum (Purseglove, 1972; Hobhouse, 1985). *Molasses*, the dark brown viscous liquid residue left behind after the centrifugal process has ended and no more sucrose can be extracted, is one of the most important by-products (contains 50% fermentable sugars) from the manufacture of cane sugar. It is used as a raw material in industry. *Rum* is produced by the fermentation of molasses, followed by distillation. Other products include industrial ethyl alcohol (*ethanol*), which is manufactured from molasses, and *bagasse*, the fibrous residue left after the extraction of juice from the cane (used for fuel in the sugar factory, as well as in various manufacturing processes). The *pith* from the bagasse is used as a stockfeed. Not much of the plant is wasted.

Sugar cane, a C_4 carbon fixation pathway species, is adapted to a range of tropical and subtropical climates, and is grown from southern Spain (37°N) to South Africa (30°S), and from sea level up to 1700 m near the equator. The optimum air temperature for growth is in the range 28–30 °C while the base temperature varies with the development stage from 12 to 19 °C (Liu *et al.*, 1998). Sugar cane can be grown on a diverse range of soils (Jones *et al.*, 1990).[1]

---

[1] By M.K.V. Carr and J.W. Knox.

Figure 9.1 Sugar cane. Irrigated estates – Swaziland (MKVC). (See also colour plate.)

In 2007, the top 10 countries in terms of value and total annual production were Brazil, India (by far the two biggest), China, Thailand, Mexico, Pakistan, Australia, Colombia, USA and Guatemala (FAO, 2009a). In 2007, the total harvested area of sugar cane in the world was about 22.7 million ha (FAO, 2009a) of which about 10.2 million ha (45%) were irrigated (Portmann *et al.*, 2008).

In Brazil alone, the planted area of sugar cane is 8 million ha producing 650 million tonnes of cane (fresh weight) in 2008 of which about 45% is used for ethanol production. Only about 1% of the total area is currently irrigated, but this is liable to increase (Laclau and Laclau, 2009). In many places, irrigation water management is a major component of the production system. In some countries (e.g. Swaziland; Figure 9.1) irrigation is essential. In others it is supplementary to variable rainfall. In South Africa, for example, about 40% of the crop is irrigated (Inman-Bamber and Smith, 2005), whereas about 60% of the sugar produced in Australia depends on irrigation to some extent. Regional water supplies are becoming increasingly limited and there is rising pressure on growers in Australia to improve their on-farm water management practices (Inman-Bamber, 2004).

Much research has been reported on the water relations and irrigation requirements of sugar cane, most recently from Australia and South Africa. Various aspects of this topic have previously been reviewed by Finkel (1983), Yates (1984), Jones *et al.* (1990), Inman-Bamber and Smith (2005) and Martin *et al.* (2007).

Figure 9.2 Sugar cane. The 'germination' of setts requires moist soil surrounding the stem – Swaziland (MKVC).

Crop development

Sugar cane is a perennial crop in which flowering is undesirable. Up to the end of the nineteenth century, only a few clones of *S. officinarum* had been used to establish the major portion of the world's sugar cane industry (Bull and Glasziou, 1976). Almost all the commercial cultivars grown today are interspecific hybrids of *Saccharum* species specially bred during the twentieth century, mainly for disease and pest resistance (Purseglove, 1972).

The crop is produced from stalk cuttings called *setts*. Each node has an axillary bud, and a band of root primordia, and is capable of giving rise to a new plant. The 'germination' of setts requires moist soil surrounding the stem (Figure 9.2). The developing bud is initially dependent on the sett for nutrients and water, but it develops its own root system after about three weeks. Once the new plant is established, roots arise from underground nodes and the axillary buds at these nodes give rise to *tillers* (Bull and Glasziou, 1976). As the crop develops there is an overproduction of tillers (*stalks*), with peak numbers (up to 25 m^{-2}) attained three to five months from planting, but 50% of these can die (as a result of shading) before a stable stalk population is reached after about nine months (Figure 9.3). Tiller senescence begins when about 70% of incident radiation is intercepted by the leaf canopy (Inman-Bamber, 1994).

Figure 9.3 Sugar cane. As the crop develops there is an overproduction of tillers (*stalks*), with peak numbers attained three to five months from planting, but 50% of these can die (as a result of shading) before a stable stalk population is reached after about nine months – Swaziland (MKVC).

The effects of water stress on crop development processes are considered here in the following sequence: leaf canopy, yield accumulation (grand growth), ripening/drying off, flowering, and ratoons, ending with a summary of factors influencing the development of root systems.

Leaf canopy

In a review, Inman-Bamber and Smith (2005) summarised the stages of vegetative growth and the influence of water stress as follows.

- Leaf initiation continues even when leaf appearance is impeded by dry soil.
- Leaf extension is sensitive to water stress declining, for example, from 40 mm d^{-1}, when the leaf water potential at midday is −0.5 MPa, to zero when the leaf water potential falls to −1.3 MPa (cvs. NCo376 and N11).
- Stem extension rates are more sensitive to water stress than leaf extension rates.

- Compensatory growth can occur when water stress is relieved, for example, the relative leaf extension rate[2] can exceed 1.0 within three days of rain.

Subsequently, Smit and Singels (2006) reported the results of a study in South Africa (29° 42′ S 31° 02′ E; alt. 96 m) of the effects of controlled water stress on leaf canopy development. They found that leaf senescence was affected most by drought, followed by leaf appearance, and then tiller senescence.

There were differences in response between cultivars. For example, cv. NCo376 was able to maintain canopy development processes, particularly slower rates of tiller and leaf senescence, for longer than cv. N22 as the soil dried, at least initially. At the same time, cv. NCo376 was also able to maintain stomatal conductances and leaf water potentials at higher levels than cv. N22 (a cultivar known commercially as being drought sensitive).

Similarly, in a preliminary pot experiment, cultivar N11 appeared to be better adapted to water stress than cv. NCo376 since it could adjust its leaf area more rapidly and tended to elongate at a slightly lower leaf water potential than cv. NCo376. This concurred with the results of field trials in South Africa in which cv. N11 produced higher sucrose yields than cv. NCo376 in dry conditions, but not under irrigation (Inman-Bamber and De Jager, 1986).

Yield accumulation/grand growth

In sugar cane, sucrose is stored in the stalk parenchyma cells. Accumulation is a continuous process throughout the life of the plant. In commercial varieties, sucrose concentration increases from 10% (dry mass) in young plants to about 50% as they mature (Julien et al., 1989). In commercial production, sucrose yield is commonly expressed as the product of the fresh weight of stalks (cane) and sucrose concentration (%).

For well-adapted cultivars grown in Australia and South Africa, partitioning of plant biomass to above-ground organs is similar for a wide range of climatic conditions, with the trash component (dead leaves and stalks) varying most (Inman-Bamber et al., 2002). For crops yielding >60 t ha^{-1} dry mass (green biomass = biomass less the trash), the stalk component reached a maximum value of about 0.85 regardless of cultivar or extremes of water regime. Seasonal and age effects on whole-stalk sucrose content are due to varying proportions of young segments (low sucrose content) and older segments (high). From a comparison of high and low sucrose content clones, Inman-Bamber et al. (2009) concluded that there is little direct genetic control on the maximum amount of sucrose that can accumulate in the stalk. Rather, differences between cultivars reside more in the morphology of the plant and responses to ripening stimuli such as mild water stress.

In Ayr, Queensland (19° 32′ S 147° 25′ E; alt. 15 m), Inman-Bamber (2004) reported the responses of two cultivars (Q96 and Q124), both with well-developed leaf canopies, to increasing soil water deficits. Yield-forming processes responded

Figure 9.4 Sugar cane. Prior to harvest and after 'drying off' the cane may be burnt to reduce the amount of leaf 'trash' – Swaziland (MKVC). (See also colour plate.)

in the following sequence as stress levels increased. Under high evaporating conditions (ET_c up to 7.5 mm d^{-1}).

- leaf and stalk extension rates declined when the measured soil water deficit (SWD) was c. 60 mm, reaching very low levels at a SWD of 130 mm;
- the number of green leaves per stalk was reduced at a SWD of 80 mm;
- biomass accumulation declined at a SWD of 130 mm; and
- sucrose yield declined at a SWD of 150 mm.

For comparison, the total available water content in the root zone was >230 mm.

Ripening/drying off

This relative sensitivity of expansive growth to water stress compared with photo-synthesis means that sucrose is diverted from growth to storage in the stem, a phenomenon exploited by the practice of 'drying off' before harvest (Figure 9.4) (Inman-Bamber and Smith, 2005).

In irrigated sugar cane production water is usually withheld prior to harvest (1) to dry the field and (2) to raise the sucrose concentration of the cane. Because past research had given conflicting results on the optimum duration of the drying-off period, Robertson and Donaldson (1998) undertook a detailed analysis of pooled data from 37 experiments (mainly cv. NCo376) conducted in Southern

Africa. In only 22% of the drying-off treatments (total 174) was there a significant increase in the yield of sucrose, averaging 8% (maximum 15%), over the well-watered control treatment. This increase occurred when the reduction in cane yield (dry mass) was no greater than about 10%. In 61% of the drying-off treatments there was a significant increase in the sucrose concentration (% fresh mass), as a result of an increase in the soluble solids together with dehydration. However, sucrose yields only increase if water stress reduces stalk biomass by less than 4% (Donaldson and Bezuidenhout, 2000). Using this information, the CANEGRO model (ICSM, 2008) was used to simulate the optimum drying-off period (for crops harvested annually) for different soil types, locations and month of harvest in South Africa. A set of tables was produced, with adjustments depending on anticipated rainfall. As a general rule, this equated to the time it would take for the cumulative total evaporation from a pan to equal twice the water-holding capacity of the soil (depth not specified) in which the crop was grown.

The results of a one-year duration experiment in Swaziland with a ratoon crop (cv. NCo376), confirmed that there was no reduction in the yield of cane if water was withheld for either seven or 10 weeks. This was equivalent to cumulative evaporation prior to harvest 1.5 and three times the total available water in the root zone (80 mm), respectively. There was a minor improvement in cane quality through an increase in sucrose content (Ellis and Lankford, 1990).

Flowering

Although nearly all sugar cane species and varieties flower ('arrow'), producing large panicles of tiny flowers and fertile seed, selection programmes are biased against flowering. Most varieties will not flower at day-lengths longer than about 13 h or shorter than 12 h (Bull and Glasziou, 1976). Water stress can delay flowering if it occurs before flower inductive conditions arise. The effect of flowering on yield is complex and dependent on several factors and is not always deleterious (Julien *et al.*, 1989).

Ratoons

A ratoon refers to the regrowth of a cane crop after harvesting. New shoots develop from the axillary buds of the stubble piece. In Swaziland, Ellis and Lankford (1990) found that, providing the plant crop was irrigated immediately after harvest, the ratoon crop did not need to be irrigated again until the onset of rapid shoot elongation. The plant crop is usually followed by up to eight ratoons, sometimes more depending on the rate of yield decline.

In tropical Queensland, Australia (18.7° S 146.2° E; alt. 150 m) ratoon crops accumulated biomass faster than plant crops (both irrigated) during the first 100 days due to higher stalk number, faster canopy development and more radiation interception (Robertson *et al.*, 1996). These differences became negligible after 220 days because maximum radiation-use efficiency was larger in the plant crop (1.72 g MJ^{-1}, excluding trash) than in the first ratoon (1.59 g MJ^{-1}). Biomass accumulation reached a plateau (53–58 t ha^{-1}) after 300 days from planting/ratooning, 140 days before harvest. This plateau was associated with the loss of live millable stalks, and not with a cessation of growth of individual stalks. Over the 15-month season the crops intercepted about 70% of the incident radiation. This study emphasised the point that maximising early radiation interception does not necessarily lead to higher yields.

Roots

An excellent review of the growth and function of the sugar cane root system was published by Smith *et al.* (2005). It focused on physical (soil compaction, high water tables) and genetic (differences between cultivars) factors influencing the capacity of roots to access water and nutrients particularly at depth, the likelihood that root water status may influence assimilation through its effect on stomatal conductance (by chemical signals), and the possibility that yield improvements by breeding may have resulted from a focus on above-ground components of yield at the expense of roots. They concluded that, because the underground carbon budget for sugar cane is poorly understood, more research is justified in order to improve further the access and utilisation of resources by roots.

Factors affecting the growth and function of the roots of sugar cane, including their sensitivity to soil compaction and waterlogging, had previously been reviewed in detail by Humbert (1968). Cultivars differ in the size of their root system and in the shoot to root ratios, which increase with increasing productivity. The maximum depths of rooting, as listed, varied with soil type, ranging from 0.9 to 2.2 m.

The root system can be divided into three types of roots: superficial, buttress and rope. *Superficial* roots are thin and branched with numerous rootlets with root hairs; they exploit the upper layers of the soil. *Buttress* roots provide anchorage while *rope* roots grow down to considerable depths (Julien *et al.*, 1989). But, as Smith *et al.* (2005) pointed out, it is unclear how common rope systems are in modern cultivars. Roots of sugar cane are not perennial. Each new stem produces its own root system from its basal nodes. The new root system cannot be formed until the soil around the base of the new stem has been moistened (Yates, 1984).

Sett roots grow at a rate of up to 24 mm d^{-1}. These stop elongating when they are 150–250 mm in length and quickly produce a much branched network of thin subroots, but these die within eight weeks of planting as roots originating from basal nodes of developing shoots take over. Primary shoot roots grow faster at up to 75 mm d^{-1} (Glover, 1967, cited by Thompson, 1976, and Jones *et al.*, 1990), and reach depths of 1.0 m in about 120 days, 1.5 m in 160 d and 2.0 m in 190 d (Wood and Wood, 1967 cited by Jones *et al.*, 1990). Maximum rooting depths for sugar cane in South Africa of between 0.8 m and 4.0 m were reported by Thompson (1976) depending largely on soil type. Following observations made in a rhizotron, Van Antwerpen (1999) recorded roots (cv. NCo376) descending in depth at average rates of 22 mm d^{-1} in sandy soils and reaching maximum depths of about 2 m in 87 d. Rates of root penetration were less in a sandy clay loam taking 176 d to reach the same depth. Smith *et al.* (2005) confirmed that root density (biomass and length) declines exponentially with depth with roots sometimes reaching depths >6 m.

A detailed study of the development and distribution of roots of a plant crop (cv. RB72454) grown on a Xanthic Ferralsol soil with and without supplementary irrigation in Piracicaba, Brazil (22° 42′ S 47° 33′ W; alt. 570 m) was recently reported by Laclau and Laclau (2009). For the first four months after planting the 'root front' extended at a rate of about 5 mm d^{-1}, and afterwards until harvest (322 days after planting) at about 18 mm d^{-1}, almost independent of the water regime. Roots reached maximum depths of 4.25 m with irrigation and 4.70 m when rain fed. About 50% of the total number of root intersects were below a depth of 1 m in both treatments, as observed on a trench wall.

The root systems of ratoon crops are less well developed than those of the plant crop, the roots of which can remain active for a considerable period after harvest (up to 60 days) as the new root system develops on the developing shoot of the ratoon crop. Partial survival of the root system of the plant crop appears to provide protection against drought during the early stages of growth of the ratoon crop (Smith *et al.*, 2005).

Summary: crop development

1. Moist soil is needed to establish a sett.
2. Stem/leaf extension is particularly sensitive to water stress.
3. An excess of tillers is produced; some die due to shading.
4. Maximising early radiation interception is not necessary.
5. Compensatory growth occurs when dry soil is rewetted.
6. Water can be withheld prior to harvest without loss of yield.
7. For a ratoon crop, water needs to be applied immediately after harvest.
8. No further water is needed until rapid stem elongation begins.
9. After planting in wet soil, the sett produces short-lived, thin roots.
10. Their function is taken over by roots that develop from basal nodes of developing shoots.
11. Each stem produces its own root system.
12. The soil needs to be moist close to the stem for roots to develop.
13. Shoot roots extend in depth at rates varying between 5 and 22 mm d^{-1}, reaching depths of 1 m in about 140–150 d, 2 m in 190–200 d, and 4 m in 300–310 d.
14. Providing that there are no physical restrictions (e.g. compact soil, waterlogging) roots can be expected to reach depths of 4 m.
15. Irrigation has a relatively small effect on root depth and distribution.
16. After harvest, the root system of the plant crop remains active for up to 60 d.
17. A new root system develops from the developing stem of the ratoon crop.

Plant water relations

Stomata are more abundant (about twice as many) on the lower (abaxial) surface of the leaf lamina than on the upper (adaxial) surface (Julien *et al.*, 1989). In India, stomata on the adaxial surface of leaves of cultivars subjected to water stress were observed to be more sensitive (closed earlier) than those on the abaxial surface (Venkataramana *et al.*, 1986).

In a field study in Hawaii (21° N 158° W; alt. 100 m), Meinzer and Grantz (1989) measured simultaneously the stomatal conductances of single leaves (steady-state porometer; cv. H65–7052) and transpiration from a developing canopy (Bowen ratio method). Because the aerodynamic resistances were large, transpiration was not under direct stomatal control (uncoupled) and, as a result, small changes in stomatal openings had little effect initially on transpiration.

Some cultivars maintain strict control over stomatal conductance during drought, so controlling water loss and maintaining green leaf area (Bull and Glasziou, 1976). With one drought-resistant Hawaiian cultivar (H69–8235), Inman-Bamber and Smith (2005) described how the stomata closed rapidly as the soil dried (ψ_m declined from 0 to −40 kPa). The stomata of the same cultivar were also sensitive to small changes in leaf water potential. Its reputation for drought resistance was considered to be a result in part of its capacity to conserve water through early stomatal closure. Similarly, the reputation that cv. N12 has in South Africa for drought resistance/avoidance was also thought to be a result of early stomatal closure.

In Mauritius, Roberts *et al.* (1990) measured diurnal changes in stomatal conductances with a portable infrared gas analyser. The maximum values recorded were around 400 µmols m^{-2} s^{-1} (measured across both leaf surfaces) in well-irrigated cane but less than this on days when the saturation deficit of the air (SD) was high (*c.* 1.7 kPa). Conductances were very low at leaf water potentials of −1.7 MPa. The sensitivity of stomatal conductance to changes in leaf water potential was illustrated under both clear sky and cloudy conditions with full stomatal closure occurring at about −1.8 MPa (Turner, 1990).

In the same series of drip irrigation experiments in Mauritius, leaf extension rate was found to be more sensitive to water stress than either stomatal conductance or photosynthesis. Daytime depressions of leaf extension rates were even observed in well-irrigated treatments during the middle of the day when evaporation rates were high. Recovery from water stress was always rapid (within a few days of water being applied) with rates reaching levels in excess of those in the well-watered control treatments (Roberts *et al.*, 1990).

In South Africa, Inman-Bamber and De Jager (1986) were able to relate observed changes in growth processes to the decline in the midday leaf water potential (ψ_l) as the soil dried:

- Stem/leaf extension rate is reduced and the youngest unfurled leaves begin to roll at $\psi_l = -0.8$ MPa.
- Stomatal conductance starts to fall at $\psi_l = -0.8$ to −1.0 MPa.

- Stem extension ceases and stomatal conductance reaches a minimum at $\psi_l = -1.3$ to -1.7 MPa.
- The youngest unfurled leaves became fully rolled at $\psi_l = -2.0$ MPa.

Osmotic regulation has been observed in Hawaii by Koehler *et al.* (1982) when, during drought, the concentrations of potassium and reducing sugars in the leaf blade increased (cv. H62–4271), but only after stem elongation rates had declined. Similarly, in Mauritius, an osmotic adjustment at full turgor of about 0.4–0.6 MPa between drip-irrigated and rain-fed sugar cane (cv. R570) was recorded by Roberts *et al.* (1990). After repeated periods of water stress, osmoregulation has also been observed in pot-grown plants, but to a lesser extent (Inman-Bamber and De Jager, 1986).

Summary: plant water relations

1. When the aerodynamic resistance is relatively large, changes in stomatal opening do not have an immediate, direct effect on transpiration.
2. Stomatal conductance is sensitive to changes in leaf water potential (possible role of root signals?).
3. Some (limited) evidence of sensitivity of stomata to dry air (SD ≥ 1.7 kPa).
4. Conductance is very low at leaf water potentials of -1.7 MPa.
5. Leaf extension rate (and leaf rolling) is a more sensitive indicator of water stress than stomatal conductance or photosynthesis.
6. Drought resistance is associated with early stomatal closure.
7. Growth processes are linked to decline in midday leaf water potential as soil dries.
8. Some (limited) evidence shows that osmotic regulation can occur.

Crop water requirements

Probably the first attempt to quantify the actual water use of sugar cane based on rational physical processes was that reported by Cowan and Innes (1956) in Jamaica. They related monthly water use by a full canopy of cane (ET_c) measured with 24 drainage lysimeters to the Penman equation (1948 and 1951 versions) estimate of evaporation from an open water surface (E_o) using standard meteorological data. The value of the ET_c/E_o ratio (f) so obtained was 0.57–0.58. They also found that leaf elongation rates were linearly related to the accumulated soil water deficit.

In South Africa (29° 26′ S 31° 12′ E), Thompson and Boyce (1967) measured daily ET_c using four hydraulic lysimeters. They confirmed the large effects advection could have along the Natal coast on ET_c rates on individual days. As a result the Penman equation (1963 version) estimate of potential ET_c was sometimes less than the measured value. The most consistent relationship was between measured ET_c and USWB Class A evaporation pan data with a mean ratio of 1.0. Later,

Thompson and Boyce (1971), in a comparison of large and small lysimeters, observed that ET rates declined by about 30% after crops lodged (plant and ratoons), an effect that lasted 2–3 months. In a follow-up study, Thompson and Boyce (1972) compared four models of estimating ET_c using standard weather data together with estimates of the aerodynamic and canopy (stomatal) resistances, values of which were derived from field measurements. The Penman–Monteith equation (1965 version) gave the best (i.e. closest to weekly lysimeter measurements) estimate of ET_c. Actual ET rates fell below potential ET_c rates when the soil water deficit exceeded 50 mm. Because of the complexity of the Penman–Monteith model, and the close relationship between ET_c and E_{pan} (effectively 1:1), USWB Class A pans were recommended for use by irrigation water managers. This confirmed advice given earlier by Thompson et al. (1963).

Subsequently, Thompson (1976) reviewed the research undertaken in South Africa, but in the context of related work reported from Argentina, Australia, Hawaii and Mauritius. Water use was considered under five stages of crop development: bare soil, partial crop cover, complete crop cover, lodged crop and during the drying-off period before harvest. Under bare soil conditions, the effects of weeds, trash, frequency of wetting, and soil type were considered at two sites in South Africa and total water use estimated for each situation. Similarly ET_c figures were tabulated for different degrees of crop cover. The ET_c/E_{pan} ratio reached 1.0 when the crop cover (viewed vertically) exceeded 80–90%.When/if the crop lodged, the ratio fell to about 0.7 for up to three months. Water use during the drying-off period fell below potential ET_c rates as irrigation water was withheld. On the basis of neutron probe measurements, Thompson (1976) considered it to be realistic to assume that 50% of the total available water (ψ_m between −0.01 and −1.5 MPa) in the root zone (maximum depth) was 'freely available' to the plant. During the drying-off phase, the actual rate of water use was then assumed to decline at a progressive rate until all the available water had been used.

Total water use varies considerably depending in part on the duration of the crop (Thompson, 1976). On an annual basis, it can range from 1100 to 1800 mm depending on the location (based on lysimeter data from different parts of the world). Similarly, peak rates of water use for sugar cane between 6 and 15 mm d^{-1} have been reported internationally. There were indications that the ET_c/E_{pan} ratio was (1) less than 1.0 in the winter months and (2) lower in the ratoon crop than the plant crop.

In a detailed study conducted at two sites (Kalamia Estate, north-east Australia 19° 6′ S 14° 4′ E, and Simunye Estate, Swaziland, 26° 12′ S 31° 55′ E; alt. 250 m), Inman-Bamber and McGlinchey (2003) measured daily ET_c rates using the Bowen ratio energy balance method. They confirmed that a realistic mean K_c value (ET_c/ET_0) during the initial stage of crop development was 0.4, and during the mid-season 1.25 (when >80% of the incoming radiation was intercepted by the canopy). For a crop that continued to be well watered, $K_c = 1.25$ was considered to be appropriate until harvest but, in order to impose water stress ahead of harvest, a K_c value of 0.7 may be desirable. (For this stage Allen et al., 1998, suggest a K_c

value of 0.75). Actual ET_c rates in mid-season averaged 5.48 ± 0.13 mm d^{-1} in Australia and 5.19 ± 0.26 mm d^{-1} in Swaziland. The corresponding ET_0 (reference crop) values were 4.44 ± 0.07 mm d^{-1} and 3.98 ± 0.16 mm d^{-1}.

An attempt was made by Chabot *et al.* (2005) to measure transpiration of sugar cane in the field (Gharb plain, Morocco, 34.67°N 8.75°W), using the sap flow technique. Taking measurements on 14 individual stems they found that estimates of mid-season ET_c (*c.* 8 mm d^{-1}) were more than 30% above those predicted from the Penman–Monteith equation (ET_0) using the appropriate mid-season crop factor ($ET_c = 1.23ET_0$). Although 8 mm d^{-1} is not an unrealistic value for August in that location, the authors believed that the sap flow technique was an inappropriate method for determining transpiration rates from a heterogeneous canopy like that of sugar cane because of uncertainties in the methodology.

A novel way of measuring actual water use of sugar cane was described by Omary and Izuno (1995) in the Everglades, south Florida, USA. They monitored daily changes in the height of the water table over a two-year period (plant crop and first ratoon). Diurnal changes in ET rates were discernable. Minimum daily values occurred in December through to February (0.7–1.5 mm d^{-1}), and maximum rates during June to September (4.5–4.6 mm d^{-1}). The total annual ET averaged 1060 mm. Crop coefficients were derived for the Penman E_o estimate (1948 version), with a peak mid-season value of 1.27 in September.

Summary: crop water requirements

1. On an annual basis, the water use of sugar cane is in the range 1100–1800 mm, depending on location.
2. Peak daily ET_c rates of 6 to 15 mm d^{-1} have been reported.
3. Different ways of calculating ET_0, E_o or E_{pan} will result in different K_c (or equivalent) values.
4. The Penman–Monteith equation gives the best estimate of ET_0.
5. Evaporation from a USWB Class A pan can give a good approximation of this estimate of ET_0 (1:1).
6. The generally accepted values for the crop coefficient ($K_c = ET_c/ET_0$) are as follows: initial stage = 0.4 (depends on wetting interval); peak season = 1.25 (when crop cover >80%); drying-off phase (if practised) = 0.75 (otherwise 1.25 continues).
7. Lodging can reduce ET_c by 30%.

Water productivity

In Hawaii, Chang *et al.* (1963) were among the first people to recognise the relationship that existed between the ratios of actual to potential yield of sugar cane and actual to potential evapotranspiration, and to develop a general equation to predict yield.

Later, in Natal, South Africa (c. 30° S), the responses of sugar cane (cv. NCo376) to a range of supplementary irrigation treatments on two soil types, sand and clay, were compared (Thompson et al., 1967). The yield response to *irrigation water applied* was similar at both sites averaging 84 kg ha^{-1} mm^{-1}. Rain-fed crops grown on the sand extracted water from depths of 2 m, while those on the clay extracted water to only 1.2 m, reflecting the relative differences in the depth of rooting on the two soils. On both soils there was a linear relationship between the rate of increase in height of the cane and actual daily water use. In three follow-up experiments on the same two soil types, Thompson and De Robillard (1968) described a linear relationship between cane yield and 'effective' water application (rain plus irrigation, range 750–1800 mm, $n = 14$) with a slope of 120 kg ha^{-1} mm^{-1}.

Data from irrigation experiments conducted across the world (Australia, Hawaii, Mauritius as well as South Africa) were collated by Thompson (1976) and significant linear relations between the yield of cane (Y, range = <100 to >300 t ha^{-1}, fresh weight) and total water use (ET, range = <1000 to >3000 mm) were derived:

$$Y(\pm15.1) = 0.0969ET - 2.4 \quad (r = 0.95; n = 91).$$

For sucrose, the corresponding relationship was:

$$Y(\pm3.43) = 0.0135ET - 1.32 \quad (r = 0.75; n = 85).$$

On the basis of their international experience, Yates and Taylor (1986) urged caution when using water-use efficiency values alone to justify investment in supplementary irrigation in areas receiving 1200 mm or more rainfall. Other factors play a role in determining yield responses to water, including the depth of rooting, soil type, climatic conditions and standards of management. An analysis of commercial yields (unirrigated) in upland Kenya, for both smallholders and estates, gave water-use efficiencies (based on estimated values of actual ET) of only 50–60 kg ha^{-1} mm^{-1}, less than the generally accepted realistic commercial value (100 kg ha^{-1} mm^{-1}) (Bull and Glasziou, 1976; Julien et al., 1989). An example was also given of how advection of hot, dry air on crop water use (in semi-arid Somalia) can reduce water-use efficiencies.

In the FAO Irrigation and Drainage Paper 'Yield response to water' (Doorenbos and Kassam, 1979), typical 'water-use efficiencies' for cane yield (80% moisture content), were presented as 50–80 kg ha^{-1} mm^{-1}, and for sucrose (dry) 6–10 kg ha^{-1} mm^{-1}. These are both of the same orders of magnitude as those summarised here.

Several researchers have looked at the effect of water stress at different stages of growth on yield. In semi-arid north-east Australia (20° S 147° E), Robertson et al. (1999) reported (for cvs. Q96 and Q117) that deficits imposed during tillering, although having large effects on leaf area, tillering and biomass production at the time, had no effect on final yield, due in part to compensatory growth when watering began again. By contrast, deficits imposed when the canopy was well established (leaf area index >2) reduced biomass and sugar production. Unfortunately, the levels of stress imposed were not well defined so it is difficult to extrapolate these detailed findings to other situations.

In a five-year field study in Texas (27° N 98° W), Wiedenfeld (2000) found that withholding (supplementary) irrigation during one of four individual six-week periods during the grand-growth stage resulted in only relatively small reductions in cane or sugar yields, depending on the level of stress. The maximum yield loss, with no rain or irrigation for six weeks in mid-summer, was predicted to be 8–15% for cane yield and 12–19% for sugar. Yield responses to nitrogen fertiliser were not affected by the irrigation treatments.

In Swaziland, Ellis and Lankford (1990), on the basis of the results of a one-year trial and previous work in Zimbabwe (Ellis *et al.*, 1985), suggested that water could be saved by scheduling irrigation using an ET_c/E_{pan} ratio of 0.8 rather than 1.0 during the period of stem elongation, without loss of yield. This treatment, when combined with no irrigation prior to rapid stem elongation and a drying-off period prior to harvest resulted in a water saving compared with local practices of about 20%.

Doorenbos and Kassam (1979) identified four growth stages: establishment, vegetative, yield formation and ripening. The corresponding values for the 'yield response factor' (K_y), which is a measure of the relative sensitivity to drought or response to irrigation, were given as 0.75, 0.75, 0.5 and 0.1, respectively. For the whole growing period K_y had a value of 1.2 (values of 1.0 or above imply sensitivity to water stress) which indicates a relative yield loss greater than any corresponding reduction in evapotranspiration. It is not explained how these values were derived and, following this review, it is not possible to verify them or to suggest alternatives.

Summary: water productivity

Despite difficulties in extrapolation, it is possible to come to the following broad conclusions from the information presented.

1. There is a linear relationship between cane yield and actual evapotranspiration (*ET*).
2. The slope of this relationship (benchmark) is of the order of 100 kg ha^{-1} mm^{-1}, but variable.
3. The corresponding value for sucrose is about 13 kg ha^{-1} mm^{-1}.
4. Water stress during tillering need not reduce final yields because of compensatory growth on rewatering.
5. Water stress during 'grand growth' does not necessarily lead to a large (>15%) loss in yield.
6. Water can be withheld prior to harvest for periods equivalent to up to twice the total depth of available water in the root zone.
7. Opportunities exist to reduce the depth of irrigation water applied without loss in yield, depending on soil type, at all growth stages.
8. It is not possible to reconcile the FAO figures for the 'yield response factor' with this information.

Irrigation systems

The choice of method for irrigating sugar cane depends on many factors that are site and context specific. The merits and limitations of different systems (not crop specific) have been described by Jones *et al.* (1990) and Kay (1990) and for sugar cane cultivation by James (2004). This review is restricted to those irrigation systems evaluated on sugar cane and particularly the drivers forcing growers to switch technology to improve system performance and efficiency.

For many crops, including sugar cane, the capacity of a system to apply water uniformly and efficiently is a major factor influencing the agronomic and economic viability of production (Qureshi *et al.*, 2001). But switching from traditional gravity-fed schemes to modern pressurised systems does not necessarily lead to better irrigation performance unless the management skills and experience of the irrigator allow the equipment to be used effectively. For sugar cane, methods that are widely used include surface- (furrow-) based systems, overhead (sprinklers, centre-pivots and rain-guns) and micro (drip) irrigation.

Furrow irrigation

Furrow irrigation was favoured in many early sugar cane developments, due to its low capital costs, suitability for land with gentle slopes and operational simplicity (Figures 9.5 and 9.6) (Holden, 1998). However, rising costs for energy (pumping) and labour (furrow is labour intensive) has led to existing schemes in Swaziland being replaced with overhead or micro-irrigation systems (Merry, 2003). There are also increasing concerns regarding the environmental impacts of large drainage flows and deep percolation losses that can occur from furrow-irrigated fields (Mhlanga *et al.*, 2006). For example, in north Queensland (Australia) the long-term use of furrow irrigation is reported to be contributing to a rise in the water table and an increase in salinity (Tilley and Chapman, 1999).

In Burdekin, Australia, Qureshi *et al.* (2001) used a bio-economic modelling approach to assess the viability of switching from furrow to either centre pivot or drip irrigation. They concluded that furrow irrigation remains the most attractive option when water charges are low. Only when volumetric water charging is considered as a policy option does centre pivot irrigation become the preferred option.

Overhead irrigation

The main overhead systems used for sugar cane are semi-permanent sprinklers and centre pivots, and to a much lesser extent high-pressure rain guns.

Figure 9.5 Sugar cane. Priming syphons for furrow irrigation – Swaziland (MKVC). (See also colour plate.)

Figure 9.6 Sugar cane. Field evaluation of furrow irrigation – Swaziland (MKVC).

Sprinklers – dragline

In dragline irrigation a rotary impact sprinkler is attached to a riser and connected to a quick-release valve via a flexible hose (Figure 9.7). Compared with conventional sprinkler systems with portable pipes, there are fewer pipe moves (as only

Figure 9.7 Sugar cane. In dragline irrigation a rotary impact sprinkler is attached to a riser and connected to a quick release valve via a flexible hose – Swaziland (MKVC).

the hose and sprinkler are moved) with consequential labour savings, for a modest increase in capital cost (installation of underground laterals). They are robust and flexible and can be designed to cope with most soil types, small and odd-shaped fields, obstructions, and even different crops (ratoons) within one field. They are simple to operate and highly visible, making faults (blocked sprinklers) easy to identify and remedy (Merry, 2003). For these reasons, they have proved popular especially in Southern Africa (Zadrazil, 1990). However, the system is susceptible to wind drift (particularly when the cane is young) as the sprinklers are mounted on tall (2.5 m) risers to cope with irrigating a full cane canopy (Figure 9.8).

Centre pivots

In a number of regions including Africa, Brazil and Australia there has been a steady uptake in the use of centre pivots in sugar cane. The reasons include low running costs (compared to furrow), lower labour and energy requirements, ease of operation, and the potential to achieve high application uniformities even under windy conditions (Teeluck, 1997).

Figure 9.8 Sugar cane. High-rise sprinklers suitable for sugar cane – Swaziland (MKVC).

Although originally adopted by the large-scale commercial growers, centre pivots have also proved popular with small-scale farmers in organised associations in Swaziland. The conversion from surface-irrigated rectangular fields can create problems dealing with field corners which then require a separate irrigation system, usually dragline sprinklers or drip. Centre pivots can cope with undulating land and awkward field boundaries (e.g. drainage ditches), but a disadvantage is the relatively high capital cost while fields generally need to be at least 40 ha to make investment worthwhile.

Rain guns

Due to their robustness and versatility, high-pressure, high-volume sprinklers (rain guns) were widely used from the 1950s to irrigate sugar cane in Mauritius, Zambia, South Africa, Swaziland and Australia. However, the large water droplets can cause damage to young sugar cane and create capping problems on sensitive soils (Figure 9.9). Since 1990, however, rising energy costs coupled with increasing demands for improved water application and crop uniformity have resulted in these systems being replaced by draglines and centre pivots (Teeluck, 1997).

Drip irrigation

With drip irrigation small quantities of water are applied at frequent intervals directly to the soil. In Hawaii (USA), drip irrigation has been the principal method

Figure 9.9 Sugar cane. High-intensity water application from rain guns can lead to run-off – Nigeria (MKVC).

used to irrigate sugar cane since the 1970s (Koehler *et al.*, 1982), while it has been evaluated elsewhere in the world. For example, Pollok *et al.* (1990) described the installation and operation of a fully automated, commercial, 40-ha drip-irrigation scheme in Swaziland. Although this pioneering initiative was considered to be a success, such a sophisticated system, it was judged, should only be contemplated where there are well-organised management systems in place. Malfunctions of equipment and components require constant monitoring and rectification as consequential crop losses are large. When well managed, drip-irrigation systems have the potential to achieve high (>90%) application efficiencies (Tilley and Chapman, 1999), while there are also potential water savings when compared with the losses associated with surface or overhead methods. Research has concentrated on design issues that relate to the siting and spacing of the lateral pipes and drippers, quantifying the amount of water to apply relative to other irrigation methods, and the amelioration of problem soils.

In a detailed experiment in Mauritius (20°S 57°E; alt. 70 m) reported by Batchelor *et al.* (1990), a treatment with buried (0.2 m) subrow drip lines out-yielded an alternate inter-row drip line treatment over three years (plant crop plus two ratoons, cv. Saipan). Applying water during tillering at $0.5ET_c$ resulted in similar yields to those obtained from applying more water at that time (1.0 or $1.5ET_c$). Despite a large number of supporting measurements, it is difficult to extrapolate the results, which are not easy to interpret, beyond the locality where these experiments were conducted. More roots grew in the inter-row areas when a drip-irrigation emitter was present. Drip irrigation also altered the relative distribution of roots vertically compared with a rain-fed crop (Soopramanien and Batchelor, 1987, also described by Gregory, 1990).

On a poorly structured saline/sodic soil in Swaziland (26° S 32° E; alt. 200 m), drip irrigation was compared with furrow irrigation to see if yields and the number of ratoons could be increased (Dodsworth et al., 1990). In the plant crop, drip irrigation gave a small (7.5%), but non-significant, yield benefit but there was no advantage in the first or second ratoon crops. Despite operational problems, there were significant improvements each year in water-use efficiencies with drip irrigation (cane fresh weight, average 108 cf. 93 kg ha^{-1} mm^{-1}; +16%). Previously, Nixon and Workman (1987) had described the results of a field observation trial in which drip irrigation was compared with furrow irrigation on a similar poorly draining saline/sodic soil. Again, there was no benefit from drip irrigation except in the plant crop and the first ratoon. Indeed, in later ratoons, there was a negative yield response as the soil structure deteriorated further.

Following a cost–benefit analysis of seven different irrigation options, a subsurface drip system replaced ageing dragline sprinklers on a large commercial estate in Swaziland (Merry, 2003). A post-investment audit confirmed that there was a resultant 15% sucrose yield increase and a water saving of 22% compared with the sprinkler system. Although these increases were above those originally envisaged, there is a problem in distinguishing the water saving due to better management from that due to the use of drip irrigation. Whether reported savings will persist once the drip system is no longer a closely monitored novelty remains to be seen.

In Mauritius, Ng Kee Kwong et al. (1999) showed how nitrogen fertiliser inputs could be reduced with drip irrigation (fertigation). Similarly, a study in Australia found that increased crop yields and sugar contents were possible, with a 25% reduction in nitrogen input relative to industry standards (Dart et al., 2000). Part of this gain arises from adjustments to nitrogen management to minimise the loss of nitrogen in wet periods, compared with other systems. However, high installation costs and problems associated with rodents and low water quality causing iron deposits in the laterals remain major barriers to the adoption of drip irrigation.

Irrigation scheduling

Irrigation scheduling is the process of deciding when to irrigate and how much water to apply. The objective is to maintain optimum soil water conditions for growth in order to meet yield and quality targets with the minimum amount of water.

In the introduction to a paper describing the results of a survey of sugar cane irrigators in South Africa, Olivier and Singels (2004) explained how irrigation schedules can be defined as *fixed* (amount and cycle are kept constant for the entire growing season), *semi-fixed* (amount and cycle are changed a few times to accommodate rainfall and significant seasonal and crop age induced changes in water demand) or *flexible* (amount and timing are changed daily or weekly according to a calculated water balance based on recent crop and weather conditions).

Although many approaches have been promoted over the years for sugar cane (and other crops), it remains the case that only a minority of farmers use an objective (scientific) method of scheduling irrigation, and most still rely solely on their judgement based on intuition and/or crop appearance.

Visible symptoms of water stress include the following:

- The lamina of the upper leaves curl inward, reducing the exposed leaf area.
- Young tillers roll their leaves before those on older stems.
- Senescence of the lower leaves begins.

Irrigation method constraints

In the four major irrigation areas in South Africa surveyed by Olivier and Singels (2004) scheduling practices were found to be highly dependent on the irrigation method being used. Dragline systems were mainly operated on a semi-fixed schedule, due to labour constraints and design limitations, with allowance being made for rainfall and seasonal effects (winter and summer) and to a lesser extent crop age. By contrast, centre pivots and drip systems were mainly operated on a flexible schedule, with direct measurements of the soil water content (using a neutron probe) being the preferred method of assessment. Water budgeting using a water balance model was perceived as being 'too much trouble' in part due to lack of availability (at that time) of appropriate crop coefficient values. It was recommended that attempts should be made to persuade users of dragline systems to change from fixed to semi-fixed scheduling, and users of centre pivots and drip systems to switch to flexible schedules. By so doing, it was estimated that the water-use efficiency of 18 000 ha of irrigated sugar cane in South Africa could be improved. Although based on a relatively small sample size (40), these are interesting findings that are probably typical of many other areas of the world. This is perhaps due in part to the fact that for many growers scheduling is perceived to be unnecessary since it simply reinforces their existing knowledge.

Olivier and Singels (2004) concluded from their study that the main barriers to the uptake of objective irrigation scheduling techniques in South Africa were (1) the complexity of the technology and the difficulty of applying it in practice on a farm, and (2) the perception that accurate scheduling provides little benefit.

Leaf/stem extension

In theory, it is possible to exploit the sensitivity of leaf extension to water stress to schedule irrigation of sugar cane. In South Africa, Inman-Bamber (1995) monitored diurnal changes in extension rates of the youngest visible leaf under different degrees of water stress. At low stress levels leaf extension rates were above 2 mm h^{-1} throughout the day, reaching 3–4 mm h^{-1} by early evening. Under moderate stress, daytime extension was minimal and was exceeded by night-time growth. Under severe stress, daytime extension was zero or negative (shrinkage), and at night it was reduced considerably. Recovery was very rapid (within a few days) on rewetting of the soil.

In Australia, relative shoot extension rate (RSER) is actually recommended as a criterion for judging when to irrigate (RSER < 0.5), but farmers rarely have a control, well-watered treatment for comparison (Inman-Bamber and Smith, 2005). It may be possible, particularly under conditions of low evaporative demand, to reduce the critical RSER value from 0.5 to 0.3 without loss of yield. If water is limiting, it was suggested that irrigation could be delayed until leaf senescence began (Inman-Bamber, 2004).

Shoot extension rate (SER) or leaf extension rate (LER) alone is confounded with temperature effects, although an attempt to allow for this, by developing an index based on a comparison of early morning and daytime LERs, has been proposed (Inman-Bamber and Spillman, 2002).

Nevertheless, the rate of stem elongation and final internode lengths are convenient ways of assessing the effects of drought stress and, when compared with well-watered plants, provide a record of the timing, length and severity of a drought.

'Simple' scheduling aids

To address the problem of complexity, attempts have been made to develop 'simple' scheduling aids. For example, a device based on the water balance approach was promoted in South Africa for sugar cane growers by George (1988). Using long-term average evaporation data, and a measure of crop cover, together with an estimate of the allowable soil water deficit, it provided a visual indication, using coloured pegs placed in a 'pegboard', of when individual fields next needed to be irrigated.

Again with the same aim of simplicity, Torres (1998) described how, in Colombia, a suitably calibrated cylindrical plastic bucket (0.3 m × 0.4 m deep) could be used to schedule irrigation of sugar cane. Based on a simple water balance, with a specified allowable depletion of the available water, it served as a visual aid of when to irrigate and how much water to apply. Its cheapness and simplicity made it appropriate for use by small holders.

Tensiometers

At another level of complexity, Bell et al. (1990) used an array of tensiometers in Mauritius for characterising the soil water status beneath a drip-irrigated row crop. This provided a means of describing numerically the soil water potential distribution within a given volume of soil to be used by designers, researchers and managers of commercial plantations. It is not known if this approach has been applied in the ways suggested, but Hodnett *et al.* (1990) compared a method of scheduling irrigation based on this concept (using 'index' tensiometers) with the water balance method with some success.

Again with drip-irrigated sugar cane, Wiedenfeld (2004) compared four different methods of scheduling over three successive seasons in south Texas. Two were based on a water balance approach (pan evaporation and ET_0–Penman–Monteith), and two on tensiometers (automatic and manual). All four methods were judged to be effective, although automatic tensiometers were found to be unreliable, with each prescribing similar quantities of water. The two water balance methods required crop coefficients appropriate to the location.

Simulation models

In an excellent review, Lisson *et al.* (2005) compared and contrasted the two main dynamic sugar cane simulation models, APSIM-Sugarcane and CANEGRO, and highlighted the role that modelling can/could play in the management of sugar cane production systems. This included: irrigation scheduling, optimising the allocation of limited water supplies, and assessing water storage options. Full details of the scientific basis of CANEGRO can be found on the web through ICSM (2008). More recently, Singels *et al.* (2010) proposed ways in which the water uptake component of the model could be improved.

Irrigation scheduling services

In South Africa, Singels and Smith (2006) described an irrigation scheduling service consisting of a web-based simulation model[3] (CANESIM) that estimates the recent, current and future water balance, crop status and yield from field information and real-time weather data. The system automatically generates and distributes simple, user specific, irrigation advice by a 'short message service' direct to farmers' cellular phones. An initial evaluation of the service on a small-scale sugar cane sprinkler irrigation scheme in South Africa indicated that, by following this advice, large reductions in water applied (33%) and deep drainage (64%) were possible. This was the result of reduced irrigation when the crop was young and during the winter months. Reliable feedback from the farmers was necessary for the service to be truly effective.

Recently, a web-based irrigation management service has been introduced in Queensland, Australia. Known as WaterSense, it is designed to assist growers who practise supplementary irrigation in areas where rainfall is variable, where water abstractions are at risk from restrictions, and whose sugar cane fields are at different stages of development. It provides 'real time' advice on how to minimise yield losses when water is limited for individual fields (Inman-Bamber *et al.*, 2008; Haines *et al.*, 2010). Data from a network of automatic weather stations are used to calculate reference crop evapotranspiration whilst the APSIM-Sugarcane model simulates canopy development and soil water processes. An irrigation schedule is derived and accessed on a dedicated web page on a centrally located server that stores the database for participating growers.

Conclusions

Robertson *et al.* (1997) used a systems modelling approach to evaluate the opportunities that existed for improving the use of limited water by sugar cane farmers with case studies in Australia and South Africa. The importance of trying to ensure that the maximum quantity of water (rain and irrigation) is stored in the root zone was emphasised, together with the problems of managing variability in the field. The paper questioned the widespread use of a common value for the water-use efficiency since it presupposes that it is relatively stable across production systems and environments (the APSIM-Sugarcane model suggested that it could vary from 50 to 150 kg ha^{-1} mm^{-1}). The authors agreed that concepts such as effective rainfall and water-use efficiency still have a useful role to play as benchmarks against which to judge performance.

Many approaches to irrigation scheduling have been proposed to suit all levels of complexity, but few are widely used by irrigators of sugar cane. This will probably remain the case until the availability and cost of water becomes a real constraint. Useful practical bulletins on the irrigation of sugar cane have been published amongst others in South Africa (SASA, 1977) and in Australia (Holden, 1998).

Drivers for change

For many row crops, including sugar cane, the adoption of modern water-saving irrigation technologies is often cited as key to increasing water-use efficiency while maintaining current levels of production (Green *et al.*, 1996). However, new technology requires greater capital investment, so irrigators are often reluctant to adopt new systems unless they can be convinced of the likely benefits. Where water costs are low, sugar cane growers have little incentive to switch technology to improve efficiency unless there are other externalities that might influence their ability to maximise net crop return. But rising energy, labour and water costs, the need to increase water productivity, less water available for abstraction due to expansion of cropped areas, increasing competition for limited resources, climate change risks (Knox *et al.*, 2010) and demands for greater environmental protection are now the driving forces influencing technology choice in irrigated sugar cane production. In this context, better scheduling may prove to be a useful adaptation strategy.

The complexity of justifying a new investment in irrigation in areas where irrigation is supplementary to variable rainfall was demonstrated by Inman-Bamber *et al.* (1999). Basing their analysis on two case studies in Queensland, Australia, they compared measured yield responses to irrigation with simulated values using the APSIM-Sugarcane model. They then predicted the likely benefits in cash terms of an investment in different irrigation systems for a range of soils and scheduling options over a sequence of years. A positive return on an investment in supplementary irrigation was by no means a foregone conclusion despite favourable yield responses. Irrigation application efficiency was, for example, an important determinant of profitability.

Conclusions

For irrigation of sugar cane to be sustainable, the issues of concern that still needed to be addressed by the sugar industry (in Australia) were highlighted by Meyer (1997). Of overriding importance was drainage management for effluent control (excess water, salts, agrochemicals and nutrients), the need to maximise water productivity and *to learn from history*.

By contrast, in the conclusion to his review paper, Thompson (1976) argued that 'the degree of sophistication of irrigated sugar cane farming ... is such that little will be gained from further research into the water requirements of the crop in the immediate future'. Thirty-five years later how justified was that statement?

On the basis of this review, progress of generic importance since 1976 appears to have been made in the following areas:

1. The sequential responses to drought have been quantified in terms of changes in leaf water potential.
2. Mechanisms responsible for drought tolerance are better understood.
3. Factors influencing the partitioning of dry matter and sucrose within the plant are better understood.
4. Factors influencing root growth rates and distribution, and soil water availability, are better understood.
5. Ways of assessing crop water requirements, including realistic values of the crop coefficient, have been ratified.
6. Opportunities to save water by withholding irrigation at different growth stages have been identified.
7. Irrigation systems have been developed that allow water to be applied at the right time with improved precision (but at a cost).
8. New ways of scheduling irrigation have been developed and promoted but little evidence is yet available regarding levels of uptake.

How much of this knowledge has led to improvements in water management and increases in water productivity is not known. Probably not enough! The 'drivers of change' may not allow this state of affairs to continue much longer.

Summary

The results of research on the water relations and irrigation need of sugar cane are collated and summarised in an attempt to link fundamental studies on crop physiology to irrigation practices. Background information on the centres of production of sugar cane is followed by reviews of crop development, including roots, plant water relations, crop water requirements, water productivity, irrigation systems, and irrigation scheduling. The majority of the recent research published in the international literature has been conducted in Australia and southern Africa. Leaf/stem extension is a more sensitive indicator of the onset of water stress than stomatal conductance or photosynthesis. Possible mechanisms by which cultivars differ in their responses to drought have been described. Roots extend in depth at rates of 5–18 mm d^{-1} reaching maximum depths of >4 m in c. 300 d providing there are no physical restrictions. The Penman–Monteith equation and the USWB Class A pan both give good estimates of reference crop evapotranspiration (ET_0). The corresponding values for the crop coefficient (K_c) are 0.4 (initial stage), 1.25 (peak season) and 0.75 (drying off phase). On an annual basis, the total water use (ET_c) is in the range 1100–1800 mm, with peak daily rates of 6–15 mm d^{-1}. There is a linear relationship between cane/sucrose yields and actual evapotranspiration (ET) over the season, with slopes of about 100 (cane) and 13 (sugar) kg ha^{-1} mm^{-1} (but variable). Water stress during tillering need not result in a loss in yield because of compensatory growth on rewatering. Water can be withheld prior to harvest for periods of time up to the equivalent of twice the depth of available water in the root zone. As alternatives to traditional furrow irrigation, dragline sprinklers and centre pivots have several advantages, such as allowing the application of small quantities of water at frequent intervals. Drip irrigation should only be contemplated when there are well-organised management systems in place. Methods for scheduling irrigation are summarised, and the reasons for their limited uptake considered. In conclusion, the 'drivers for change', including the need for improved environmental protection, influencing technology choice if irrigated sugar cane production is to be sustainable are summarised.

Endnotes

1 The reader is referred to a review paper by Sumner (1997), focusing on sugar cane, which describes the chemical and physical degradation of soils resulting from organic matter depletion, crust formation, acidification, salinisation and sodification. Available strategies for amelioration and case studies with successful outcomes are discussed.
2 The ratio of the actual leaf extension rate to that of the control well-watered treatment.
3 http://sasex.sasa.org.za/irricane/index.htm

10 Tea

Introduction

Tea (*Camellia sinensis* L.) is believed to have originated within the fan-shaped area extending from the Assam/Burma border in the west to China in the east (*c.* 26° N), and south from this line through Burma and Thailand to Vietnam (*c.* 14° N) (Kingdom-Ward, 1950; Mair and Hoh, 2009). This is an area of monsoon climates with a warm, wet summer and a cool, dry (or less wet) winter. From the main centres of cultivation in South-east Asia tea has been introduced into many other areas of the world and is now grown in conditions that range from Mediterranean-type climates to the hot, humid tropics, from Georgia in the north (42° N)[1] to Argentina (27° S) and New Zealand (37° S) in the south, and from sea level to 2700 m altitude (Carr, 1972). By far the largest producer of tea is China (estimated planted area by 2008 = 1.4 million ha; annual production of processed tea = 1.3 million t) followed by India (474 000 ha; 800 000 t) and then Kenya (158 000 ha; 345 000 t) and Sri Lanka (222 000 ha; 319 000 t) (FAO, 2010b).

In this chapter, the contribution of research into the water relations and irrigation requirements of tea over the last 40 years to commercial crop management, and to our understanding of the physiology of the crop, is summarised and reviewed. The focus is on work done in eastern Africa where the Tea Research Foundation of Kenya (TRFK; 0° 22c′ S; alt. 2200 m), the Tea Research Institute of Tanzania (TRIT, 8° 32′ S; alt. 1840 m), and the Tea Research Foundation of Central Africa (TRFCA, 16° 05′ S; alt. 630 m), based in Malawi, have all undertaken fundamental research on these topics. Prior to 1978, the Tea Research Institute of East Africa (TRIEA) served three countries – Kenya, Tanzania and Uganda. Some of the major private tea companies also undertake their own research.

Commercial tea production in Africa started in the late nineteenth century when missionaries planted seedlings in Malawi. Afterwards (1920s/1930s), tea industries began to be established in Kenya (planted area by 2008 = 158 000 ha, annual production of processed tea = 345 000 t), Uganda (21 000 ha, 43 000 t) and Tanzania (19 000 ha, 35 000 t). These three countries, together with Malawi (19 500 ha, 48 000 t), are still the leading producers in Africa followed by Rwanda (12 000 ha, 20 000 t), Burundi (9000 ha, 7700 t), Zimbabwe (6700 ha, 22 300 t) and Mozambique 8500 ha, 16 900 t). Smaller industries exist in the Democratic

Figure 10.1 Tea. In Africa away from the equator (e.g. in southern Tanzania, Malawi and countries to the south) there is a single dry season that can last from four to six months. In these areas, tea is often irrigated during the dry season – Malawi (MKVC). (See also colour plate.)

Republic of the Congo, Ethiopia, South Africa and Zambia and in West Africa, Cameroon and Nigeria. In 2008, Africa produced 548 000 t of processed tea, which is equivalent to about 11% of the total world production (4.9 million t; FAO, 2010b).

In the equatorial areas of eastern Africa where tea is grown (Kenya, Uganda, northern Tanzania, Rwanda and Burundi) rainfall distribution is bimodal with two dry, or less wet, seasons (Stephens *et al.*, 1992). Although some tea in these areas may be irrigated, drought mitigation is usually more appropriate and cost effective. Away from the equator, in southern Tanzania, Malawi and countries to the south, there is a single dry season that can last from four to six months. In these areas, tea is often irrigated during the dry season (Figure 10.1) (Carr and Stephens, 1992). The rainfall amount and distribution in many tea locations in Africa is often considerably modified by local physical features such as altitude (tea is found from 600 to 2700 m a.s.l.), proximity to a large body of open water (e.g. Lake Victoria) or to mountains (e.g. the Rwenzoris in western Uganda, or Mulanje mountain in southern Malawi). Research priorities at the principal tea research organisations are influenced by these regional differences in rainfall, availability of water for irrigation, expected yield increases and likely financial benefits.

The results of research undertaken at the Ngwazi Tea Research Unit/Station (NTRS, now the headquarters of TRIT) since 1967 are reviewed in detail, but in the context of other research undertaken in eastern Africa and elsewhere in the world. This chapter extends a previous evaluation of the developmental and scientific impacts of research on irrigated tea carried out intermittently over a 30-year period in Tanzania (Carr, 1999). Other reviews of aspects of the ecophysiology of tea include those by Carr (1972), Fordham (1977), Squire and Callander (1981), Carr and Stephens (1992), Tanton (1992) and, most recently, De Costa et al. (2007).

Crop development[2]

It is generally believed that existing populations of tea plants are largely derived from two original taxa, which are given varietal status within the species Camellia sinensis by Sealy (1958) but specific status by Wight (1962). Wight refers to these species as C. sinensis (the China plant, which is a shrub, with small leaves, thought to have originally grown in the open) and C. assamica (the Assam plant, which is a small tree, with large leaves, thought to have originally grown in forest). The two species can be distinguished by leaf form and floral characters (Wight, 1959; Banerjee, 1992). Tea is a highly heterogeneous outbreeder, which results in a cline extending from extreme 'China type' plants to those of distinct 'Assam' origin (Wight, 1959). Generally all teas are classified under the name C. sinensis (L.) O. Kuntze irrespective of taxonomic variation (Paul et al., 1997).

In Africa, tea plantations were originally established with heterogeneous seedlings derived from seeds imported from India (although in Malawi some seeds were of Chinese origin). From the 1960s onwards, most new plantings were with vegetatively propagated clones selected for superior yield and/or tea-making qualities (Ellis and Nyirenda, 1995). Recently, AFLP markers have been successfully employed to detect diversity and genetic differentiation among Indian and Kenyan populations of tea. Principal coordinate analysis showed that the Assam genotypes/clones selected in Kenya were indeed of Indian origin (Paul et al., 1997).

Vegetative

In its natural state, the tea plant grows to become a tree of moderate size (or a shrub); under cultivation, however, it is pruned horizontally (typically at two to five year intervals) to form and maintain a low spreading bush (Figure 10.2). This increases the number of young tender shoots, which supply the produce, and allows them to be removed at relatively frequent intervals. The growth of a plant is therefore continually curtailed by cultural operations, which although familiar to present-day tea culture are alien to the normal growth processes of the tea plant (Carr, 1970).

After planting container-grown rooted plants in the field at commercial densities, usually in the range 10 000–15 000 ha^{-1}, the young plants are encouraged to develop a low spreading frame by a range of techniques, a process known as

Figure 10.2 Tea. In its natural state, the tea plant grows to become a tree of moderate size; under cultivation it is pruned horizontally (typically at two to five year intervals) to form and maintain a low spreading bush –Tanzania (MKVC).

bringing into bearing. These include the removal (or bending) of the dominant shoot (*decentreing*) to encourage branching, the continuous removal of harvestable shoots above a specified height (*tipping*), pruning of woody branches at one or more fixed heights above the ground (*formative pruning*), and spreading of lateral branches by *pegging*. The aim is to encourage the canopy to cover the ground, to intercept radiation and to come into profitable, commercial production as quickly as possible at least cost. Each method, or combination of practices, has its proponents, but all can have an effect on the shoot or canopy to root ratio and hence the susceptibility to drought of immature tea plants.

In Malawi, the most common clones (referred to locally as 'superior cultivars') are SFS150, SFS204 (both field selections, released in the 1970s), PC81, PC105, PC108 and PC110 (selections from crosses in a breeding programme, mostly released in 1981), followed by smaller areas of a large number of more recent releases, including PC122 and PC123 (released in 1994). These clones (and related composites/grafted plants) have all been selected for good field performance and high quality for planting in southern Africa (from 11°S to 32°S) (Ellis and Nyirenda, 1995). Their properties (including observations on drought tolerance) have been described in a catalogue (TRFCA, 2000).

In eastern Africa, the most popular clones include: S15/10, 6/8, 6/10, 31/8, BBK35, TN14/3, BBT207 (all long-established field selections), SC12/28 and the 303 series (selected from crosses, and released in 1987 and 1994 respectively).

These clones were selected and evaluated by commercial companies and/or research institutes, usually at single sites. In Kericho, Kenya, four of these clones with contrasting characteristics (S15/10, 6/8, BBK35, TN14/3) were compared at a range of altitudes (1800–2200 m) and significant genotype × environment interactions were demonstrated (Ng'etich and Stephens, 2001a) emphasising the complexity of selection for the diverse geographic locations where tea is grown (Wachira *et al.*, 2002).

In both Malawi and Kenya, grafting of clones onto suitable rootstocks to create higher yielding, high-quality composite plants is being promoted as a commercial practice (Kayange *et al.*, 1981; Tuwei *et al.*, 2008a) and to improve drought tolerance (Tuwei *et al.*, 2008b). Trials on grafting have also been conducted in Tanzania (Mizambwa, 2002a).

Components of yield

There are three principal contributors to the annual yield in tea: (1) the mean *number of shoots* harvested per unit area (N, m^{-2}); (2) the mean shoot *dry mass* at harvest (M, g); and (3) the *number of harvests* in a year. When a shoot is harvested, the uppermost axillary bud on the residual shoot is released from apical dominance and begins to develop into a new shoot. The time taken for this shoot to reach a harvestable size and stage of development (two or three leaves and a terminal bud) is known as the *shoot replacement cycle* (S, d). The annual yield of tea (Y, g m^{-2}) can therefore be expressed in the following way:

$$Y = N \times M \times 365/S.$$

However, when tested with experimental data from Tanzania, this simple model consistently overestimated yields in low input plots for no obvious reason, although there was a very good linear relation between predicted and actual yields ($r^2 = 0.83$, $n = 9$) (Stephens and Carr, 1994). It is possible that the number of shoots that reached harvestable size in the low-input plots was overestimated. There is the additional complication that more than one shoot can sometimes develop from a residual shoot: that is, from secondary and tertiary leaf axils. Each component of yield is considered in turn below.

Number of shoots

This is probably the main determinant of yield. It varies with the clone, inputs such as nitrogen and water, temperature and stage in the pruning cycle. We can define the number of *harvested shoots* (usually those with two or three unfolded leaves and an unopened terminal bud) together with the number of *basal shoots* (small shoots remaining after harvest) as the *total shoot population*. It is also sometimes necessary to differentiate between shoots with terminal buds that are active and those with buds that are dormant, or *banjhi* (Stephens and Carr, 1990, 1994). For example, in southern Tanzania, the mean annual, basal shoot population density increased from 310 to 560 m^{-2}, averaged across all fertiliser and

Figure 10.3 Tea. A shoot with three leaves and a bud can weigh up to 50% more than a shoot with two leaves and a bud (clone 6/8) – Tanzania (MKVC).

irrigation treatments, over a three-year period, with peaks of 850 m^{-2} in well-fertilised, well-watered tea (clone 6/8). Applying fertiliser increased the number of shoots harvested in all treatments compared with tea without fertiliser, and increased the proportion of shoots that were actively growing. Irrigation had no effect on the annual mean basal shoot population density, but within seasons there were considerable differences. The main effect of water stress was to delay the peak basal shoot population density from the warm dry season to the early rains, without affecting the annual mean total. Over the period of the experiment, the *shoot replacement ratio* increased from 1:1.1 to 1:1.6 in the high input plots. That is, each harvested shoot was replaced by 1.6 new shoots (Stephens and Carr, 1994).

Shoot mass

The *fresh mass* of an individual shoot increases linearly with the number of leaves on the shoot. The slope of the relationship varies with the clone, season and irrigation (0.16–0.35 g $leaf^{-1}$) but not with fertiliser level (Stephens and Carr, 1994; Burgess *et al.*, 2006). A shoot with three leaves and a bud can weigh up to 50% more than a shoot with two leaves and a bud (Figure 10.3). The *dry matter content* of a shoot varies with the season and also with fertiliser and irrigation (range 0.19–0.30; Burgess, 1992b). The product of fresh mass times dry matter content gives the *shoot dry mass*. Processed (or made) teas leaving the factory usually have water contents of 2–4%: that is, their mass is 2–4% more than the dry mass.

Shoot replacement cycle

Shoot growth can be considered as a three-stage process. Stage 1 is a long lag phase as the axillary bud slowly expands, and as the leaf primordia within it develop. Stage 2 is a period of rapid shoot extension and leaf development (the unfolding of leaves). Stage 3 is when the terminal bud becomes dormant. The duration of the shoot replacement cycle (i.e. stage 1 and part of stage 2) is mainly a function of temperature, but it is also influenced by water stress, nutrition, and the dryness of the air, and varies between individual clones. To aid our understanding of the processes involved, it is necessary to distinguish between shoot *extension* (increase in length) and shoot *development* (change in appearance). Rates of shoot extension and development are both temperature dependent. The base air temperature for shoot extension (T_{be}) is in the range 8–13°C, depending on the clone. The optimum mean temperature is about 24–26°C, with growth rates declining at temperatures above 30–35°C. For comparison, the base temperature for shoot development is about 2–3°C less than that for extension. For this reason, internodes are shorter during the 'winter' months (or at high altitude) than in the 'summer' (or low altitudes), and tea is then more difficult to harvest by hand (Stephens and Carr, 1990, 1993; Burgess and Carr, 1997).

This approach to understanding the effects of climate and weather on crop developmental processes has been reviewed by Carr and Stephens (1992). In summary, Squire (1979) demonstrated a linear relationship between shoot extension rates and mean air temperature over the range 17–25°C. By extrapolation backwards the base temperature for shoot extension was identified (about 12.5°C for clone SFS204). Squire (1979) and afterwards Tanton (1982a) then applied the concept of thermal time to predict the time taken for a bud released from apical dominance to reach a harvestable size. Subsequently, a total of 475 day degrees became the accepted value (although unconfirmed) in Malawi, summed above a base temperature of 12.5°C. This has allowed the seasonal effects of temperature on the duration of the shoot replacement cycle to be quantified and geographical sites compared. A limitation to this approach occurs when temperatures exceed the optimum for growth (>30–35°C), and/or when the saturation deficit of the air exceeds a value of about 2.0–2.3 kPa (Squire, 1979; Tanton, 1982b). As a practical example of its importance, Carr *et al.* (1987) demonstrated how dry air could limit the responses to irrigation in the hot, dry season in Malawi (at a relatively low altitude, 650 m), but rarely in southern Tanzania (at a relatively high altitude, 1800 m). In other words, irrigation cannot substitute entirely for rainfall.

Subsequently, Stephens and Carr (1990) showed how the apparent base temperature for shoot extension varies between clones, from 14–15°C (BBT1) to 10.5 °C (clones BBT28 and BBT36) to as low as 7.5°C (S15/10; Squire *et al.*, 1993). On the basis of measurements on seven clones in Malawi, Smith *et al.* (1993c) urged caution about taking too simplistic an approach to shoot extension and overreliance on the view that there is an inherent base temperature characteristic of a particular clone. The analysis by Stephens and Carr (1990) also showed how the results of studies like these could be biased by the shoot selection technique employed. If shoots already 30–50 mm long and actively growing are selected for measurement, there is a risk that the growth rates of only inherently fast-growing shoots are recorded. To obtain a representative sample it is essential to select axillary buds at random immediately after harvesting for subsequent measurement.

Using this approach, Burgess and Carr (1997), in a comparison of six clones at NTRS, found that an exponential function with two constants, an initial shoot length and a relative extension rate (r), provided a realistic description of the length of axillary shoots at the end of the period from the release from apical dominance to the unfurling of three true leaves. Differences between clones in the values of T_{be} (8.9°C; SFS150 to 11.3°C; BBK35) and in thermal extension rates (r, derived from linear relations between r and the mean air temperature) could be used to explain seasonal differences in yield in southern Tanzania. The apparent base temperatures for shoot development (T_{bd}, the unfolding of leaves) were consistently 1.7 (BBT207) to 3.4°C (6/8) below those for extension; an observation supported by the effects of temperature on the length of the internodes. The value of T_{be} for young clone 6/8 plants (9.5°C) is close to that identified for mature plants of the same clone (10.0°C), which was also some 2–3°C above T_{bd} (Stephens and Carr, 1993). Both sets of results apply to well-fertilised, well-watered plants, but nutrition and water supply also influence how shoots respond to the environment. For example, shoots of well-fertilised tea were always longer (at a given stage of development) than those from unfertilised plants, while the length of shoots with three leaves and a bud ranged from 15 mm in unirrigated plots at the end of the dry season to 130 mm in high-input plots at the start of the rains. Similarly, the duration of the shoot replacement cycle varied from 65 d (warm, wet season) to 95 d (cool, dry but irrigated) for high-input plots, and from 75–180 d for unirrigated, unfertilised tea, results that all have commercial implications in terms of the choice of clones, harvesting policies and planning, and yield distribution (Stephens and Carr, 1993).

Yield distribution

Crop yield is not uniform during the year. Large peaks in production can occur after a limiting factor, such as low temperature or drought, has ended and allowed the accumulated buds of many ages to develop together. This large peak, such as that which follows the start of the rains in southern Malawi, or for irrigated crops the rise in temperatures after the cool winter period in southern Tanzania, can cause major logistical problems for farmers and factory managers. There is then a decline in production until the next generation of shoots has developed. A second, but smaller, peak in production is followed by a third and sometimes a fourth until once again drought and/or low temperatures reduce rates of development and the cycle is repeated. The scale of these oscillations is proportional to the degree of synchronisation induced during the period of stress (Stephens and Carr, 1990).

The initial yield peak followed by the subsequent decline is often referred to as the '*Fordham effect*' after the scientist who first described and modelled the process in Malawi. It must be emphasised that the period between peaks is one of active growth and not dormancy (Fordham, 1970, 1977; Fordham and Palmer-Jones, 1977).

To assist growers in Tanzania to plan a harvesting schedule, relationships were derived for six contrasting clones between the 'leaf appearance rate' (leaf d^{-1}, equivalent to the reciprocal of a phyllochron[3]) and the mean daily air temperature (range 13–20 °C). It was assumed that the most advanced cohort of shoots remaining on the bush after a harvest has, on average, one unfolded leaf, and that these will form the basis for the next harvest. If the target shoot for harvesting is then three leaves and a bud, the time to the next harvest will be two phyllochrons. A simple procedure has been described for determining harvest intervals from mean air temperature. Allowances can also be made for the effects of drought using estimates of the potential soil water deficit (Burgess and Carr, 1998).

Burgess (1994a) has highlighted the importance of restricting the increase in height of the plucking surface, if potential yields are to be achieved. Each unnecessary 10-mm increase represents an annual yield loss of 70–150 kg ha^{-1}. Growers in Tanzania are recommended to set limits on the maximum allowable annual increases in crop height (not more than 80 mm for Assam-type clones, or 100 mm for China-type clones). This is also very important in research, where yield responses to inputs like irrigation and fertiliser can be greatly underestimated if harvesting is not strictly controlled and crop heights regularly monitored. It also influences the frequency of pruning, allowing pruning intervals to be extended from the conventional three to four years to five or six, or even more, as has been possible with many of the irrigation experiments in Tanzania reported below.

The impact of harvesting method and timing on yields of contrasting clones is illustrated in the results of a long-term experiment in Tanzania (Burgess *et al.*, 2006), while Mouli *et al.* (2007) have emphasised the importance of specifying the 'intensity of harvest' (IoH) if rational comparisons are to be made between different harvesting methods. The IoH is defined in terms of the number of leaves (or axillary buds) left behind after a shoot is harvested. Since yields increase with more intense (harder) harvesting, Mouli *et al.* (2007) recommended that IoH should be measured routinely in experiments with tea. For example, specifying only the duration of the interval between harvests limited the value of a long-term (10 year), multi-site, single clone (BBK35), harvesting frequency experiment in Kenya (Owuor *et al.*, 2009).

Roots

Root depth

In Malawi, roots of mature seedling tea have been found at depths of at least 5.5 m, with evidence of water extraction during the dry season at that depth (Laycock and Wood, 1963). At TRFCA, the soils are deep (>5 m), well-drained latosols, classed as clays to sandy-clays with a water-holding capacity of about 130 mm m^{-1} (Willatt, 1970; Carr *et al.*, 1987). Similarly in Kenya, Kerfoot (1961, 1962) observed roots of seven-year-old seedling tea at depths of 3 m. He also reported that roots of tea had been exposed at depths of more than 6 m elsewhere in East and Central Africa. Later in Kericho, Kenya, Cooper (1979) traced roots of 17-year-old seedling tea to depths below 6 m. The soil at this site is a deep and red, freely draining friable clay (Humic Ferralsol) with an available water holding capacity of about 215 mm m^{-1}. In southern Tanzania, roots of eight-year-old seedling tea were found at 4.5 m (Carr, 1969), and those of a 23-year-old clone (6/8) at >5 m, both irrigated (Nixon *et al.*, 2001). Here the soil is classified as a Xanthic Ferralsol, with a brown, medium to fine textured topsoil over a deep, very light coloured, unmottled clay subsoil. The available water-holding capacity averages about 100 mm m^{-1} (Baillie and Burton, 1993; Carr, 1974). In Africa, tea is definitely not a shallow-rooted crop.

In Malawi, Willatt (1970) illustrated the beneficial effects of irrigation on root depth and distribution of four clones in the first year after field planting. By contrast, grass mulches reduced the dry root mass (by 20%) and rooting depth (from 1.20 m in the control to 0.60–0.90 m depending on the mulch used) and influenced root distribution in three-year-old plants (clone 6/8) in Kenya (Othieno and Ahn, 1980). These and other examples are summarised in Table 10.1 (Figure 10.4). They include observations of the root systems of two rain-fed, three-year-old clones brought into bearing by pegging or pruning at TRFK, Kericho, Kenya. Pruning appeared to reduce the size of the root system compared with pegging (Carr, 1976).

Detailed observations made at NTRS in southern Tanzania on irrigated tea are also summarised in Table 10.1. For example, Burgess and Carr (1996b) showed how the maximum rooting depth of four irrigated (and mulched) clones (BBT1, 6/8, SFS150 and S15/10) increased linearly with time after field planting (from 12 to 48 months) at similar rates (averaging 2.0 ± 0.11 mm d^{-1}), reaching 2.8 m depths within four years (clone S15/10). By contrast, comparisons of four clones at four rain-fed sites in Kericho, Kenya, showed roots reaching depths of 1.0 to 1.5 m three years after planting (averaging 1.0 to 1.2 mm d^{-1}), depending on site (Ng'etich and Stephens, 2001b). The relationship between rooting depth and time from planting was, however, in this case best described by an exponential equation. Clone S15/10 had the shallowest roots at each site and TN14/3 the deepest: clones 6/8 and BBK35 were intermediate. Maximum rooting depths for each clone increased linearly with the mean air temperature over the range 16 to 19.5°C. Previously, Othieno and Ahn (1980) had demonstrated the positive influence of soil temperature (16–22°C) on root mass.

Root distribution
In order to help to explain differences in response to irrigation by young and mature tea, a detailed comparison of root distribution of 'young' (six years after

Table 10.1 Examples of factors influencing the maximum rooting depth of tea at different locations in Africa

Site	Cultivar	Age[a] (years)	Depth (m)	Treatments	Source
Kericho, Kenya	Clone 12/12	3	0.72	Pegged; rain-fed	Carr (1976)
			0.94	Pruned; rain-fed	
	Clone 6/8		1.43	Pegged; rain-fed	
			1.17	Pruned; rain-fed	
	Seedlings	9	1.4–1.6	Rain-fed	
	Clones 6/8, 6/10, 6/11	7	3.0	Rain-fed	Carr (1977a)
	Clone 6/8	3	0.6–1.2 m	Depending on the mulch used; rain-fed	Othieno (1977)
	Seedlings	17	>6.0	Rain-fed	Cooper (1979)
	Clones 6/8, S15/10, TN14/3, BBK35	<3	1.2–1.5	Rain-fed	Ng'etich and Stephens (2001b)
Ngwazi, Tanzania	Clones BBT1, 6/8, SFS150, S15/10	2.25	1.4–1.7	Irrigated and part irrigated	Burgess and Carr (1996b)
	Clone S15/10	4	2.8	Irrigated	
	Clone 6/8	23	>5.0	Irrigated	Nixon et al. (2001)
		4–5	>3.0		
	Seedlings	8	4.5	Part irrigated	Carr (1974)
	Clone S15/10	3	2.3	Irrigated	Kigalu (1997)
	Clone BBK35		1.7		
	Clone S15/10	6	3.6–4.3	Plant density/ water variables	Kigalu (2002)
	Clone BBK35		2.5–3.0		
Marikitanda, Tanzania	Clone BBK35	4	1.5	Rain-fed	Sanga and Kigalu (2006)
Mambilla, Nigeria	Clone 31/8	2	0.6–0.8	Rain-fed	Carr, personal observation (1978)
Mulanje, Malawi	Clones MT12, SFS204, SFS371	0.75	0.54–0.60	Irrigated	Willatt (1970)
			0.45	Rain-fed	

[a] Time from planting in the field.

field planting) and 'mature' (23 years) clone 6/8 plants was undertaken at NTRS, Tanzania, in 1994 (Nixon et al., 2001). Both crops had been irrigated since planting. Sampling was restricted to the top 3.0 m, although roots of both crops went deeper than this. The total dry mass of structural roots (>1.0 mm diameter) to 3.0 m depth was four times greater in the mature crop (5.82 kg plant^{-1}) than the young crop (1.56 kg plant^{-1}), and for fine roots (<1.0 mm) eight times greater (1.86 and 0.24 kg plant^{-1} respectively). For both crops, over 85% of the structural roots (dry weight, g l^{-1}) were found in the top 0.40 m of soil, but the absolute weights again differed by a factor of four. The corresponding shoot:root ratios (dry mass) were about 1:1 and 2:1 for 'old' and 'young' plants respectively.

Figure 10.4 Tea. Clones differ in the depth and distribution of roots, but not consistently across sites – Kericho, Kenya (MKVC).

By contrast, the concentration of fine roots (sampled towards the end of the dry season) was much lower in the corresponding mature clone 6/8 plants that had not been irrigated for the previous nine years (0.13 g l^{-1}, to 3 m depth) compared with the equivalent for well-irrigated plants (0.53 g l^{-1}). The weight and distribution of the structural roots, and the shoot (or canopy):root ratios (1:1), were, however, similar under both water regimes (Nixon and Sanga, 1995).

Root depth and distribution were monitored in selected treatment combinations in a plant density × drought × clone experiment at NTRS. There were six spacings, giving densities ranging from 8333 (D1) to 83 333 (D6) plants ha^{-1} (Kigalu, 1997) of which three only are considered here. By August 1995, 32 months after planting (and mulching), roots of well-watered clone BBK35 had reached depths of 0.9 m (D1) to 1.1 m (intermediate density, D3 = 16 667 ha^{-1}), and those of clone S15/10 from 1.5 m (D1) to 2.1 m (D3). Roots in treatment D6 went less deep than those in D3. The largest effect of plant density on root distribution occurred in the top 0.2 m where root density increased with plant density for both clones. Below this depth the largest root density occurred with treatment D3, followed by D6 (S15/10 only), and then D1. S15/10 had more fine roots below 0.2 m than K35 at all three densities. The corresponding rates of increase in root depth equated to about 0.9–1.1 mm d^{-1} (BBK35) and 1.6–2.2 mm d^{-1} (S15/10).

It is of interest to note that relative differences between clones in rooting depth were not consistent between sites. Thus, roots of three-year-old BBK35 plants were shallower than those of S15/10 in Ngwazi, but deeper in Kericho.

The treatment effects on root depth and distribution were again investigated at NTRS in early 1999, six years after field planting and four years after differential irrigation/drought treatments had first been imposed. This time the two extreme densities were compared (high, D6, and low, D1) at each of the two extreme drought treatments (I0, most droughted, a total of 40 weeks without irrigation/rain, and I6, well-irrigated since planting) for both clones. Three plants from each treatment combination were excavated at 0.20-m-depth increments. The results were not easy to interpret. Although on average, roots of clone S15/10 reached depths (4.0 m, equivalent to about 1.8 mm d^{-1}) substantially greater than those of BBK35 (2.8 m, 1.3 mm d^{-1}), the effect of water regime was not consistent between the two clones. For S15/10, dry conditions (I0) resulted in deeper rooting (4.3 m, 2.0 mm d^{-1}) than wet (3.6 m, 1.6 mm d^{-1}). For clone BBK35 the situation was reversed, with droughted plants rooting less deeply (2.5 m, 1.1 mm d^{-1}) than those well-watered since planting (3.0 m, 1.4 mm d^{-1}). The effect of plant density was more complicated: for S15/10 roots of high-density plants were about 0.2–0.4 m shallower than those grown at low density; by contrast, for BBK35, although roots of droughted plants were 0.8 m shallower at low density, they were 0.2 m deeper when irrigated (Kigalu, 2002).

In terms of root distribution, there were more fine roots (<1.0 mm) at the low plant density than at the high density (expressed as g l^{-1} soil). For droughted plants this difference extended to depths of 2.0 m with both clones, but to only 0.5–1.0 m for irrigated bushes (Kigalu, 2002). In all four treatment combinations, BBK35 had fewer fine roots than S15/10, low-density plants had more than high-density plants and, with one exception (S15/10, high density), droughted plants had more than those well irrigated. The largest differences (averaged over 1.0 m depth only) in fine root densities for (a) BBK35 were between low plant density, well irrigated (0.4 g l^{-1}) and low density, droughted (1.5 g l^{-1}), and for (b) S15/10 between high density both droughted and well irrigated (0.9 g l^{-1}) and low density, droughted (2.0 g l^{-1}).

The canopy:root ratios (dry mass), as calculated from the slope of the linear relationship between the cumulative totals recorded in sequential whole plant harvests over the three years following planting, were independent of plant density but were consistently less for BBK35 (3:1) than for S15/10 (5:1) (Kigalu, 1997).

Root extension

The growth of tea (white, unsuberised) feeder roots was observed and monitored against glass in several simple underground root chambers over a 21-month period 1968–70 at Ngwazi, Tanzania (Carr, 1969, 1971a). The principal observations were:

- Roots of irrigated China-type BBT1 grew throughout the 'winter' months (June–September) when shoot growth was negligible.
- Roots of previously unirrigated BBT1 began to grow only after the first 'flush' of shoot growth was coming to an end, six to eight weeks after water was first added to the soil.

- Roots of two Assam-type clones (BBT28 and BBT36) grew only slowly during the 'winter', although some shoots continued to grow, albeit slowly.

In a similar study in southern Malawi, Fordham (1972b) observed roots of both young clones (SFS204, MT12 and MFS76) and mature seedling tea under irrigated field conditions in simple root observation trenches. For the clones, periods of maximum shoot growth were associated with minimal root growth. For mature tea, there was a similar reduction of root growth during a period of intense shoot growth. Pruning caused roots to stop growing for approximately three months. Herd and Squire (1976) also observed a stimulation of root growth in the 'winter' months in Malawi.

Summary: crop development

1. There are three principal contributors to the annual yield in tea: the mean number of shoots, the dry mass of a shoot; and the number of harvests in a year.
2. Water stress affects all three components but especially the number of harvests (by extending the duration of the shoot replacement cycle).
3. Rates of shoot extension and development are both temperature dependent; the base temperature for shoot development is about 2–3 °C less than that for shoot extension.
4. Thermal time can be used to estimate the time taken for a bud released from apical dominance to reach harvestable size.
5. Dry air (saturation deficit >2.0 kPa) can restrict shoot extension rates.
6. Peaks of shoot production followed by troughs are the result of synchronised shoot development following the relief of water- or temperature-induced stress (the Fordham effect).
7. The roots of tea can extend to considerable depths (5–6 m), providing that there are no physical restrictions.
8. Roots extend in depth at rates of between 1 and 2 mm d^{-1} (temperature dependent), reaching 1.0 m in about 18–36 months after planting in the field.
9. About 85% of the structural roots (by mass) occur in the top 0.4 m of soil, regardless of the plant age.
10. The density of roots continues to increase at all depths with age from planting.
11. The canopy:root ratio (by mass) decreases with plant age.
12. Clones differ in the depth and distribution of roots, but not consistently across sites.
13. Methods of bringing young tea into bearing can influence the form of the resultant root system in young tea. Pruning results in the cessation of root growth.
14. There is no consistent evidence that irrigation reduces the size or depth of a root system (indeed there is some evidence for the opposite).
15. Grass mulches can reduce the depth and mass of root systems of young plants.

16. Planting density can also influence root depth and distribution, but the effects are complex, and appear to vary with clone and watering regime.
17. Roots can continue growing during 'winter' months when shoot extension rates are slow.
18. There is evidence of periodicity of root growth, with root growth alternating with active shoot growth.

Plant water relations

The water relations of tea have been studied at all three east African research sites. Emphasis has been on measurements of stomatal behaviour, photosynthesis, transpiration and xylem water potential in attempts to understand better how the crop responds to its environment (in order to inform field management practices), and to find ways of identifying drought-tolerant cultivars at an early stage in the selection process.

Stomata

Stomata occur only on the lower (abaxial) surface of the leaf at an average density of about 190 mm^{-2} (Fordham, 1971), 130 mm^{-2} (Squire, 1976), 150–200 mm^{-2} (Samson *et al.*, 2000; Olyslaegers *et al.*, 2002). By contrast, Ng'etich and Wachira (2003), in a comparison of common cultivars grown in Kenya (all diploid), reported densities in the range 240–312 mm^{-2} with significant differences between individual clones (triploid and tetraploid cultivars had smaller densities).

A range of techniques has been used to monitor the behaviour of tea stomata. These include the infiltration score technique, the pressure drop porometer, the steady-state constant flow porometer, the non-ventilated transient (or dynamic) porometer, and direct observation. Unfortunately, they have not always given results that are consistent or directly comparable.

In early work in Tanzania and Kenya, the infiltration technique, based on isopropyl alcohol, was used to study stomatal behaviour (i.e. a measure of the degree of opening of the widest stomatal pores) in relation to environmental variables, and specifically to water stress. Preliminary measurements showed that the most consistent estimates were obtained from healthy, fully grown, yet still supple leaves that were fully exposed to the sunlight (Carr, 1971b). In Tanzania, partial stomatal closure was observed during the middle of the day on nearly all occasions when diurnal assessments were made (Carr, 1968). Similar diurnal patterns of stomatal opening were observed during dry weather for a selection of clones in Kenya. Stomata were always wider open in the rains than during the dry season in both (previously) irrigated and unirrigated plants. Differences in the dryness of the air (saturation deficit) were believed to be largely responsible for both the diurnal and seasonal differences in the degrees of stomatal opening observed (Carr, 1977a; Othieno, 1978b).

In Malawi, Fordham (1971) used a pressure drop porometer as well as the infiltration method to measure both diurnal and seasonal changes in stomatal opening of mature Assam-type seedling plants. Progressive closure of the stomata was observed from midday onwards in both irrigated and unirrigated tea, and there were marked seasonal changes associated with the dry season. Also in Malawi, Squire (1976), using a diffusion porometer in the cool dry season, showed leaf conductances (measured on the second leaf of an actively growing shoot of several clones) increasing from low values in the morning, and peaking during the middle of the day, before declining throughout the afternoon. Later, Squire (1977) used silicone rubber impressions of leaves to estimate stomatal opening in the wet season. Again, the stomata were wider open at midday (3.5–4 mm) than at dawn or dusk (2 mm).

By contrast to these observations in Malawi, diurnal changes in stomatal conductances recorded in Sri Lanka (6° 55c' N; alt. 1382 m) were similar to those summarised above using the infiltration technique. In the early morning, conductances (clone TRI 2025, measured with a portable infrared gas analyser on recently mature leaves) were large; they decreased towards midday and increased again in the afternoon (Mohotti and Lawlor, 2002). The pattern was the same regardless of the degree of shade or nitrogen level. There were negative linear relations between conductance and leaf temperature (range 16–33°C), saturation deficit of the air (range 0.5–3.8 kPa) and illuminance (range 0–2000 mmol m^{-2} s^{-1}).

Burgess (1992a) attempted to make direct comparisons of diurnal patterns of stomatal behaviour using steady-state and transient porometers and the infiltration technique. The absolute and relative measures of stomatal conductances of six irrigated clones were also compared with the corresponding infiltration score. Despite taking all the precautions possible (for example, in the calibration of the porometers), it was difficult to reconcile the results obtained. Relationships between conductance (transient porometer) and the infiltration score were linear but varied with the clone. As a result of this experience, Smith et al. (1993a) urged caution when using the results of porometry to identify drought-tolerant clones.

In an interesting short-term pot experiment in Colombia, Hernandez et al. (1989) demonstrated convincingly the relative sensitivity of tea stomata (and also coffee and cacao) to the dryness of the air. Conductances of all three species (shaded) declined rapidly as the saturation deficit increased from 0.5 to 4.0 kPa, while transpiration rates were reduced when the saturation deficit exceeded 1.0–1.5 kPa. By comparison, sunflower stomata (not shaded) were less sensitive to the dryness of the air, and transpiration continued to increase over the range 0 to 4.0 kPa.

Photosynthesis

Using a portable gas exchange system, photosynthetic rates (A) of individual mature leaves (clone 6/8) at the surface of the canopy were monitored during the warm dry season in southern Tanzania (Smith et al., 1993b). Rates increased up to an illuminance (photon flux density) of about 1000 mmol m^{-2} s^{-1} but above this value they remained relatively constant. There was a quadratic relationship between A and stomatal conductance (g), with A increasing with g over the range 8 to 100 mmol m^{-2} s^{-1}, but with minimal changes in A when g exceeded 100 mmol m^{-2} s^{-1}. Irrigation and fertiliser increased photosynthetic rates, both by increasing A per unit leaf area and by increasing the proportion of light intercepted by photosynthetically efficient leaves. Although there was a broad temperature (air and leaf) optimum for photosynthesis (range 20–36°C), irrigation-induced increases in A could be accounted for by increases in g and associated reductions in leaf temperature. The effects of fertiliser were more complicated.

By comparison, Barman *et al.* (2008) at Tocklai in north-east India (26°47c′ N; alt. 97 m) identified for a selection of tea clones an optimum illuminance for A of about 1200 mmol m^{-2} s^{-1} (possible range 1000–1400), while the optimum leaf temperature was considered to be about 26°C (range 25–30), values comparable to those reported by Smith *et al.* (1993b). Similarly, in Sri Lanka Mohotti and Lawlor (2002) found, at a high elevation site (1400 m), that A increased rapidly from zero at dawn to a maximum between 0800 h and 0900 h followed by a progressive decrease during the remainder of the (bright and sunny) day, even when the environmental conditions were improving during the late afternoon (irradiance, temperature and saturation deficit declined). Photoinhibition was implicated in the (complicated) explanation put forward to explain this diurnal pattern in A, linked to the corresponding diurnal pattern in stomatal conductances summarised above (Figure 10.5).

Previously, Squire (1977) had monitored representative diurnal changes in photosynthesis in the wet, cool and dry seasons in southern Malawi using the ^{14}C technique. Photosynthesis of leaves on the bush surface was light saturated when irradiance reached 350–400 W m^{-2} (equivalent to an illuminance of about 800–900 mmol m^{-2} s^{-1}). Declines in photosynthesis observed in the afternoon appeared, in large part, to be closely related to falls in the xylem water potential (range −0.4 to −1.5 MPa, measured with a pressure chamber). Photosynthesis did not decrease at the start of the cool season when the yield of tea declined, while irrigation in the dry season had little immediate effect on shoot growth, but increased photosynthesis. These observations are important to our understanding of the allocation of assimilates within the tea plant and seasonal yield distribution.

As De Costa *et al.* (2007) concluded in their detailed review of the photosynthetic process: 'the photosynthetic apparatus and partial processes of tea show specific adaptations to shade. Maximum light saturated photosynthetic rates are below the average for C$_3$ plants and photoinhibition occurs at high light intensities. These processes restrict the source capacity of tea.'

Transpiration

Although portable gas analysers have been used to monitor instantaneous rates of transpiration (Smith *et al.*, 1994), they do not allow whole plant water use to be

Figure 10.5 Tea. Shade trees (in this case *Grevillea robusta*) are still used in some parts of the world, but rarely in Africa. Note the peppers growing up the trunk of the trees – South India (MKVC).

determined. Accurate estimates of transpiration are needed when scheduling irrigation for immature tea or, for example, comparing the water use of different clones. Kigalu (2007a) has described the successful use of sap flow meters, based on the stem heat flow method, for measuring transpiration rates of individual plants in a plant density experiment (three years after field planting) in Tanzania. There were differences in water use (on a per unit leaf area basis) between the two well-watered clones, with S15/10 (with a spreading habit) transpiring faster for most of the day than BBK35 (more upright). Similarly, there were differences in diurnal patterns of water use between plants (both clones) grown at low density (8333 ha^{-1}) or at very high density (83 333 ha^{-1}). The same method was used by Samson *et al.* (2000) to compare the responses of four clones to atmospheric water stress in South Africa with some success.

Xylem water potential

The Scholander pressure chamber has proved to be a useful way of measuring the water status of tea in the field. It was first used by Carr (1971a, 1971b, 1976, 1977a, 1977b) in Tanzania and Kenya, and by Williams (1971) and Fordham (1977) in Malawi, and subsequently by Squire (1977, 1979) in Malawi, Othieno (1978b, 1980) and Odhiambo *et al.* (1993) in Kenya, Renard *et al.* (1979) in Burundi and Olyslaegers *et al.* (2002) in South Africa. The technique has been variously used to

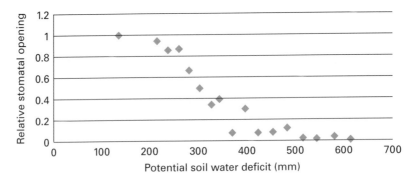

Figure 10.6 Tea. Relation between relative stomatal opening (ratio infiltration score unirrigated to irrigated tea, clone BBT1) and potential soil water deficit.

identify the critical values of xylem water potential (ψ_x) that limit the growth of shoots and yield (−0.7 to −0.8 MPa) in relation to potential soil water deficits (SWDs) and saturation deficits of the air, to explain the seasonality of yield, to understand the effects of shelter from wind and mulches on crop water use, to determine the relationship to stomatal behaviour, to compare clones for drought tolerance, to assess the relative yield potential of clones, to evaluate the effects of methods of bringing tea into bearing on crop water status, and to compare the responses of seedling tea bushes and their clones to water stress. These issues are considered in more detail, under the appropriate headings, below.

Drought tolerance

In Tanzania, differential stomatal closure between well-irrigated plants (BBT1) and unirrigated plants was first observed when the potential SWD exceeded about 200–300 mm (Carr, 1968, 1969). There was then a progressive decline in relative opening until the potential SWD reached about 500 mm, followed by all-day closure. BBT1, a China-type cultivar, is unusual in many respects, being very drought tolerant on an annual yield basis, but responsive to irrigation in the short term. Because of these properties, it is worthy of study (Figure 10.6).

The slope of the linear relationship between ψ_x and the infiltration score is a possible indicator of drought tolerance. For example, compared with heterogeneous, mature seedling tea, the stomata of BBT1 are relatively insensitive to large changes in the shoot water potential (range −0.1 to −1.5 MPa). Thus, there was an eightfold difference in the slopes of the regression line for seedling tea (−0.71) and BBT1 (−0.09) (Carr, 1971b). A follow-up study during dry weather in Kenya confirmed that there were differences between clones in the slope of this relationship, with clone 6/127 being classified as relatively drought tolerant and a seedling population as drought sensitive, but with none as extreme as BBT1 (Carr, 1977a). As the soil dried (to a maximum potential SWD of 300 mm), the daily minimum shoot water potentials declined (to −2.0 MPa), more in some clones than others. There was some evidence of a genotype × environment interaction for both

variables. The infiltration score and the xylem water potential were both negatively correlated with air temperature (range 13–28 °C) and the saturation deficit
(0.06–2.5 kPa).

In a related study again in Kenya during the same dry season (1971), similar
comparisons were made between the responses of seedling plants and the clones
derived from them (Carr, 1977b). There was no evidence to suggest that the
response of a clone to drought could be predicted from that of its seedling 'parent'
(ortet). It was postulated that, since the original seedlings had been identified in a
field on account of their comparative vigour, they must have been able to compete
effectively with their neighbours for water. This advantage would not be maintained when clonal plants were competing with neighbouring plants that were
genetically identical. Using the same criteria as described above (the ψ_x/infiltration
score relationship), clone 7/4 was considered to be relatively *drought tolerant*. The
stomata of this clone remained open during the dry season, and the shoot water
potential remained high, suggesting a very efficient root and water transport
system throughout the plant (or possibly changes in osmotic potential, as suggested by Karunaratne *et al.*, 1999). By contrast the stomata of clone 7/14 began
to close early in the dry season and were slow to reopen when the rains began (an
example of *drought avoidance*).

In Malawi, Squire (1976) found that the xylem water potential (ψ_x), measured in
the cool and rainy seasons, was least in clones that gave the largest yields
(recorded over a full year), and postulated that measurements of ψ_x with a
pressure chamber might be used to screen new clones for high productivity at an
early stage of growth, although the physiological link between ψ_x and yield was
not obvious. In a comparison of six irrigated clones in Tanzania, Smith *et al.*
(1993a) were not able to confirm this relationship.

In a detailed comparison of six, young (<3 years), contrasting clones in
southern Tanzania, Smith *et al.* (1994), using a portable gas analysis system, found
that stomatal conductances (*g*) of clones S15/10 and 6/8 were consistently at least
10% higher than those of clones BBT1 and BBT207 (both China-type). The rates
of photosynthesis (*A*) in clones BBT1, SFS150, S15/10 and 6/8 were always greater
than those in BBT207. Irrigation increased *g*, *A* and the *A/g* ratio (a measure of
instantaneous water-use efficiency) in all six clones. Clones also differed in the
relationship between leaf temperature and *A*. By contrast to the results summarised above (under photosynthesis) for 'mature' clone 6/8 plants, irrigation also
increased the temperature optimum for photosynthesis, and reduced photoinhibition at high illuminance (Smith *et al.*, 1993b).

In the same experiment, clones BBT1 and SFS150 were classified as 'drought
resistant' (based on the relative annual yield loss) although different mechanisms
were involved (see below). Drought resistance for both these clones, and also for
the high yielding S15/10, was strongly related to high ψ_x values (less negative) in
the dry season. Smith *et al.* (1993a) were of the view that 'there is now sufficient
accumulated evidence for the relationship between drought resistance and ψ_x
(measured with a pressure bomb) to be used to help to identify drought-resistant

clones during a dry season.' They also believed that further investigations were justified into establishing the relationships between A and A/g and an appropriate measure of drought resistance. Subsequently, in an evaluation of the effects of grafting in Kericho, Kenya, on drought tolerance (based on an index derived from the ratio of yields in a drought year with those in the previous or subsequent year, or equivalent period, with only a mild drought) Tuwei *et al.* (2008b) considered that xylem water potential measurements (i.e. the least negative values) of a clone under well-watered conditions may be related to its drought tolerance as a rootstock, and could be helpful in a selection programme. By contrast, the performance of clones as drought-resistant scions, based on visual symptoms, was correlated with stomatal conductance values.

Among the morphological leaf traits studied by Olyslaegers *et al.* (2002) in South Africa, stomatal density, pore diameter and pore depth were not linked consistently to stress tolerance. Cuticle thickness was also not a good indicator. In contrast, leaf conductances were greater and leaf water potentials lower in two clones considered to be sensitive to very dry air (PC113 and SFS204) compared with two clones thought to be tolerant (PC114 and SFS150), but this observation was site dependent. Previously, Samson *et al.* (2000), in a comparison of the same four clones under controlled conditions, considered leaf-related sap flow measurements (which are related to transpiration rate) to be a possible discriminator for identifying clones sensitive to drought stress induced by dry air conditions (considered to be an important limiting factor to tea production in South Africa), in contrast to daytime stomatal conductance. Similarly, Nijs *et al.* (2000) compared PC113 and PC114 using a canopy-level energy balance approach together with measurements of stomatal conductance and leaf water potential of individual leaves. Neither approach could distinguish clearly between the two clones, which differed (visually) considerably in their response to water stress. PC114 (tolerant) did though exhibit greater stomatal control in young leaves, and associated higher (less negative) water potentials, than (susceptible) PC113.

Summary: plant water relations

1. Stomata only occur on the abaxial surface of the leaf, at densities between 130 and 300 mm^{-2}.
2. A range of techniques has been used to monitor stomatal behaviour but results are not always directly comparable.
3. In particular, evidence for diurnal changes in stomatal opening is inconsistent, partly depending on the technique used.
4. Nevertheless, it appears that stomata are sensitive to temperature and/or dry air.
5. Rates of photosynthesis increase up to an illuminance of about 1000 mmol $m^{-2} s^{-1}$, and then remain relatively constant.
6. The optimum leaf temperature for photosynthesis is about 25–30°C.
7. Rates of photosynthesis can vary between clones but are not linked directly with yield.

8. Sap flow meters have been used with apparent success to monitor actual transpiration by young clones.

9. Measurements of xylem water potential with a pressure chamber have proved to be a useful way of quantifying plant water status, and possibly drought resistance, in the field.

10. The sensitivity of stomata to changes in xylem water potential varies between clones, and this may also offer a drought tolerance/avoidance selection procedure.

11. There is no apparent direct relationship between the responses of a clone to water stress and that of its ortet.

12. Caution is urged when using porometry for identifying drought-resistant clones.

Crop water requirements

Early work

In many areas where tea is grown in Africa, evapotranspiration from a crop with complete ground cover (ET_c) ranges between 3 and 6 mm d^{-1}, equivalent to 90–180 mm month^{-1}, depending on the season. A selection of diverse experimental techniques has been used to develop the crop factor (f) relating ET_c to evaporation from an open water surface (E_o), which has traditionally been calculated for tea with the McCulloch (1965) version of the Penman equation. These include a long-term, large-scale catchment water balance study (Blackie, 1979), a weighing lysimeter (Dagg, 1970), both in Kenya, together with changes in soil water content using gypsum blocks (Laycock and Wood, 1963), gravimetrically (Willatt, 1973), both in Malawi, and with a neutron probe (Cooper, 1979) in Kenya. The usually accepted value for f for mature, well-watered tea is 0.85 (Laycock, 1964; Dagg, 1970).

At NTRS, unirrigated, mature seedling tea, rooting to depths of 4.3 m, extracted water from a depth of 3.6 m by the end of the six-month-long dry season (based on observations made with gypsum blocks) (Carr, 1974). This corresponded to an actual soil water deficit (SWD) of about 330 mm (equivalent to 80% of the estimated total available water), and to a potential SWD of about 1000 mm ($ET_c = 0.85E_o$). In Malawi, the actual SWD under unirrigated mature seedling tea reached 380 mm (gravimetric sampling) by the end of the dry season, with water extraction down to 3.0–3.5 m depth (Willatt, 1973).

Mature clones

Subsequently, at NTRS, unirrigated, mature clonal tea (6/8), rooting to depths >5 m, extracted water from depths below 4 m. Based on a water balance model developed from neutron probe data, the estimated annual actual water use (ET)

ranged from 800 mm (unirrigated tea) to 1200 mm (fully irrigated). When the soil water deficit exceeded 60 mm, *ET* declined linearly with increases in the SWD until, by the end of the dry season, the estimated *actual* SWD for unirrigated tea reached about 330–350 mm, representing 95% depletion of the extractable water (Stephens and Carr, 1991b). These results are similar to the earlier findings for seedling tea in Tanzania and Malawi.

At NTRS, Burgess (1994b) compared several methods for estimating the water requirements of tea. Using 41 months' data, the Penman–Monteith equation, with appropriate crop and aerodynamic resistance values, was recommended for estimating daily crop water use during the dry season, without the need to include a crop factor ($ET_c = 0.84(\pm 0.005) E_o$; $r^2 = 97\%$; $n = 41$). Where weather data are not available, Burgess (1994a) confirmed that a sunken, 1.83 m square, 0.6 m deep, screened 'British' evaporation pan could be used with confidence for scheduling the irrigation of tea (with suitable correction factors to allow for the colour of the pan and its siting). Previously, Stephens and Carr (1991a) had shown that evaporation from a British pan matches closely ET_c ($\beta = 0.99$; $r^2 = 93\%$; $n = 20$), calculated using the McCulloch method for estimating E_o with a crop factor of 0.85.

Immature clones

The water use of young tea with partial crop cover can be estimated with the equation developed by Dagg (1970) using a large hydraulic lysimeter in Kericho, Kenya:

$$ET = E_o[0.9a + (1 - a)0.9n]$$

where *ET* is monthly evapotranspiration, E_o is the Penman estimate of open water evaporation, *a* is the fraction of the soil covered by crop at noon, and *n* is the fractional number of rain days per month.

On the basis of measurements with a neutron probe made under young tea at NTRS, Burgess (1992a) developed a simple model for estimating the cumulative water loss. Given the degree of ground cover, soil available water capacity, maximum rooting depth and ET_c, the model partitions crop water use into transpiration, soil evaporation and drainage. Using this approach, Burgess (1993a) described a simple method for scheduling irrigation of immature tea. For each of three pan evaporation rates (3, 4 and 5 mm d^{-1}) graphs were presented from which the potential SWD could be predicted for different degrees of ground cover, given the number of days since the soil profile was last at field capacity.

Summary: crop water requirements

1. The generally accepted crop factor (*f*) for tea with complete ground cover is 0.85 ($ET_c = 0.85E_o$).

2. Mature tea can extract water from depths greater than 4 m; the depth of extractable water at NTRS was 330–350 mm.
3. Actual annual water use (*ET*) was about 800 mm for rain-fed tea, and 1200 mm for well-irrigated tea. *ET* declined linearly with increase in the SWD after it exceeded 60 mm.
4. The Penman–Monteith equation gives a direct estimate of ET_c.
5. A sunken and screened 'British' pan can also provide an estimate of ET_c in the dry season.
6. For young tea, *ET* can be estimated with a model that includes crop cover and the number of rain days as variables.
7. A simple method for scheduling irrigation of young tea on the basis of these measures has been devised.

Water productivity

There are several ways of expressing the yield responses to irrigation, or crop water productivity, and it is necessary to differentiate between them to minimise confusion. They include: yield (absolute totals or increments) as a function of actual water use (*ET*), potential water use (ET_c) (these can be numerically equivalent to actual or potential SWDs respectively), transpiration (*T*), and water applied (irrigation alone and/or rainfall). Similarly the time period represented can be annual totals, or dry season values only. The results of experiments synthesised below include all these descriptors. Yields can also be expressed as 'dry' tea (no moisture) or as 'processed' (or 'made') tea, implying water contents of 2–4%. The results for mature (>7 years from field planting) and immature (<7 years) tea are considered separately.

Mature tea

Early work
One of the first tea irrigation experiments reported from Africa was undertaken at NTRS over the period 1967–70 (Carr, 1974). In this high-altitude, seasonal location, there is an extended dry season alternating with one rainy season. Frequent irrigation doubled the annual yields (from about 1000 to 2000 kg ha^{-1} processed tea) obtained from mature (planted 1959), heterogeneous seedling tea and improved within-year crop distribution to some extent. These are low yields by today's expectations. Irrigation was applied at potential SWDs ranging from 25 to 150 mm with the soil profile wetted to field capacity at each irrigation. The limiting SWD for both *annual* and *dry season* yields for this deep-rooting tea (4.3 m) was in the range 100–150 mm, equivalent to about 25–35% depletion of the available water. There was some evidence of 'compensatory' shoot growth following the relief of water stress in the less frequently irrigated plots, and especially in the 'dry' unirrigated plots after the start of the rains. This was the result, in part, of seasonal

differences in shoot population densities following the synchronisation of shoot growth and development during the period of water stress. In each of the irrigated treatments, an average of about 700 mm of water was applied during the dry season. This corresponded to a *yield response to irrigation* totalled over the three-year period of about 1.4 kg ha^{-1} mm^{-1}. For the one treatment, in which only half the calculated quantity of water required was applied at each irrigation (75 mm at a potential SWD of 150 mm), the corresponding yield response was about 1.8 kg ha^{-1} mm^{-1}. Part of the benefit from irrigation was due to the control of the disease *Phomopsis theae*.

The principal objective of another irrigation experiment at that time was to determine whether it is possible to adjust the frequency of irrigation during the dry season, and particularly in the 'winter' months when shoot extension rates are slow (Carr, 1971a; Nixon and Carr, 1995). The results clearly showed that there were no adverse effects of reduced irrigation frequency on total *annual* yields. However, crop distribution was dependent on the timing of irrigation. For clone BBT1 (which has a high base temperature for shoot extension, 14°C, and consequently low winter yields) there were no adverse effects from irrigating less frequently in the cooler dry months (full replacement at a potential SWD = 150 mm) compared with more frequent irrigation (SWD = 50 mm). Indeed, when these treatment combinations were reversed in the warm dry season, yields were greater in the tea that had previously been irrigated at SWD = 150 mm than in the control tea (irrigated at a SWD = 50 mm throughout the dry season). Similarly, compensatory growth during the rains made up for loss in yield during the warm dry season in the treatment irrigated less frequently at that time (SWD = 150 mm). The timing of nitrogen applications modified these responses. The large peaks in production that occurred after the relief of stress (low-temperature or dry-soil induced) were the result of synchronisation of shoot growth, which is more extreme in this clone than most others. This is one of the contributory explanations for the lack of response of BBT1 to drought/irrigation, in terms of the total *annual* yield, reported below (Stephens and Carr, 1990). In addition, after the start of the rains, the extension rate of shoots of previously unirrigated BBT1 was 86% greater than that of previously irrigated plants, the rate of increase in leaf area was 75% greater, and the final area of an expanding leaf was 74% larger. These exceptional rates of growth did not, however, last long (Carr, 1969).

At about the same time, similar experiments were being undertaken independently in southern Malawi (16°05′ S; alt. 650 m) where there is an equivalent seasonal climate to that found in southern Tanzania (Dale, 1971). Although most conditions were similar at both sites, there were marked differences in the dry season yield responses to irrigation. In particular, there was variability between years in Malawi despite similar quantities of water being applied. In some years, yields in the dry season were limited by the hot (air temperatures >30°C for 30 consecutive days in October), dry air (saturation deficits >2.0 kPa) conditions experienced at that time in this relatively low altitude site. In such a year, *yield responses to irrigation (plus rainfall) in the dry season* were only 0.3 kg ha^{-1} mm^{-1},

Figure 10.7 The yield responses of mature clonal tea to irrigation and fertiliser were studied in a long-term field experiment based on the line-source technique. To obtain the required variable water application across the experiment irrigation must be applied under very low wind speeds, which usually occur at dawn or dusk – Ngwazi Tea Research Station, southern Tanzania (MKVC).

about one third of those obtained in less hot years (Carr *et al.*, 1987). Later, an analysis of commercial yields in Malawi suggested a *dry season* yield loss corresponding to about 1.2 kg ha^{-1} mm^{-1} for increases in the *potential SWD* over the range 100–650 mm. This compared to an equivalent value of 1.4 kg ha^{-1} mm^{-1} for Kericho, Kenya (Carr and Stephens, 1992).

Irrigation × fertiliser

More recently at NTRS, the yield responses of mature clonal tea (6/8) to irrigation and fertiliser were studied in a long-term (1986–2004) field experiment (labelled N9), based on the line source technique[4] (Figure 10.7). Stephens and Carr (1991a) reported the results for the first three years of the experiment (*phase 1*) in terms of *yield loss for each millimetre increase in the potential SWD*. In all three years there were significant linear or slightly curvilinear reductions in yields during the *dry season* with increasing potential SWD (up to *c.* 700 mm). At relatively high fertiliser levels (300 kg N ha^{-1}), the rate of yield loss (dry tea) in the dry season was equivalent to 1.9–2.7 kg ha^{-1} (mm SWD)$^{-1}$ and, for unfertilised tea, 0.9–1.4 kg ha^{-1} (mm SWD)$^{-1}$. In the first two years the benefits from irrigation continued into the rains, but in the third year there was evidence of some yield compensation in the dry plots, with yields from the unirrigated tea exceeding those from irrigated tea during the early part of the rainy season (Figure 10.8). *Annual* yields from

Figure 10.8 Tea. The line-source technique allows differential quantities of water to be applied across an experiment, enabling crop-yield/water-use response functions to be developed. Dry (rain-only) plots can be seen in the foreground; these merge into partially and fully irrigated plots in the background (experiment N9) – Ngwazi Tea Research Station, southern Tanzania (MKVC).

unirrigated tea in well-fertilised plots increased from 2000 to 3200 kg ha^{-1} over the three years and, for the equivalent irrigated tea, from 3400 to 4900 kg ha^{-1}, reflecting in part cumulative benefits. Irrigation increased the proportion of the annual crop harvested in the dry season to up to 45%.

Following the approach used by Doorenbos and Kassam (1979), relative yield loss during the dry season was plotted against the corresponding relative reduction in *actual water use* for each of the first four years of the experiment (1986–89). The slopes of the response curves (K_y) were similar for each year (except in the year of prune), and independent of fertiliser level, with a pooled value of 1.3, meaning that the reductions in yield *during the dry seasons* were proportionally more than the reductions in crop water use (Stephens and Carr, 1991b).

After pruning in November 1990 (*phase 2*), annual yields continued to increase in the following three years, reaching 6000 kg ha^{-1} in well irrigated and fertilised plots by 1991/92 (Burgess, 1993b). *Annual yield responses to increases in the maximum potential SWD* were now consistently curvilinear at fertiliser inputs corresponding to 300 kg N ha^{-1} and above but, apart from in the year of prune (1990/91), linear at nitrogen levels less than these (Figure 10.9). It is not clear what caused these differences in the shape of the response curves. For example, the yield loss for each millimetre increase in the potential SWD at 225 kg N ha^{-1} was between 1.7 and 2.5 kg ha^{-1}. By contrast, at 300 kg N ha^{-1}, the yield loss

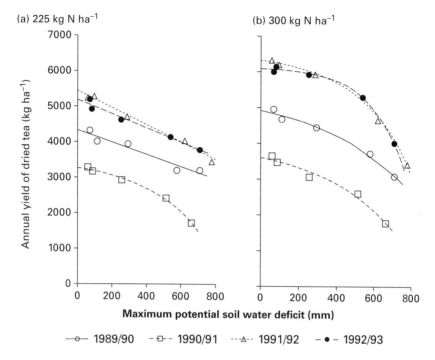

(a) 225 kg N ha^{-1} (b) 300 kg N ha^{-1}

Annual yield of dried tea (kg ha^{-1})

Maximum potential soil water deficit (mm)

—○— 1989/90 - □ - 1990/91 ···△··· 1991/92 - ● - 1992/93

Figure 10.9 Tea. Relations between the yield of dry tea and the maximum potential soil water deficit for clone 6/8 receiving (a) 225 and (b) 300 kg N ha^{-1} (as NPK) for each of four years from June 1989. The tea was pruned in November 1990. (Redrawn from Burgess, 1993b.)

increased from 0.75–1.8 kg ha^{-1} mm^{-1} at SWD = 200 mm to 2.3–3.0 kg ha^{-1} mm^{-1} at a SWD = 400 mm, and 4.9–11.8 kg ha^{-1} mm^{-1} at a SWD = 600 mm.

Afterwards, Stephens *et al.* (1994) analysed the *annual and seasonal yields and actual water use* (*ET*) for the first eight years of the experiment (1986–93). For plots receiving annual applications of 225 kg N ha^{-1} (the commercial level applied at that time), yields increased over this period from 1800 to 3700 kg ha^{-1} in the unirrigated tea and from 3200 to 5700 kg ha^{-1} in well-irrigated tea. *ET* did not vary greatly between years and averaged 700 mm and 1200 mm respectively. There was therefore an upward trend in water productivity over time from 2.3 to 5.3 kg ha^{-1} mm^{-1} and from 2.7 to 4.6 kg ha^{-1} mm^{-1} for unirrigated and irrigated tea respectively. In each year, *annual yields* declined linearly with increases in the maximum estimated *actual soil water deficit* (SWD$_a$) at rates of up to 6.3 kg ha^{-1} for each mm increase in SWD$_a$. The slope and shape of the relationships between yields and water use were strongly dependent on fertiliser application levels (range 0–450 kg N ha^{-1}).

In the cropping year 1995/96, to see if water productivity could be increased, irrigation was withheld for seven weeks prior to the start of the rains to allow a 230-mm potential SWD to accumulate in the normally well-irrigated treatments. Although some crop may have been lost at this time, there was no obvious compensation in yield following the start of the rains. At the 300 kg N ha^{-1} input

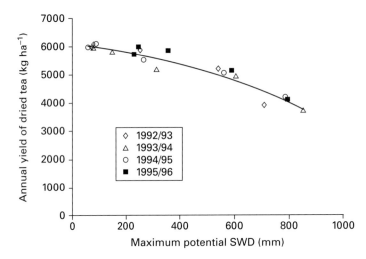

Figure 10.10 Tea. The relationship between the annual yield of dry tea from mature clone 6/8 (with 300 kg N ha^{-1}) and the maximum potential soil water deficit (SWD) for four successive years. The curve is for the first three years only (open symbols). In 1995/96, water was withheld at the end of the dry season to allow a 230-mm SWD to develop in the 'fully' irrigated treatment (solid symbols) before the rains began. (Redrawn from Nixon, 1996a.)

level, the average *annual yield response to water (irrigation plus annual rainfall)* received during the year (1700 mm) was 3.4 kg ha^{-1} mm^{-1}, close to the values obtained in the previous three years (3.1–3.4 kg) ha^{-1} mm^{-1}. Similarly, the yield response to drought over these four years showed a consistent curvilinear relationship between the *annual yield* (range 6000 to 4000 kg ha^{-1}) and the *potential maximum SWD* (range 50 to 800 mm). Yields declined at an increasing rate, from 1 to >6 kg ha^{-1} (mm SWD)$^{-1}$, as the SWD increased; for comparison, the corresponding linear mean was 3.2 kg ha^{-1} (mm SWD)$^{-1}$ (Figure 10.10) (Nixon, 1995, 1996a).

The weight of prunings recorded in June 1996 (all foliage more than 0.40 m above the ground removed) reflected the cumulative effect of treatments on dry matter production in the five and a half years since the previous prune (in November 1990). At the 300 kg N ha^{-1} level, dry weights increased from about 20 to 27 t ha^{-1} with increase in water applied, and from 7 to 30 t ha^{-1} with increase in fertiliser inputs (as NPK) from zero to 450 kg N ha^{-1} in the well-irrigated treatment (Mizambwa, 1997).

After pruning in June 1996 (*phase 3*), the experiment was severely damaged by frost. The tea was pruned at ground level to aid recovery, and at the same time plastic barriers were installed to 1.0 m depth between the fertiliser treatments in order to restrict movement of nutrients between adjacent plots. The fertiliser treatments were also modified, with nitrogen alone becoming the principal variable, not NPK. For the next two years, the whole experiment was irrigated uniformly during the dry season until complete ground cover (>95%) was again achieved. Afterwards, for the five years from 1999/2000 to 2003/2004 differential

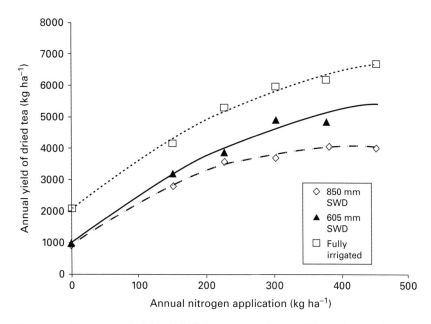

Figure 10.11 Tea. Annual yield (1993/94) response of clone 6/8 to nitrogen fertiliser (applied as NPK 20:10:10) at three potential maximum soil water deficits (SWD). (Redrawn from Burgess, 1994c.)

irrigation treatments were reinstated. At the 300 kg N ha^{-1} level (which is close to the optimum for well-irrigated tea; Figure 10.11) *annual* yields increased linearly ($r^2 = 64\%$; $n = 25$) with the *depth of irrigation water* applied (range 10–820 mm) during the dry season from about 4300 kg ha^{-1} (rain-fed) to 6900 kg ha^{-1} (fully irrigated) at an average rate of about 3.1 ($\pm$ 0.49) kg ha^{-1} mm^{-1} (Kimambo, 2005a).

Summary: mature tea

1. For low yielding (<2000 kg ha^{-1}) seedling tea, the *annual* yield benefit from effective irrigation equates to about 1.3 kg ha^{-1} mm^{-1}, increasing to 1.8 kg ha^{-1} mm^{-1} with deficit irrigation. Where high temperatures or dry air occurs this may be reduced to 0.3–1.0 kg ha^{-1} mm^{-1}. Commercial yield losses during the dry season in Malawi are about 1.3 kg ha^{-1} (mm SWD)$^{-1}$.
2. For reasonably high yielding (up to 6000 kg ha^{-1}) clone 6/8, over the first three years of treatments, reductions in *dry season* yields for each millimetre increase in the potential SWD went from 0.9 to 1.4 kg ha^{-1} mm^{-1} for unfertilised tea, and from 1.9 to 2.7 kg ha^{-1} mm^{-1} at 300 kg N ha^{-1}. Over the same period, the K_y factor for dry season yields averaged 1.3, independent of fertiliser level, implying drought sensitivity at this time.
3. After pruning, the *annual* yield losses equated to 1.7 to 2.5 kg ha^{-1} mm^{-1} at 225 kg N ha^{-1}, over three years, and from 0.8 to 12 kg ha^{-1} mm^{-1} at 300 kg N ha^{-1}, depending on the potential SWD.

4. Later, once full crop cover was achieved, over a four-year period, the corresponding *annual* values at 300 kg N ha^{-1} were between 1 and >6 kg ha^{-1} mm^{-1}, again increasing with the potential SWD, with a linear mean of 3.2 kg ha^{-1} mm $^{-1}$.

5. The *annual* yield responses to actual water use (*ET*) for unirrigated clone 6/8 increased over an eight-year period from 2.3 to 5.3 kg ha^{-1} mm $^{-1}$ and for irrigated tea from 2.7 to 4.6 kg ha^{-1} mm^{-1}, as absolute yields increased (at 225 kg N ha^{-1}).

6. Over the same period, *annual* yields declined with increase in the actual SWD at rates up to 6.3 kg ha^{-1} mm^{-1}, depending on the fertiliser level.

7. The *annual* yield response to irrigation plus total annual rainfall averaged 3.1 to 3.4 kg ha^{-1} mm $^{-1}$ at 300 kg N ha^{-1}.

8. The critical potential SWD above which *annual* yields declined was in the range 200–300 mm.

9. During the *dry season*, the critical SWD for yield was in the range 50–100 mm.

Conclusion

For planning purposes, an annual yield response of 3–5 kg ha^{-1} for each millimetre of effective irrigation for well-managed clonal tea, after allowing for a residual potential SWD at the end of the dry season of 200 mm, is probably realistic, providing there are no other limiting factors (Figure 10.12). For low input/low yielding farming systems, the yield response is proportionally less.

Immature tea

Early work

Irrigation of young tea is intended to increase plant survival and to improve early growth leading to cumulative yield benefits over time (Dale, 1971). In Malawi, Willatt (1970) showed that irrigation increased survival of 18-month-old rooted cuttings, and improved rooting depth and lateral spread. Similarly, in Kericho, Kenya, Othieno (1978a) compared the responses of five young contrasting clones to supplementary irrigation during the dry season. As a result of better survival and early growth, cumulative benefits were recorded over the first two years. There was, however, little improvement in yield distribution, with low yields from both irrigated and unirrigated plants during the dry season when other factors were probably limiting shoot growth, such as dry air, or low night temperatures (Othieno, 1978b).

Clone × irrigation

At NTRS, Burgess and Carr (1996a) described the yield responses to irrigation/drought of six contrasting clones, all commercially and/or scientifically important in eastern Africa (experiment N10; Figure 10.13), for the first four years (*phase 1*) after field planting (in 1988). Based on the line-source experimental design,

Figure 10.12 Tea. In general, the sensitivity of tea to drought declines six or seven years after planting – southern Malawi (MKVC). (See also colour plate.)

Figure 10.13 Tea. The shape of the annual yield/water-use or irrigation response functions changes with time from planting, and varies between clones (experiment N10) – Ngwazi Tea Research Station, Tanzania (MKVC). (See also colour plate.)

differential drought treatments were imposed towards the end of the dry seasons for 16 weeks in 1990 and for 13 weeks in 1991. Although *annual* yields for all clones decreased curvilinearly as the *maximum SWD* (simulated using a water balance model) increased, single values for the drought sensitivity of each clone could be derived using stress time (summed above a derived value for the critical SWD) as an index of drought. The derived SWD values were 70 mm in 1990 and 90 mm in 1991. Using this approach, clones S15/10 and BBT207 (probably because of its sensitivity to a stem canker, *Phomopsis theae*) were identified over the two years as drought sensitive, and SFS150 and BBT1 as drought tolerant; clones BBK35 and 6/8 were intermediate in response.

A similar analysis was conducted for *dry season* yields. Using the mean yields for all six clones, the derived critical SWDs were 40 mm in 1990 and 50 mm in 1991, both less than the estimated values for annual yields. Clone S15/10 was again identified as drought sensitive in both years, but not BBT207. Overall, S15/10 was the highest yielding clone, reaching 5600 kg ha^{-1} in the fourth year (1991/92) after planting in the field when well irrigated.

Later, reporting the same experiment (*phase 2*), Nixon (1996b) demonstrated how responses to drought changed as the clones matured. Figure 10.14 shows how the shape of the *annual* yield responses curves to the *maximum potential soil water deficit* varied considerably between clones, with BBT1 again being particularly tolerant of drought, and 6/8 and S15/10 both relatively sensitive. Annual yields declined rapidly once the SWD exceeded about 250–300 mm in 1993/94, 5–6 years after planting (the year before pruning), but less rapidly in 1995/96, 7–8 years after planting (the year after pruning). There was no change in the shape of the response curve for BBT1 (an extreme China-type clone), which remained flat. Clones SFS150, BBK35 and BBT207 were similar in response to 6/8. The annual yields of dry tea under well-irrigated conditions in 1995/96 were all in the range 5300–6300 kg ha^{-1}, with S15/10 yielding the most, and BBT1 the least. Thus, when the potential SWD exceeded about 300 mm, the *annual* yield loss (or yield gain from irrigation) ranged from 1 (BBT1) to 15 (BBT207) kg ha^{-1} mm^{-1} in 1993/94 to 2 (BBT1) and 7 (BBT207) kg ha^{-1} mm^{-1} in 1995/96 (Nixon, 1996b). *Yield distribution* during the year also varied between clones. For example, weekly yields declined most rapidly as the SWD increased for BBT207, while shoot growth continued for longer with SFS150.

By 2002/03, the year before the experiment closed (*phase 3*), there were further changes in the *annual yield* responses to drought (*potential SWD*). For clones 6/8 and BBT207 responses appeared to be linear with an average loss for both clones of 6.6 kg ha^{-1} mm^{-1}; for S15/10, BBK35 and SFS150, it was two-step linear with yields declining by 4.4 kg ha^{-1} mm^{-1} once the potential SWD exceeded about 370 mm; for BBT1 the relationship had not changed (Mizambwa, 2004a).

Ranking clones on the basis of *cumulative yields* over the first two pruning cycles (1990–94 and 1994–99) identified clone S15/10 as the highest yielding under well-irrigated conditions with 5000–8000 kg ha^{-1} (12–17%) more than each of the other five clones in the order SFS150 (next highest), BBT207, BBK35, 6/8 and

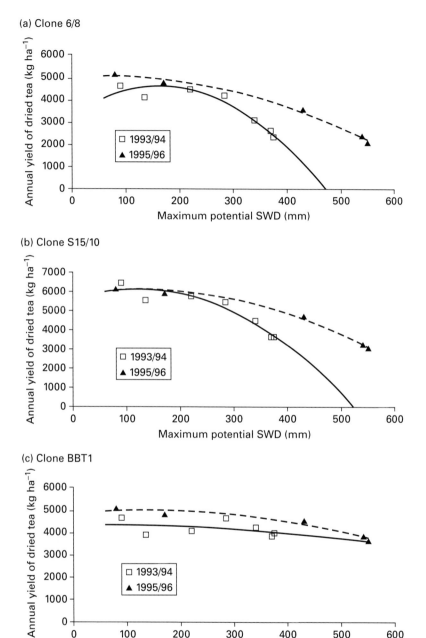

Figure 10.14 Tea. The effects of the maximum potential soil water deficit (SWD) on the annual yields of dry tea for each of three clones over two years 1993/94 and 1995/96, five and seven years after the year of planting, the years before and after the year of pruning respectively. (Redrawn from Nixon, 1996b.)

BBT1 (TRIT, 2000). Under droughted conditions, BBT1 yielded 3000 kg ha^{-1} (11%) more than S15/10 and SFS150, followed by BBK35, 6/8 and BBT207. In terms of yield loss due to drought, BBT207 suffered most with a reduction in cumulative yield of 24 500 kg ha^{-1} (-58%), followed by 6/8 (22 000; -54%), BBK35 (20 000; -48%), S15/10 (20 000; -41%), SFS150 (15 000; -34%) and lastly BBT1 (9000; -24%). On the basis of absolute yields under dry conditions, as well as on both relative and absolute yield loss, BBT1, S15/10 and SFS150 could be identified as the three most drought tolerant, with BBT207 and 6/8 the most drought susceptible.

Based on visual assessments of the severity of drought symptoms in 2001, the clones were ranked in the order 6/8 (worst), BBT207, S15/10 = BBK35 (but variable), SFS150, and BBT1 (best).

A similar experiment (labelled N14) to the one reported above was established at NTRS in 1997, with six different clones, three from Malawi and three from Kenya. The results for the first pruning cycle were summarised by Mizambwa (2004b). By 2002/03, yields from the best clones (12/28, PC81 and 31/8) had exceeded 7000 kg ha^{-1} when well irrigated. In that year, *annual* yields from these three clones declined linearly with increases in the *potential SWD* (up to 410 mm) at a rate of 5.5 kg ha^{-1} mm^{-1} from an average yield of 7500 kg ha^{-1}. From a lower maximum yield (5250 kg ha^{-1}) yields from clones PC105 and TN14/3 also declined linearly (at a rate of 6.2 kg ha^{-1} mm^{-1}). By contrast, the response of clone PC105 (intermediate in yield) was curvilinear.

Cumulative yields under well-watered conditions were similar for clones PC81, 12/28 and 31/8, totalling 26 000 kg ha^{-1} over the five years 1998/99 to 2002/03. These were followed by PC113 (22 000), 14/3 (19 000) and PC105 (18 000). When only partially irrigated, the clones were ranked in a similar order, with yield losses ranging from 9000 kg ha^{-1} (-35%) for 12/28 to 7000 kg ha^{-1} for PC81 (-24%) and 14/3 (-37%). On the basis of relative yield loss, 12/28 (and also to a lesser extent TN14/3, PC113 and especially PC105) would be described as drought sensitive, but in absolute terms 12/28 was the third highest yielding under dry conditions and second highest under wet conditions. PC81 was the 'best' clone using any of the three criteria (yield when well irrigated, when droughted, and relative yield loss), and PC105 the 'worst'.

Composite plants × irrigation

A similar experiment (labelled N13), again based on the line-source technique, was undertaken to evaluate the responses to drought/irrigation of selected composite plants at NTRS. There were two scion clones (BBK35 and S15/10) each grafted on to each of three rootstocks (6/8, SFS150 and PC81), as well as on to each other. In the sixth and final year of the experiment (2001/02) rootstocks 6/8 and SFS150, as in previous years, increased the *annual* yields of both scions by 9% and 7% respectively in the two driest treatments (but this increase was not statistically significant). At the end of the dry season (potential SWD ≥ 700 mm), there were clear visual differences, with S15/10 on 6/8 and BBK35 on SFS150 both surviving the drought better than the corresponding ungrafted scion control plants. Visually, scion BBK35 showed fewer drought symptoms than S15/10 on all rootstocks, an observation confirmed by measurements of the xylem water potential (Mizambwa, 2002b, 2003). The *annual* yield responses to water for each scion (averaged across all treatment combinations) differed. For scion S15/10 there was a linear decline in yield over the whole *potential SWD* range (50–770 mm) with a slope of -6.0 kg ha^{-1} mm^{-1} ($r^2 = 0.96$). For scion BBK35, the relationship was curvilinear, with annual yields declining rapidly (*c.* 10 kg ha^{-1} mm^{-1}) once the potential SWD exceeded about 200–300 mm. Maximum yields for scion S15/10 were in excess of 7 000 kg ha^{-1}, and for BBK35 around 6 500 kg ha^{-1}.

Trials in Kericho, Kenya, have shown that yields of clones can be increased significantly by grafting onto suitable rootstocks, with the best combinations giving profitable returns compared with straight cuttings. Grafting increased the number of shoots harvested, and had little or no effect on the quality of the tea produced (Tuwei *et al.*, 2008a). One of the benefits of grafting was to improve the drought tolerance (defined by a drought tolerance index) of susceptible scions (Tuwei *et al.*, 2008b). Useful rootstocks included TN14/3, China 3, BBT1 and PC87. The most productive scion was 31/8. Yield increases averaged 400–500 kg ha^{-1} (*c.* 10%) for 31/8 on TN14/3. Unilever Tea Kenya Ltd. has planted this combination on a commercial scale (Tuwei *et al.*, 2008a). In Malawi, rootstocks are selected for, among other attributes, high degrees of drought tolerance and compatibility with a wide range of scions. TRFCA-recommended rootstocks include MFS87, PC87, 7/7, 14/22, PC138 and PC141 (TRFCA, 2000).

Plant density × irrigation

In order to assess the effect of plant population on water productivity, the responses of two contrasting clones to plant density and drought were evaluated in a field experiment at NTRS (labelled N12), again based on the line-source irrigation method (Kigalu and Nixon, 1997). Kigalu (2007b) has reported some of the results for the first six years from planting in January 1993 (*phase 1*). The two clones that were compared differ in their growth habit: BBK35 is upright in nature, whilst S15/10 is spreading. The six densities that were compared ranged from 8333 ha^{-1} (1.2 × 1.0 m), below the current commercial practice, to 83 333 ha^{-1} (0.6 × 0.2 m), which is considerably in excess of that density. In addition, there were seven irrigation/drought treatments: these were first imposed in the third year after planting (1995/96). Prior to this all the treatments were uniformly irrigated during the dry season. After planting the soil was mulched, and the young plants were brought into production by decentreing and tipping. As expected, clone S15/10 covered the ground much more quickly than BBK35, especially at the low densities. At the two highest densities, both clones had virtually complete ground cover within 24 months of planting.

Yields in the first cropping year (1993/94) increased linearly with plant density, but the slope of the relationship was three times greater for S15/10 than BBK35 (6.9 cf. 2.3 kg plant^{-1}). In the following four years, yields from well-irrigated treatments increased asymptopically with plant density, but with the magnitude of the responses decreasing year on year, particularly with S15/10. By year six (1998/99), there were no longer any yield benefits from increases in plant population. Annual yields peaked at about 6700 kg ha^{-1} (S15/10) and 6000 kg ha^{-1} (BBK35).

The yield responses to drought at different densities have only been reported for one *dry season* (1995). Yields declined linearly in all cases as the *potential SWD* increased from 40 to about 270 mm. For S15/10, the lines depicting these declines were approximately parallel for each density treatment (-3.6 kg ha^{-1} mm^{-1}). For BBK35 there were differences between densities; at low densities the lines were virtually flat (-0.05 kg ha^{-1} mm^{-1}) with dry season yields remaining constant regardless of the level of drought stress. That is, for this clone, crops planted at low densities (8300 and 12 500 ha^{-1}) withstood drought better than those at a higher density (21 000 ha^{-1}). For S15/10, all densities were affected equally.

Inspection of the *cumulative yields* extracted from the TRIT annual reports and summed for the period from when regular harvesting began (1994/95) to the first prune five years later (1998/99) suggests that a plant density close to 40 000 ha^{-1} gives the maximum yield advantage for both clones, when fully irrigated as well as when droughted. When *averaged across all six irrigation treatments*, the cumulative yields for S15/10 were 33% (7000 kg ha^{-1}) more at this density (D5) than for the corresponding low (current commercial) density (D2, 12 500 ha^{-1}), namely 28 500 kg ha^{-1} compared with 21 500 kg ha^{-1}. The benefits from high-density planting were similar for BBK35, with a 40% (6000 kg ha^{-1}) increase (21 000 compared with 15 000 kg ha^{-1}).

There were further yield benefits from high-density planting in the year following the first prune while full crop cover was being reestablished (1999/2000, *phase 2*). On a base yield of 1700 kg ha^{-1}, this equated to about 500 kg ha^{-1} for BBK35 when averaged across watering treatments, and for S15/10 about 300 kg ha^{-1} (base yield only 1000 kg ha^{-1}). Clone S15/10 is well known for its slow recovery from pruning. In succeeding years, there was virtually no response to density beyond D2 (12 500 ha^{-1}).

These large cumulative yield benefits, which appear to be consistent between clones and between drought treatments, demand a detailed analysis to identify the economic conditions that justify high-density planting (20 000–40 000 ha^{-1}), including savings in the cost of weeding.

Over the first four years of differential irrigation, a total of 1540 mm more water was applied to the fully irrigated crop (I6) compared with the most droughted treatment (I0). This equates to a *cumulative yield response to irrigation* ranging from a consistent 3.0–3.3 kg ha^{-1} mm^{-1} (for low-density S15/10 and BBK35, and also high-density, D5, S15/10) up to 6.0 kg ha^{-1} mm^{-1} for high-density BBK35. In the five years after pruning, *yield responses to irrigation* water applied again averaged about 3–4 kg ha^{-1} mm^{-1} for S15/10, once full crop cover was achieved. For BBK35 the shape of the response curve continued to be curvilinear, with maximum

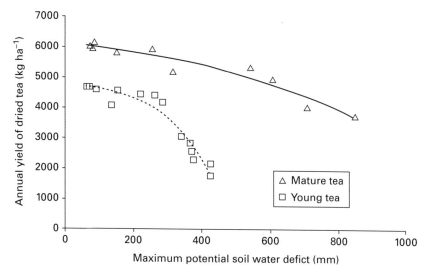

Figure 10.15 Tea. A comparison between the annual (1993/94) yield response to drought of young (four years after the year of planting) and mature (21 years) clone 6/8. (Redrawn from Nixon *et al.*, 2001.)

responses reaching 7–9 kg ha^{-1} mm^{-1}, regardless of plant density. Visual observations indicated that BBK35 suffered more in dry conditions, especially at high densities, than S15/10. Perversely, this was the opposite ranking to that observed for BBK35 as a scion, in the neighbouring composite experiment (N13).

Mature versus immature comparisons

In an attempt to understand better the processes leading to reduced drought sensitivity with age, Nixon *et al.* (2001) compared the responses of mature and young clone 6/8 plants in two experiments reported above (N9 and N10, which at the time were about 22 and 5 years respectively after field planting). For the mature crop, *annual yields* (1993/94) did not decline until the maximum potential SWD exceeded about 300–400 mm, and for the young tea 200–250 mm (Figure 10.15). At deficits greater than these, yields declined relatively slowly in mature tea (up to 6.5 kg ha^{-1} mm^{-1}) but rapidly in young tea (up to 22 kg ha^{-1} mm^{-1}). The apparent insensitivity of the mature crop to drought was principally due to compensation during the rains for yield lost during the dry season. Differences in dry matter distribution and shoot:root ratios also contributed to these contrasting responses. Thus the total above-ground dry mass of well-irrigated, mature plants was about twice that for the corresponding young plants, but the dry mass of structural roots was four times greater and for fine roots eight times greater. In addition, each unit area of leaf in the canopy had six times (by weight) more fine roots to extract and supply water than did young plants.

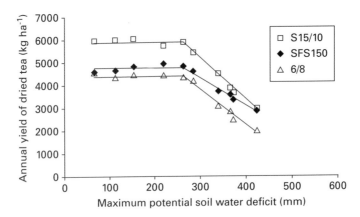

Figure 10.16 Tea. Examples of two-step linear relationships between the annual yields of dry tea (for 1992/93 and 1993/94, years four and five following the year of planting) and the maximum potential soil water deficit for each of three immature tea clones. The relative yield decline was greatest for clone S15/10 and least for SFS150. (Redrawn from Nixon, 1995.)

Summary: water productivity immature tea

General

1. The line-source irrigation system provided a successful way for evaluating a selection of contrasting clones to drought/irrigation and for assessing water productivity. It also served as an excellent visual demonstration of treatment effects for visitors.
2. With strictly controlled harvesting, the annual yield potential of some clones (e.g. S15/10, PC81, 12/28 and 31/8) exceeded 7000 kg ha^{-1} (dry weight) within six years of planting in this part of southern Tanzania.
3. Water productivity increases as yields rise.

Response curves

4. The shape of the annual yield/water use or irrigation response functions changed with time from planting, and varied between the clones tested.
5. For some clones the response function was sometimes linear (e.g. S15/10), for others it was curvilinear (e.g. K35), and for these clones and others it could sometimes be two-step linear (Figure 10.16). But it was not always consistent for individual clones between years.

Yield responses

6. Critical potential SWD for *annual* yields varied between about 50 mm and 200–300 mm, depending mainly on plant age. During the dry season, it was <50 mm for immature tea.
7. For 5–7 year-old plants (the year before the first prune), *annual* yield losses due to drought/gains from irrigation varied between 5–6 (N14) and 10–15 kg ha^{-1} mm^{-1} (experiment N10) once the critical SWD was exceeded (Table 10.2).

Table 10.2 Average *annual* yield losses to drought/gains from irrigation water applied for a selection of well-fertilised clones at different times from planting when potential soil water deficits exceeded *c.* 200–300 mm

Experiment	Clone	Year 5–6[a]	Year 5.5–6.5[a]	Year 7–8[b]	Year >10
		Water productivity [kg (ha mm)$^{-1}$]			
N9					
	6/8				6.5
N10					
	BBT1	1		2	0
	BBT207	15		7	6.6
	6/8	13 (up to 22)		6	6.6
	K35	14		6	4.5
	SFS150	10		4	4.5
	S15/10	14		6	4.5
N13					
	S15/10 scion	6			
	K35 scion	10			
N14					
	12/28		5.5		
	PC81		5.5		
	31/8		5.5		
	TN14/3		6.2		
	PC113		6.2		
	PC105		6.2		

[a] Year before first prune.
[b] Year after first prune.

8. Clone BBT1 behaved differently; its (lack of) *annual* yield response to irrigation/drought was totally different from any other clone. This was mainly the result of yield compensation, more than in other clones, when drought stress was relieved by rain or irrigation.

9. In general, the sensitivity to drought declined after six or seven years from planting (the year after the first prune) when it tended to stabilise at around 3–6 kg ha^{-1} mm $^{-1}$ (experiments N9, N10, N12, N13, N14).

10. Clones 6/8 and BBT207 were both drought sensitive when young, but clone 6/8 in particular became more drought tolerant with age. Care has to be taken when defining what is meant by drought tolerance (experiments N9, N10, N14).

11. Increases in drought tolerance as plants mature were associated with reductions in the shoot:root ratios with time from planting (experiments N9, N10).

12. Based on cumulative yields under dry conditions, clones BBT1, SFS150 and S15/10 could be considered to be drought tolerant. SFS150 also had a low base temperature for shoot growth, while S15/10 recovered slowly from pruning (experiment N10). PC81, 12/28 and 31/8 also performed relatively well (experiment N14).

Composite plants

13. There is some evidence for worthwhile (*c.* 10%) yield benefits from selected scion:rootstock combinations (e.g. 31/8 on TN14/3), including improved drought tolerance.

Plant density

14. Large cumulative yield benefits (up to about 7000 kg ha^{-1}) were achieved from high-density planting (20 000–40 000 ha^{-1}) in the first six years after planting, and also in the year following the first prune (experiment N12). A full cost/benefit analysis is needed.

Conclusion

These complex findings, only partly summarised here, have implications in terms of clone selection for improved water productivity under rain-fed or irrigated conditions, changes in the economics of irrigation as crops mature, and prioritising the allocation of water between clones and plants of different ages.

Dry matter production and partitioning

The proportion of the total dry mass that is allocated to the economically useful structural component, the so-called harvest index, governs the yield potential of a crop. For tea the useful structure is the immature shoot. The partition of dry matter between these harvested shoots, the remainder of the leaf canopy, the woody structural stems and the roots of the tea plant is influenced by all the main environmental and cultural factors, and also varies with genotype as illustrated by the following examples.

Clone 6/8

At a high-altitude site (2178 m), close to the equator, in Kenya, Magambo and Cannell (1981) found that over a year only about 8% of the total dry matter (including roots) was allocated to the harvested shoots (two leaves and a bud; clone 6/8; 6–7 years after planting) or 11% when root weight was excluded. Harvested tea bushes produced 36% less dry mass in a year (17 t ha^{-1}) than those left to grow freely (26 t ha^{-1}), and 64% less wood.

At NTRS, the total above-ground dry matter production, and the corresponding harvest index, were determined for selected fertiliser × irrigation treatments (experiment N9; mature clone 6/8) over the first complete pruning cycle (1986–90, 5½ years) following the imposition of treatments (Burgess, 1992c). The mean annual dry matter production in the well-irrigated, relatively well fertilised treatment was 16.9 t ha^{-1}, of which the harvested shoots comprised 3.3 t ha^{-1} (20%) On the realistic assumption (then) that 25% of the total dry weight was allocated to roots, the net annual increment in the total dry weight (including roots) was estimated to be 22.5 t ha^{-1}, giving a harvest index of 15%.

Over the following four year period (1990–94) Burgess and Sanga (1994a) estimated the net annual dry matter production (including roots) of similar well-irrigated plants in the same experiment (N9) to be 21.5 t ha^{-1}, of which 5.1 t ha^{-1} (24%) were harvested as shoots. For young plants (2–6 years from planting; experiment N10), the corresponding values were 22.5 t ha^{-1} and 4.1 t ha^{-1} (18%) (Burgess and Sanga, 1994b). Both these harvest indices are substantially more than the corresponding value (8%) for the same clone derived by Magambo and Cannell (1981).

Cultivar comparisons

Clone × irrigation (experiment N10)

At NTRS, plant dry weights including roots of four contrasting clones (BBT1, 6/8, SFS150 and S15/10) were measured eight months after field planting, and subsequently at intervals of three to four months, corresponding to the different seasons, during the following two years, 1989–90. Fully irrigated plants of one clone (S15/10) were also harvested after four years in the field (1992). Clones differed in their rates of canopy development, and hence in their capacity to intercept solar radiation. Radiation-use efficiencies were similar for all four well-irrigated clones and ranged from 0.40 to 0.66 g MJ^{-1}, reducing to 0.09 g MJ^{-1} when a 16-week drought was imposed (Burgess and Carr, 1996b). Clone S15/10, a large-yielding cultivar from Kenya, partitioned a greater proportion of dry matter to leaves and harvested shoots than the other three clones, and correspondingly less to structural roots, resulting in a maximum harvest index of 24% (including roots). There were also seasonal differences in partitioning, with more dry matter being diverted to roots and less to shoots during the cool season ($T_{mean} = 14$–$15°C$). Although drought had no significant effect on root growth, the amount of dry matter diverted to leaves, stems and harvested shoots declined by 80–95%.

Clone × irrigation (experiment N14)

The dry weight of the foliage removed at pruning (above 0.35 m, seven and a half years after planting) was greatest for clone TN14/3 in both the well-irrigated (34 t ha^{-1}) and the droughted treatments (22 t ha^{-1}). By comparison, the lowest values were for clones PC81 (27 t ha^{-1}, well-irrigated) and PC105 (18 t ha^{-1}, droughted). The largest differential between wet and dry plots was for PC105 (13 t ha^{-1}) and the least for PC81 (6 t ha^{-1}). This matches the ranking based on the cumulative yields over the first pruning cycle (see above). Clone TN14/3 is notoriously difficult to harvest by hand, which is reflected in the height of the canopy at pruning (1.08 m) compared with the average for all six clones (0.94 m) (Kimambo, 2005b).

Clone × density (experiment N12)

Based on the slope of the linear relationships between the cumulative yield of harvested shoots (dry mass), over a three-year period following planting, and the corresponding cumulative total dry matter production (based on seven harvests, including roots), the harvest index (or partition factor; Squire, 1990) for clone S15/10 was estimated to average 17% (independent of plant density) and 14% for clone K35 (Kigalu, 1997).

In Kericho, Kenya, dry matter production and yields of four contrasting clones were compared at four sites differing in altitude by 400 m (range 1800–2200 m, corresponding to a difference in mean air temperature of about 3.5°C) over a 34-month period, from planting in 1991 to the final harvest in 1994 (Ng'etich and Stephens, 2001a). Total dry matter production over this time (assumed to include roots), averaged across clones, ranged from 18–23 t ha^{-1} from high to low altitude. Clone TN14/3 produced the largest mass of dry matter (31 t ha^{-1}) at the low-altitude site, while S15/10 produced the least (15 t ha^{-1}) at high altitude. There were significant clone × site interactions: clones S15/10, BBK35 and 6/8 demonstrated below-average adaptability to the different environments. Clone TN14/3, by contrast, had wide adaptation responding well, in terms of dry matter, to the higher temperatures at the low-altitude site. For the harvested yield (third year after planting), the position was reversed: TN14/3 was below average in adaptability, and S15/10 was above average. This reflected large differences in the harvest indices, with as little as 10% of the total dry matter being partitioned to yield for TN14/3, and up to 19% for S15/10. Partitioning of dry matter to coarse roots was least for S15/10 at all four sites (*c.* 10%) and greatest for 6/8 (up to 20%). Radiation-use efficiencies varied between 0.3 and 0.45 g MJ^{-1} (Ng'etich and Stephens, 2001b), while annual yields (third year after planting) increased with temperature at an average rate of 326 kg ha^{-1}°C^{-1} (range 275 (6/8) to 441 kg ha^{-1}°C^{-1} (S15/10)) on mean site yields ranging from 1900 to 3000 kg ha^{-1} (Ng'etich *et al.*, 2001).

In a complementary study by Balasuriya (2000) in Sri Lanka, partitioning of dry matter to roots increased linearly with altitude from 30 to 1860 m (corresponding to mean air temperatures, over the seven-month study period, ranging from 26 to 15°C respectively). Clone TRI 2023 was particularly responsive to temperature, while TRI 2025 was relatively stable across the four environments that were compared.

Summary: dry matter production and partitioning

1. The harvest index varied with plant age, clone and site. For mature clone 6/8 (including roots) it ranged from 8% (Kericho) to 24% (NTRS). For immature plants it reached 18% (6/8), and up to 24% (S15/10) (NTRS).
2. Total annual net dry matter production (including roots) for immature tea (6/8) reached 22.5 t ha^{-1} (NTRS).
3. Clones differed in the amount of dry matter partitioned to roots. Of those tested, S15/10 allocated least and 6/8 the most (Kericho; NTRS).
4. Partitioning of dry matter to roots increased with increase in altitude, and during the 'winter' months (Kericho, NTRS, Sri Lanka).

5. Harvested tea bushes (6/8) produced less total dry matter than those left unplucked (Kericho).
6. The ranking of clones for drought tolerance based on differential dry matter production (wet and dry) matched the ranking based on cumulative yields, with PC81 performing best and PC105 least well (NTRS; N14).
7. There was evidence of a genotype × environment (G×E) interaction in relation to altitude (temperature) in terms of total dry matter production, with TN14/3 showing adaptability across environments, but not for yield (harvest index = 10% only) (Kericho).

Drought mitigation

Run-off

Measurements taken in young tea on a uniform 10% slope in Kenya showed that once crop cover exceeds 60%, run-off (and erosion) is negligible (Figure 10.17) (Othieno, 1975; Othieno and Laycock, 1977). Prior to this stage being reached, cover crops, microcatchments and (in particular) grass mulches reduce run-off (Figure 10.18). Grass mulches also reduce evaporation from the soil surface and can conserve water to depths of 0.90 m (Othieno, 1980). The practical aim must therefore be to encourage the crop canopy to cover the ground as quickly as possible, through such practices as choosing a naturally spreading clone and/or planting at a high density. This could include planting a high-density row (plants 0.2 m apart) across the slope at specified intervals down the slope as a live crop barrier, thereby avoiding the problem of competition between tea and a cover crop, as well as gaining extra crop (Sanga and Kigalu, 2005).

Bringing into bearing

Based on measurements of plant water status made on three-year-old plants in Kenya, there is no convincing evidence that tea plants brought into bearing by pegging are necessarily more susceptible to water stress than similar pruned plants. This was despite large visual differences in the size and form of shoot and root systems of the young clones (6/8 and 12/12), and older seedlings. Differences in responses to dry conditions by individual clones appeared to be of greater significance (Carr, 1976).

Pruning

In mature tea, the time of pruning can play a major role in conserving water and minimising drought stress. The removal of most of the foliage stops transpiration immediately, and the crop residue or prunings (as long as they remain in the field) act as a mulch. If pruning is carried out at the start of the dry season, refoliation occurs using water stored in the soil (Laycock and Wood, 1963; Laycock, 1964).

Figure 10.17 Tea. Measurements taken in young tea on a uniform 10% slope showed that once crop cover exceeds 60% run-off (and erosion) is negligible – Kericho, Kenya (MKVC).

Figure 10.18 Tea. Microcatchments and (in particular) grass mulches reduce run-off in young tea. Grass mulches also reduce evaporation from the soil surface – Malawi (MKVC).

In areas south of the equator, where the dry season is initially cool this is common practice. If pruning is delayed, there are yield advantages from bringing the soil profile back to field capacity (by irrigation) immediately after pruning but, depending on how fast the crop canopy regenerates, there is usually no need to irrigate pruned tea again (Carr, 1969). In areas close to the equator, a compromise has to be reached since there is a crop peak following the rise in temperature prior to the start of the dry season. Pruning then begins once this crop has been harvested, but before the plants come under severe water stress (Othieno, 1983).

Shelter from wind

Contrary to conventional wisdom in the late 1960s, shelter from wind by *Hakea saligna* hedges does not conserve water in the dry season. Rather, research in the Mufindi District of southern Tanzania showed clearly that (young, 3–6 years) sheltered clonal tea experienced greater water stress than wind-exposed tea plants during the long dry season. Measurements of xylem water potential, stomatal opening and soil water content supported visual symptoms of drought stress and yield records (Carr, 1971b, 1985). Theory suggests that this unexpected result is possibly an effect of wind on relative changes in the ratio of the canopy resistance to the diffusion of water vapour, and the aerodynamic resistance. In other words, if the cooling effect of wind on the leaf surface, and the resultant reduction in the vapour pressure gradient between the canopy and the surrounding air, is greater than the increase in the rate of transfer of water vapour away from the crop surface, increases in wind speed (away from the shelter belt) will *reduce* transpiration. The adverse effect of shelter on crop water use is rarely observed in the field, but the seasonal climate in Mufindi facilitates its manifestation in the tea crop there.

Nevertheless, providing the adverse effects of drought in the dry season were not excessive, wind-sheltered tea outyielded wind-exposed tea during the main growing season, and during the cool weather. These responses could be explained by the influence of a 1–2°C increase in daytime air temperatures in wind-sheltered areas on rates of shoot extension at this high-altitude (1900 m) site. The results indicate the complexity of interpreting what at first appeared to be a relatively simple study, and the problem of making blanket recommendations on the value or otherwise of shelter belts in tea (Carr and Stephens, 1992).

Modelling

Perhaps the first attempt to relate the yield of tea to weather conditions was by Laycock (1964) in Malawi. Although its limitations were recognised, a linear relationship was established between an empirical weather factor (ETw, a simplified water balance) and annual yields. Subsequently, also in Malawi, a more fundamental approach to modelling yield processes in tea was developed

(Fordham, 1970; Fordham and Palmer-Jones, 1977). Following a period of axillary bud growth inhibition when evaporation rates were high in the latter part of the dry season, the model predicted the subsequent synchronous development of shoots, and the timing of the resultant yield peaks and troughs, which followed the removal of apical dominance when shoots were harvested (the so-called Fordham effect). But, it could not explain the size and conformation of the first peak or the decline in yields at the end of the rains (Matthews and Stephens, 1998a).

In southern Tanzania, where shoot growth is synchronised by a period of cold weather, a simple Stress Time Index (STI) for predicting yield loss due to drought was described by Stephens and Carr (1989). It was based on the daily summation of the difference between the potential SWD and a specified limiting value. Validation of the technique (for clone 6/8) suggested that there was a linear relationship between STI and relative yield loss due to drought during the warm dry season. As a result of changes in the composition of the shoot population at each harvest (the proportion of relatively fast growing shoots, sensitive to water stress, to slow growing ones, less sensitive), the apparent critical deficit at which shoot growth was restricted increased from below 20 mm to 300 mm as the dry season progressed. The rate of yield loss with increasing STI also varied through the dry season for the same reason.

Following a detailed review of other attempts to model the growth of tea, including those by Tanton (1981) and Cannell *et al.* (1990), Matthews and Stephens (1998a) described a model (known as CUPPA-Tea) that simulates the behaviour of a population of shoots, which develop and extend independently at different rates in response to variations in natural conditions, temperature, humidity and day-length. The model, which was developed with data from experiment N9 (clone 6/8), can be used to estimate the potential yield and seasonal distribution of yields under non-limiting conditions. Additional submodels were included describing dry matter production and partitioning, and water uptake and use (Matthews and Stephens, 1998b). Predicted yields matched observed yields well across a range of irrigation treatments, and for different years, at sites in Tanzania and Zimbabwe. To explain the occurrence of large yield peaks in September/October followed by progressively smaller peaks in well-fertilised, irrigated tea, Matthews and Stephens (1998c) found it necessary to include photoperiod as a variable (from 12.15 h, no bud dormancy, to 11.25 h, 100% bud dormancy) through its effect on shoot activity, despite the lack of experimental evidence (Herd and Squire, 1976), or limited evidence (Tanton, 1982b).

The model has been used to evaluate different irrigation options at NTRS. It showed, for example, that annual yields (mature clone 6/8) only start to decline once the actual SWD exceeds about 140 mm. It also confirmed that irrigation water applied during the cool dry season could be as effective as water applied later during the warm dry season. Where there are limited water storage facilities this is the most productive way of using water (Matthews and Stephens, 1997). Similarly, the model has been used to predict potential yields (with and without water stress) at a range of contrasting sites (rainfall and altitude) in Tanzania, and to simulate the likely benefits (5–15%) to be achieved from mist irrigation in eastern Zimbabwe where dry air can be a constraint to yield (Martin *et al.*, 2007).

The model has also been validated at two sites in north-east India (26° N; alt. 80–160 m), and used to simulate the effects of drought on yields of mature seedling tea and responses to irrigation (Panda *et al.*, 2003). At one site (clay loam soil) the mean yield loss due to drought was 1.5 kg ha^{-1} for each mm reduction in ET_c, and at the second site (loamy sand) the corresponding figure was 2.2 kg (ha mm)$^{-1}$. These were on observed annual yields of up to 4000 kg ha^{-1}. The corresponding limiting SWDs were 240 and 85 mm respectively. The model was also used to predict the frequency distribution of annual yield increases from irrigation and the corresponding irrigation water-use efficiencies (up to 2.7 kg ha^{-1} mm^{-1}).

Support for users together with further development of the model are needed to ensure that it is widely and successfully accessed and used.

Irrigation systems

In those areas of Africa where tea is irrigated, semi-portable sprinkler irrigation systems are generally used. Poor design, excessively wide sprinkler spacings, and the adverse effects of wind can lead to uneven water distribution, and water wastage (Figure 10.19) (Flowers, 1996). This has led to a commercial interest in centre-pivot (Figure 10.20) and drip irrigation as alternative ways of applying water to the crop. Möller and Weatherhead (2007) evaluated in detail the technical and financial performance of the first commercial drip irrigation scheme (55 ha) in Njombe, southern Tanzania (Kibena Tea Ltd.; 9° 12′ S ; alt. 1860 m; deep clay loam soil, Xanthic Ferralsol). They found that the uniformity of irrigation water distribution, and efficiency (related to water losses) were all considerably better than those previously reported for adjacent sprinkler irrigated areas. Scheduling irrigation (on three/four day intervals) using tensiometers offered potential water savings of 26% compared with the water balance approach conventionally used. Root systems became adapted to the spacing of the dripper lines (1.2 m, single row or 2.4 m apart, alternate rows). Gross margins were sensitive to the price and yield of processed tea, the amount and cost of fertiliser applied (fertigation was practised), and electricity costs. There were considerable savings in labour compared with sprinklers. Damage by rodents to the lateral pipes and to the emitters was a problem. Overall, drip irrigation was considered to be a technical success, although there were management problems, and maintenance was expensive.

In a supporting large-scale (9 ha) drip irrigation field experiment at the same site, Kigalu *et al.* (2008) compared different levels of water application (0.25. 0.50, 0.75 and 1.00 times the cumulative SWD, one dripper line per row; together with 0.25 and 1.00 SWD, dripper line in alternate rows). Water was applied at three/four day intervals, over four successive dry seasons (2003/04 to 2006/07). Unfortunately, shortages of water in three of these years meant that irrigation was curtailed before the end of the dry weather. Annual yields for each of four clones

Figure 10.19 Tea. Locally designed and built mechanical harvesters operating in Mufindi, southern Tanzania. The circular brown patches are caused by uneven water applications, the result of operating sprinklers at low pressures (MKVC). (See also colour plate.)

Figure 10.20 Tea. A planned new tea estate will be irrigated with water applied through centre-pivot irrigators – southern Tanzania (MKVC). (See also colour plate.)

were reported, and related to the depths of irrigation water applied. As presented, these results are not easy to explain or interpret, some of the curves suggesting that *annual* yields declined as the amount of water applied increased. The well-fertilised clones differed in their responses, with PC81 outyielding BBT207, S15/10 and BBK35 in that order. On average, irrigating alternate rows resulted in yield reductions of about 10%, on base yields of about 5000 kg ha^{-1}, compared with irrigating every row. Until confirmed by further work, these results need to be treated with caution.

Limiting factors

Altitude, through its influence on temperature, can be an important controlling factor in the yield of tea. This is illustrated by the results of an analysis of a long-term (16 year) experiment, under constant management, at the headquarters of the TRFK, close to the equator, in Kericho, Kenya ($0°22'$ S; alt. 2180 m) (Othieno *et al.*, 1992). Yields were constrained by low mean air (15.5–18°C) and soil (17–19°C) temperatures, which can restrict shoot extension rates and extend the duration of the shoot replacement cycle (Carr and Stephens, 1992; Squire *et al.*, 1993). As a result, the yield potential for seedling plants at this site was only about 2.0 t ha^{-1}. The within-year yield distribution was sometimes limited by large potential SWDs ($>$120 and up to 400 mm), which restricted annual yields of made tea by about 1.3 kg ha^{-1} mm $^{-1}$, and by the incidence of damaging hail storms, which caused yield losses of 10% of the annual yield. An analysis of commercial yields from 21 neighbouring estates over a 400-m altitude range (assuming an adiabatic lapse rate of 6°C km^{-1}) suggested a yield decline, below a base yield of about 4000 kg ha^{-1}, of 200–300 kg ha^{-1} for every 100 m increase in altitude above about 1700 m, with yields declining more rapidly above about 2200 m (Carr and Stephens, 1992). All these factors are important when attempting to extrapolate the results of experiments from one (unrepresentative) site to commercial practice. To aid this process, the climate at the TRFK has been characterised in terms of year-to-year variability, and probabilities of exceedance, in the key weather variables: temperature, solar radiation, humidity, wind, rainfall and evapotranspiration. Indices of water stress based on potential SWDs and the saturation deficit of the air have also been derived, as well as the duration of the shoot replacement cycle based on thermal time (Stephens *et al.*, 1992). Similar analyses are needed for all the major tea producing areas if the results of research and field experience are to be applied with confidence to different situations.

A relatively simple, hypothetical approach illustrates how actual yields from, say, a resource-poor farm can fall below potential yields. For example in southern Tanzania at an altitude of 1800 m, the yield potential (mature clone 6/8) is about 6000 kg ha^{-1}. Drought reduces this by up to 1500 kg ha^{-1} (3 kg ha^{-1} mm $^{-1}$ × 500 mm (the effective SWD)); nitrogen deficiency reduces the net total (4500 kg ha^{-1}) by a further 3000 kg ha^{-1} (10 kg ha^{-1} (kg N)$^{-1}$ × 300 kg N ha^{-1}); incomplete

Figure 10.21 Tea. Smallholder tea farms – Mufindi, Tanzania (MKVC). (See also colour plate.)

ground cover may reduce the new net total (1500 kg ha^{-1}) by a further 20% = 1200 kg ha^{-1}, while poor harvesting may halve this total to 600 kg ha^{-1}, a typical yield from a smallholding in Mufindi (Figure 10.21) (Carr *et al.* 1992).

General conclusions

This chapter provides a synthesis of some of the fundamental studies, mainly conducted within eastern Africa, and the practical outcomes, which contribute to three important objectives:

1. To answer questions of immediate practical importance to the industry.
2. To understand, where possible, the mechanisms responsible for the observed responses so that the results can be applied with confidence elsewhere.
3. To communicate the results of the research to the industry, and to the scientific community, in the most appropriate ways.

Considerable progress has been made in understanding the fundamentals of the water relations of the tea plant, and how it responds to its immediate environment. This, in turn, has fed back into practical advice to the industry. In particular, our understanding of the role temperature plays in determining shoot growth rates, and yield distribution, and the controlling effect of dry air on photosynthesis rates and shoot growth has contributed to ways of predicting harvest intervals, responses to irrigation and, by understanding limiting factors, site selection. In addition some progress has been made in identifying drought-tolerant cultivars at an early stage in the selection process. All this knowledge crosses national boundaries, and is not location or time specific.

The research described here on water productivity has contributed to the sustainability and profitability of the tea industry in eastern Africa in the following ways:

1. New target yields have been set for well-managed tea, currently in excess of 7000 kg ha^{-1} (location/temperature dependent).
2. Some of the more important cultivars have been evaluated in terms of their responses to drought and irrigation, and possible selection criteria identified.
3. Circumstances in which it may be justified to increase the plant population above the current recommended levels have been identified.
4. Yield responses to irrigation/drought have been quantified for immature and mature tea, and an explanation for the differences in response proposed.
5. Water productivity values can now be used to help to allocate water rationally and efficiently between tea cultivars, and between plants of different ages.
6. Opportunities to save water through deficit irrigation without a corresponding reduction in the *annual* yield have been identified.
7. Yield response curves to nitrogen fertiliser for rain-fed and irrigated tea have been developed.
8. Practical ways of estimating the water use of tea, and for scheduling irrigation, have been developed and promoted.
9. Ways of scheduling harvest intervals to maximise crop productivity, using the concepts of thermal time and phyllochrons, have been developed.
10. 'New' ways of irrigating tea have been evaluated.
11. Progress has been made in integrating our knowledge of how tea responds to its environment through the development of simulation models for use in planning new projects, and for evaluating alternative strategies (both of value in the context of climate change and competition for declining water resources).

Progressive, well-resourced tea producers will continue to make use of this knowledge and refine it for their own individual business circumstances. Resource-poor tea farmers will need support to access and interpret the information now available that is relevant to their needs. This is currently the focus of the Tea Research Institute of Tanzania's Technology Transfer Programme. Tea scientists must, though, continue to plan ahead and anticipate the questions that tea producers will be asking in 20 years' time. They are unlikely to be very different from those being asked 40 years ago but the context will not be the same, and the answers will certainly be different. These are the continuing challenges for tea research.

Summary

With a focus on eastern Africa, the results of research on the ecophysiology, water relations and irrigation requirements of tea are reviewed. In particular, work undertaken at the Ngwazi Tea Research Station (formerly Unit) in southern

Tanzania is synthesised and interpreted in relation to work reported from elsewhere in Africa (and beyond where appropriate). Topics covered include factors influencing: the components of yield, yield distribution, root growth, stomatal behaviour, photosynthesis, transpiration, xylem water potential and how cultivars vary in their responses to water stress. Also covered are crop water requirements, yield responses to water for mature and immature tea, comparisons between clones (including composite plants), plant density/water availability interactions, dry matter production and partitioning, drought mitigation, crop modelling and irrigation systems. Emphasis is placed on the practical outcomes of the work, and its relevance to the tea industry it serves.

Endnotes

1 There is even a small area of tea in southern England (50° N).
2 A glossary of terms commonly used by tea scientists and commercial producers in Africa to describe tea harvesting practices can be found in Carr (2000b).
3 The interval (d) between the unfolding of two successive leaves.
4 As originally described by Hanks *et al.* (1976), using the method of data analysis proposed by Morgan and Carr (1988).

11 Synthesis

A diverse range of crops has been covered in the preceding chapters, embracing a wide range of products, all united by the common collective title of plantation crops. An attempt is now made to synthesise the findings reported for each crop, despite the difficulties in making direct comparisons, to review the reporting of research findings, and finally to draw some collective conclusions in order to identify a way forward.

Crop comparisons

Summaries of the main findings from the review process are presented for ease of comparison in Tables 11.1 (origins and centres of production), 11.2 (stages of crop development), 11.3 (plant water relations) and 11.4 (water productivity) for each crop.

Origins and centres of production

In nearly all cases the centres of production of a crop are well away from its centre of origin or diversity (Table 11.1). Tea and coconuts are perhaps exceptional in that they are examples of crops that are still grown in quantity in areas close to their origin, although both, like others, continue to spread around the world. Most plantation crops have until relatively recently been associated with large estates but, in nearly all the examples described here, smallholders now dominate in global terms, the exceptions probably being tea and sugar cane. But that generalisation does not hold for individual countries, where the balance between the two main production systems is one of continuing change. In terms of the total harvested area of each crop, sugar cane is the largest (22.7 million ha) followed by oil palm (14.7 million ha). The smallest in area is sisal (0.4 million ha), a crop in decline, followed by tea (3 million ha, but expanding). The world total for all nine crops covered here is about 84 million ha, but the reliability of some of these area estimates is open to question. To put these figures into context there are, for example, about 225 million ha of wheat, 161 million ha of rice and 159 million ha of maize (grain) (FAO, 2011b).

Table 11.1 Origins and centres of production: summary table for all nine crops

Crop	Region of origin	Principal producing countries	Harvested area (ha × 10⁶)	Farming system	Research centres	Principal product	Shade adapted
Banana	SE Asia	India, China, Philippines	4.8	Estate/small farms/subsistence	South Africa, Australia, Brazil	Fruit	No
Cocoa	Amazon rainforest	Côte d'Ivoire, Indonesia, Ghana	8.6	Small farms	Brazil, Ghana, Malaysia, Indonesia	Fruit/seed (confectionery/beverage)	Yes
Coconut	SW Pacific/Indian Ocean islands	Indonesia, Philippines, India	11.2	Small farms/subsistence	Brazil, S. India, Sri Lanka, Côte d'Ivoire	Copra (oil) Oil	No
Coffee (Coffea arabica)	Ethiopia	Brazil, Colombia, India	8.2[a]	Small farms/estates	Brazil, Eastern Africa, Hawaii, India	Seed (beverage)	Yes
Oil palm	West Africa	Indonesia, Malaysia	14.7	Estate	Malaysia, Côte d'Ivoire, Indonesia	Oil	No
Rubber	Amazon rainforest	Thailand, Indonesia, Malaysia	8.9	Small farms (70%)/estates	Malaysia, Thailand, Sri Lanka, Côte d'Ivoire, India	Latex (rubber)	No (except when young)
Sisal	Central America/Mexico	Brazil, Tanzania	0.4	Small farms (Brazil)/estates (Tanzania)		Fibre (from leaf)	No
Sugar cane	New Guinea	Brazil, India, China	22.7	Estates/small farms	Australia, Southern Africa	Sucrose (ethanol)	No
Tea	Assam/Burma/China borders	China, India, Kenya, Sri Lanka	3.0	Estates/small farms	Eastern Africa, India, Sri Lanka	Young shoots (beverage)	Yes

[a] Includes *Coffea canephora* (robusta).

The most progress with respect to crop water-related research has not necessarily been made in the main production centres, but usually where the need is greatest and where water stress is an important constraint. Examples include research on banana and sugar cane, both in southern Africa and Australia, tea research in eastern Africa and coffee research in Brazil (Table 11.1). Since they all evolved in the understorey of forests, cocoa, coffee and tea (var. *assamica*) are all adapted to growing under shade.

Stages of crop development

Among the crops represented here, there is great diversity in crop structure and development, but some similarities (Table 11.2). For example, there are two palms (coconut and oil palm) in both of which the stem develops from a single apical meristem, and where the leaves take many months from initiation to emergence. Similarly, for both these crops, there is a long time lag (three years plus) between the initiation of the inflorescence and the harvest of the mature fruit, making it difficult to establish causal links between water stress and yield. In oil palm, the sex ratio (the proportion of female flowers) is an important determinant of yield, but not in coconut. Roots of both these species are adventitious and can reach depths of 2–4 m (coconut) or 6 m (oil palm).

The common link between coffee and cocoa is that both have a dimorphic growth habit with synchronised growth flushes (leaf and shoot) following the alleviation of drought stress by rain. In both crops flowering is also stimulated by rain following a period of water stress, and water needs to be freely available during fruit development. Roots can reach depths of 1.5–2 m (cocoa) or 2.5–3 m (coffee). Root growth is apparently continuous in coffee but rhythmic in cocoa.

Tea and rubber are both tree crops; in tea the leaves are the principal product, and in rubber it is latex. Rubber is deciduous with defoliation initiated (it is believed) by dry weather, while tea is evergreen except during a severe drought. In both crops shoots extend and leaves unfold in a series of flushes. Tea can root to depths of 5–6 m and rubber to >4 m. In both crops root growth apparently alternates with shoot growth.

Both banana and sugar cane are vegetatively propagated. Banana is a giant perennial herb, propagated by suckers, whereas sugar cane, a perennial grass, produces tillers. Each sucker or tiller/stem produces its own root system. Roots can reach depths of 1–1.5 m and 4 m respectively. In both crops, water stress reduces the rate of vegetative growth. Sisal, which is also propagated vegetatively, is included because it is an example of a succulent with a CAM photosynthetic pathway, which makes it special.

Leaf emergence, extension, unfurling, rolling and shedding are all visible indicators of water stress (Table 11.2).

Plant water relations

Among the plantation crops described here, only in banana, sugar cane (a C_4 species) and sisal (CAM) are stomata found on both leaf surfaces (Table 11.3).

Table 11.2 Stages of crop development: summary table for all nine crops

Crop	Vegetative	Inflorescence	Fruit	Roots		Root growth	Initial drought symptoms
				Root distribution			
Banana	Giant perennial herb; 'apparent' aerial shoot (pseudostem) develops on rhizome	Floral initiation occurs after 30–40 leaves produced; bunch emergence on erect aerial stem. Drought reduces flower numbers	Drought reduces fresh weight of fruit and delays maturity	Maximum depth 1.0–1.5 m; effective depth 0.4–0.6 m; spread 2–3 m		Seasonal/ temperature dependent; ceases at flowering	Rate of leaf extension declines
Cocoa	Dimorphic growth habit; leaf and shoot growth synchronised by rains after dry season alternates with period of dormancy	Flowering is inhibited by water stress; start of rains leads to synchronous flowering	Water stress during fruit development results in small pods	Maximum 1.5–2 m; effective 0.4 m		Roots grow in rhythmic pattern	Premature leaf fall
Coconut	A palm; single stem develops from apical meristem, dwarf and tall types; stem acts as water store; leaves differentiated one at a time 30 months before emergence	Single inflorescence born in axil of each leaf; inflorescence initiated 44 months before fruit is harvested	Drought causes immature fruits and later maturing nuts to be shed	Adventitious; can reach depths from 2 to 4 m; densest top 0.5–1.0 m; lateral spread >3 m		Not recorded	Drought increases rate of frond shedding and slows emergence of new leaves
Coffee	Dimorphic growth habit; water stress reduces rates of shoot extension	Period of water stress needed to prepare flower buds; (water	Water must be freely variable during rapid fruit expansion:	Tap root seedlings; maximum depth 2.5–3 m; effective 1.2 m		More or less continuous root growth even	Rate of production of new leaves declines; internodes

Crop						
	and number of nodes; synchronous growth flushes after rain (cooling)	potential <-1.2 MPa); blossoming stimulated by rain/irrigation.	unable to shed excess fruit early in cycle		during cool dry season	reduced in length; fruits go blue-green before being shed
Oil palm	A palm; single stem develops from apical meristem; leaf primordium initiated every two weeks; 2 years for leaf to fully expand; flush of leaf opening after drought	Single inflorescence in each leaf axil; male and female flowers on separate inflorescence; sex differentiation 25 months before harvest; sex ratio is important yield determinant; abortion of inflorescences	Fruiting is a continuous process; timing of water stress relative to development stage is important	Adventitious; up to 6 m depth, greatest concentration 0–0.6 m; lateral spread >25 m	Primary roots 30 mm d^{-1}	Leaflet opening is delayed; 'spears' accumulate in the crown
Rubber	A deciduous tree; single erect stem; clones on rootstocks; defoliation initiated by dry weather, refoliation after 2–4 weeks	Not an issue	Not an issue	Reach depths >4 m, spread >9 m; most roots within 0.3 m depth	Elongation depressed during leaf growth (flushes) when root branching is enhanced	Radial stem growth is reduced
Sisal	A succulent; short stem with apical meristem surrounded by many leaves; rhizomes produce suckers	Flowers once on tall stem (pole) after 5–12 years and then dies; forms plantlets	Rarely sets seed	Rarely deeper than 1.5 m; most roots within 0.4 m depth; spread up to 5 m	Roots deteriorate when flower stem emerges	Rate of unfurling of leaves declines

Table 11.2 (cont.)

Crop	Vegetative	Inflorescence	Fruit	Roots		Initial drought symptoms
				Root distribution	Root growth	
Sugar cane	A perennial grass; thick stem stores sugar; forms tillers; ratoon cropping	Flowering is undesirable	Not relevant	Each stem produces its own root system from underground nodes, including ratoon crop; reach depths >4 m in 300 d	Roots extend in depth at 5–22 mm d^{-1}	Rates of stem and leaf extension are reduced; leaf rolling
Tea	A tree pruned to form low spreading bush; shoot extension and development both temperature dependent; clones differ in base temperatures (8–13°C); high temperatures (30–35°C) and dry air (SD >2 kPa) restrict growth	Flowering is undesirable	Not relevant	Seedlings tap roots; clones adventitious; both can reach depths of 5–6 m; spread limited by adjacent plants	Extend in depth at 1–2 mm d^{-1}; temperature dependent; root growth may alternate with active shoot growth	Rate of shoot extension declines; shedding of older leaves

Table 11.3 Plant water relations: summary table for all nine crops

Crop	Stomatal density	Conductance	Leaf water potential	Photosynthesis	Transpiration	Other indicators of plant water status	Drought-resistance indicators
Banana	Abaxial up to 4× adaxial; 130–170/ 35–50 mm^{-2}	Good indicator of soil water availability and plant water status	Poor indicator of plant water status; diurnal range 0 to −0.35 MPa only	C$_3$ Soil water potential (ψ_m) not <−33 kPa at 0.2–0.3 m depth	Limited by dry air >2–2.3 kPa	Leaf extension rate; refractive index of exuded latex	Presence of B genome
Cocoa	Abaxial only 700–1100 mm^{-2}	Open at low light intensities; remain fully open in full sunlight	Good indicator of water status; partial stomatal closure at −1.5 MPa	C$_3$ Sensitive to dry air 1.0–3.5 kPa	Limited by dry air 1.0–4 kPa		Leaf water potential and stomatal conductance
Coconut	Abaxial only 200 mm^{-2}; 'talls' > 'dwarfs'?; Cl$^-$ ions play important function	Stomata close as saturation deficit of the air increases; stomata of 'dwarf' cultivars remain open longer than 'talls'	Sensitive indicator of plant water status; reach −1.3 MPa when soil is wet, −2.0 MPa if dry; declines with increase in saturation deficit of air	C$_3$ Ceases at pre-dawn leaf water potential of −1.2 MPa; delayed recovery of photosynthesis after rewatering	Instantaneous water-use efficiency increased linearly with saturation deficit of air		Several, including rate of decline in leaf water potential in excised leaves; stomatal control of water loss; epicuticular wax; accumulation of organic solutes; 'talls' more resistant than 'dwarfs'

Table 11.3 (cont.)

Crop	Stomatal density	Conductance	Leaf water potential	Photosynthesis	Transpiration	Other indicators of plant water status	Drought-resistance indicators
Coffee	Abaxial 150–330 mm^{-2}, also green fruits 30–60 mm^{-2}	Partial closure of stomata when leaf temperature >25°C and/or saturation deficit >1.5 kPa; conductance falls linearly as leaf water potential declines from −1.0 to −3.0 MPa	Leaf water potential as low as −1.5 MPa on sunny days even when soil is wet	C$_3$ Photosynthesis declines at temperatures >25°C linked to decline in leaf water potential below −1.0 MPa	Transpiration declines at saturation deficits >1.5 kPa or leaf water potentials <−1.0 MPa	Relative turgidity is *not* a useful measure	Decline in stomatal conductance is an early indicator of water stress; maintaining high leaf water potentials during dry weather; osmotic adjustment not proven
Oil palm	Abaxial 130–200 mm^{-2}	Stomata close rapidly when air temperature >32–33°C and/or saturation deficit >1.7 kPa	Leaf water potential is not a good indicator of crop water status; value falls to −1.5 MPa at midday when soil wet	C$_3$ Rate of photosynthesis declines in concert with stomatal conductance	Transpiration rates decline with saturation deficit of the air <1 to 4 kPa	Relative stomatal opening is a good indicator of soil water status	Leaf to air temperature difference; rate of spear leaf extension
Rubber					Not reported		

	Abaxial 280–700 mm^{-2}	Closure of stomata restricts cavitation	Cavitation begins at xylem water potential −1.8 to −2.0 MPa	C$_3$ Light inhibition of photosynthesis can occur, particularly young plants		Clones differ in susceptibility to cavitation	Stomatal closure slows reduction in leaf water potential; rate of girth increment
Sisal	Both surfaces, equal numbers 30:30 mm^{-2}	Stomata closed during day, open at night		CAM CO$_2$ uptake at night	Restricted due to daytime stomatal closure		Stomata closure when droughted is delayed by water storage in tissues
Sugar cane	Abaxial 2× adaxial	Sensitive to changes in leaf water potential; conductance very low at ψ_l of −1.7 MPa; limited evidence sensitive to dry air (≥1.7 kPa)	Growth processes are linked to decline in midday leaf water potential as the soil dries	C$_4$	When aerodynamic resistance is large changes in stomatal opening do not have an immediate effect on transpiration	Leaf extension and rolling are better indicators of water stress than conductance or photosynthesis	Associated with early stomatal closure; limited evidence of osmotic regulation
Tea	Abaxial surface only 130–300 mm^{-2}	Diurnal changes variable results; sensitive to high temperatures/dry air	Xylem water potential is a useful method of quantifying plant water status; yield limiting value c. −0.8 MPa	C$_3$ Varies between clones but not linked directly to yield; photoinhibition at high light intensities	Limited measurements		Sensitivity of stomata to changes in xylem water potential varies between clones; small-leafed 'China types' more resistant than large-leafed 'Assam types'

On all the other crops mentioned they occur only on the abaxial surface, at densities ranging from 130 mm^{-2} (oil palm) to >1000 mm^{-2} (cocoa). In banana, coffee and oil palm relative stomatal opening is a good indicator of soil water availability and plant water status. High air temperatures (variable > 25–30°C) and/or dry air (SD > 1.5–2 kPa) reduce stomatal conductances in most of the crops (cocoa, coconut, coffee, oil palm and tea). Leaf water potential is a good indicator of plant water status in cocoa, coconut (but not oil palm), sugar cane (but not banana) and tea. There is a paucity of information on sisal.

Rates of photosynthesis are limited by high temperatures/dry air in cocoa, coffee, oil palm and tea, as are rates of transpiration (banana, cocoa, coffee and oil palm). Tall, aerodynamically rough crops are better coupled to the atmosphere than short, field crops and a decline in the stomatal conductance of tall crops is translated into a corresponding reduction in transpiration (and to a large extent photosynthesis). Cocoa, coffee and tea are all tolerant of shade but, with good management, grow better in full sunlight. Drought-resistance indicators have been identified for a number of crops, but it is not clear whether this has led to the development of new drought-resistant cultivars or merely confirmed the drought tolerance of existing cultivars (Table 11.3).

Water productivity

As competition for water increases worldwide, quantifying water productivity will soon become an important prerequisite to justify a water abstraction licence for irrigation or to get permission to grow certain crops rather than others in a rain-fed area. As this review has shown, water productivity is not an easy thing to quantify; there are usually a range of values, often site specific, and before extrapolation of these values to new locations it is important to be sure that the comparisons are valid. For very few of the internationally traded plantation crops reviewed here has water productivity been quantified with any reasonable level of precision.

Key to specifying water productivity is a reliable estimate of actual evapotranspiration (ET). It is difficult enough to reconcile ET rates for one crop, given the range of measurement techniques used and the diversity of ecological areas represented, let alone for several crops (Table 11.4). Typical potential evapotranspiration rates (ET_c) rates in the humid tropics, for crops with a full canopy, appear to be about 3–4 mm d^{-1}, and in the subtropics up to 7–8 mm d^{-1} in the summer. The Penman–Monteith estimate of reference crop evapotranspiration (ET_0) has general support, but there is no agreement on values of the crop coefficient (K_c), apart from for tea, sugar cane and coffee. Few attempts have been made to estimate the limiting soil water deficits or the optimum irrigation intervals or indeed the yield responses to water. It is therefore still difficult to provide criteria with which to justify irrigation (or to quantify the impact of drought), and to design or to manage, on a day-to-day basis, an irrigation system. The most researched crops in this respect are probably sugar cane, banana and tea. The full responses to irrigation in the two palms are only realised in the third and subsequent years after irrigation is introduced because of the time delay from the initiation of the inflorescence to the harvest of the fruit. Mulching is a universally recommended drought-mitigation practice (Table 11.4).

Reporting research findings

Much of the research information on plantation crops is not in the form of readily available peer-reviewed papers. Although the internet makes literature searches much easier than in the past, it would not have been easy to access publications in some of the 'grey' literature without the help of colleagues with their own private collections of, for example, conference proceedings. With some crops (e.g. cocoa and oil palm) single crop, industry led, annual conferences have been the traditional way of reporting the results of research. Although this is a very important line of communication, it also means that much of the research reported was not validated by the peer-review procedures required when papers are published in international journals. Much good research also goes unreported except in rather dull and repetitive annual reports prepared by individual crop research institutes. Although these meet a statutory obligation, they avoid the need to synthesise and interpret the results collected over a number of years and to subject the work to peer review. Sponsors of research need to have some confidence that the research they are supporting meets an international standard. Research undertaken by universities is often on short-term, time-limited contracts and runs the risk of not being published because the scientists involved move on to other more secure jobs, and the supervisors of the research are busy trying to win the next contract. The work reported in many PhD theses also never gets the recognition it deserves because of similar constraints.

Working at a single crop research institute can mean that the scientist is in intellectual, as well as physical, isolation from fellow scientists working elsewhere on another crop, perhaps even a competing crop (e.g. oil palm and rubber), with similar challenges. There is considerable scope for people engaged in tropical plantation agriculture, regardless of crop, to learn from one another.

A way forward

The previous discussion highlights the importance of good and consistent research into the water relations of plantation crops, and in reporting research findings in clear, unambiguous ways. To assist in identifying a way forward, the main points that emerge from the conclusions presented at the end of each chapter are summarised in Box 11.1. Two principal components can be identified: (1) the priority technical issues related to irrigation/drought-mitigation research, and (2) the need to understand the requirements of the industry for which the outputs of the research are intended.

Table 11.4 Water productivity: summary table for all nine crops

Crop	ET_c	K_c	Limiting soil water deficit	Water productivity	Limiting factors	Irrigation interval	Drought mitigation	Irrigation method	Other
Banana	Typical rates in the tropics 3–4 mm d^{-1}, up to 8 mm d^{-1} in summer elsewhere	No consensus in tropics; varies with growth stage; seasonal differences in subtropics 0.6 (winter) 1.0 E_{pan} (summer), and in Mediterranean climates	Soil water potential (ψ_m) >−20 kPa at 0.2 m depth	Variable results; 40–80 kg fresh fruit ha^{-1} mm^{-1} (irrigation); yield response factor = 0.63(?); quality criteria improved	Dry air restricts gas exchange	In subtropics 2–3 days only in summer	Mulch	Cooling of pseudostem by undertree sprinklers can delay development and reduce yields	Ratoon crop is 'nomadic'
Cocoa	Data limited <2 mm d^{-1}	Not known	Not known	Not available	Dry air		Mulch; resistant cultivars		
Coconut	Direct comparisons difficult, range = 1.2–7.8 mm d^{-1}; typical *c.* 3.0–3.5 mm d^{-1}	Uncertain; perhaps seasonal variability; mature palms 0.5–1.02 × ET_0; working value $0.7ET_0$	Not known	Full response only in third and subsequent years after irrigation begins; ball park yield response extra 20–40 nuts per palm	Causal links difficult to establish because of time delays	Not known	Husk burial; mulching; common salt	Basins, microsprinklers, drip	Can tolerate using seawater for irrigation
Coffee	Depends on planting arrangement; typically 4 but up to 7 mm d^{-1}	Mature crop 0.7–0.8E_{pan}; 1.0ET_0; less than this if ET_0 >4(?) mm d^{-1}	Not specified	Recorded values are limited: 0.8–1.3 kg (green bean) ha^{-1} mm^{-1} (irrigation)	High temperatures 25–30°C; dry air >1.5 kPa	Uncertain	Mulching; planting arrangement (cova); ratooning	Over-tree sprinkler systems, microsprinklers, drip or basins	Irrigation can be used to control time of flowering
Oil palm	During rains 4.1 mm d^{-1} (range 3.5–5.5 mm d^{-1}); dry season 1.9 mm d^{-1} (0.6–2.9 mm d^{-1})	Averaged 0.9ET_0? (range 0.8–1.0)	Sandy clay loams/sandy soils 15–30 mm	Full response to irrigation only seen from year 3; 20–25 kg fresh fruit ha^{-1} mm^{-1}	Dry air	Not known	Bare soil along rows; mulching; reduce plant population; remove young inflorescences	Drip, microsprinklers	Irrigation increases sex ratio and reduces abortion; will also advance time of fruiting in young palms

Rubber	Low values reported <3 mm d^{-1}	Not known	Not known	Not known		Not known	Mulch	Basin, drip	In dry areas irrigation can reduce immaturity period from >10 to 6 years
Sisal	Limited data; 1.4 mm d^{-1} summer	Not known	Not known	High		Not known	Not known	Not known	
Sugar cane	Annual total 1100–1800 mm; daily 6–15 mm d^{-1}; P–M equation and USWB pan give good estimates of ET_0	Varies with growth stage 0.4 to 1.25 to 0.75ET_0; lodging reduces K_c	Not specified; variable; scheduling services available	100 kg cane (fresh wt) ha^{-1} mm^{-1}; 13 kg sucrose ha^{-1} mm^{-1}		Variable, depends on growth stage; prior to harvest can withhold water for periods up to twice the total depth of available water in the root zone	Opportunities exist to save water without loss in yield at all growth stages	Furrow, sprinklers (drag line), rain guns, centre pivots, drip	Compensatory growth occurs on rewatering after water stress
Tea	Annual up to 1200 mm; daily 3–5 mm d^{-1}	0.85E_0; 1.0ET_0	c. 60 mm for ET_c; 50–100 mm for yield	3 to 5 kg (dry mass) ha^{-1} mm^{-1} (irrigation), for well-managed clonal tea when potential SWD >200 mm	Dry air >2.3 kPa	Variable up to 30 days	Pruning	Sprinklers; centre pivot, drip	Compensatory shoot growth occurs on rewatering after water stress

---

Box 11.1 Summaries of the principal conclusions for each crop

Banana

With the banana, few experiments have been reported in which the aim was to identify at what stages in the development of the crop water applications can be reduced below the maximum without a proportional loss in marketable yields. As Fereres and Soriano (2006) stated, research linking the physiological basis of these responses (generally well understood for the banana) to the design of practical 'regulated deficit irrigation strategies' could have a significant impact in water-limited areas (or where it is expensive to deliver water to the field). In addition, micro-irrigation systems are ideally suited for controlling water applications and therefore for this form of stress management. Their design and operating criteria, usually the preserve of engineers, need to be specified with appropriate levels of precision (for specific farming systems) in order to maximise (marketable) crop water productivity, while ensuring minimum adverse effects on the water environment.

Cocoa

Since cocoa is a drought-sensitive crop, and a large proportion of the world's cocoa is grown in parts of the tropics having a distinct alternation between wet and dry seasons, it is to be expected that the water relations of cocoa would have been the subject of research. What is surprising is the limited amount of work done in the field with mature crops as compared with research on immature plants in relatively controlled conditions. Although research on immature plants has led to a good understanding of aspects of cocoa physiology, there is a paucity of information of direct practical value. For example, the lack of data on crop water use and water productivity means it is impossible to quantify yield losses due to drought or yield benefits from irrigation. With the threat of climate change leading to less, or more erratic, rainfall in the tropics, and higher temperatures and drier air, uncertainty in yield forecasting will increase and yields will decrease on average. Why has there been this emphasis on fundamental research and less on its practical application? The answer must be in part be due to the structure and nature of the industry, and the way cocoa research is organised and funded.

Coconut

Until relatively recently much of the research reported was empirical, so that the results were only of value in the immediate location of the experiments. They were time and space limited. This is understandable and is due, in part, to the difficulty of undertaking research on this fascinating crop. It is also due in part to limited funding at the relatively small research institutes with the mandate to undertake this research. There has also been, with some exceptions,

Continued

Box 11.1 (*cont.*)

a notable lack of international collaboration in research (coconuts are outside the CGIAR system) for a crop on which millions of people depend for their livelihoods (Carr and Punchihewa, 2002).

Coffee

Water availability plays a dominant role in many aspects of the growth, development and yield of the coffee crop. Despite the increasing international importance of irrigation in commercial coffee production, many of the crop/water relationships have not been quantified in commercially useful ways. This is especially surprising in view of the detailed understanding of many aspects of the water relations of the coffee plant, particularly the mechanisms controlling the development of flower buds. The notorious lack of stability in world coffee prices is often caused by fluctuations, real or imaginary, in the predicted production levels. The profitability of coffee growing changes accordingly, impacting directly onto the economies of individual countries, as well as on the livelihoods of the people involved in its production, including smallholders and the employees of large-scale commercial producers. Rainfall variability is one of the principal contributing factors to this instability, but we are still unable to quantify the effects of drought with precision, or to recommend with confidence where and when irrigation is worthwhile, and how it is best practised.

Oil palm

As Henson (2006) wrote:

Despite many trials, what constitutes an optimum or adequate water supply to maintain yield is still poorly defined. The issue has been complicated and deductions hindered by the variability of conditions under which the trials have been conducted, by poor experimental design and inadequate controls, by a lack of comprehensive monitoring of soil and atmospheric conditions, and by differences in methodology relating to definitions of soil water status.

All irrigation experiments are notoriously difficult to perform, and to interpret the data in ways that are of practical value beyond the place and time that they were undertaken. For a tree crop like oil palm, it is particularly difficult, and compromise is always necessary between what is desirable and what is practical. Nevertheless, there is scope to develop functional relations between yield and crop water use that can be used for rational irrigation planning and water management in the dry areas where oil palm is now being increasingly grown.

Rubber

The structure of the industry plays a role too in determining research priorities. Using rubber as the case study, five stages of development were identified

Continued

Box 11.1 (*cont.*)

beginning with when a plantation crop is introduced into a subsistence economy, and ending with its demise when, as the economy develops a manufacturing base, it is no longer profitable to grow a plantation tree crop in the traditional way. Rubber-producing countries are at different stages on this continuum. This therefore is the context within which research on water relations of rubber has to be considered. Compared with other plantation crops, very little research has been reported in the literature on the water relations and irrigation requirements of rubber. When the crop was confined to the humid tropics this may not have been a surprise, but with its expansion into regions with extended dry seasons one might have expected more emphasis to have been placed on this aspect of the agronomy of the crop, especially the selection of drought-resistant clones. It is not known for example what the yield losses due to drought are in the different areas where rubber is grown (or conversely the likely benefits from irrigation). This is essential information for rational planning purposes.

Sisal

Very little water-related research specific to sisal has been reported. Sisal is an example of a CAM photosynthetic pathway succulent with a high water-use efficiency.

Sugar cane

For many row crops, including sugar cane, the adoption of modern water-saving irrigation technologies is often cited as key to increasing water-use efficiency while maintaining current levels of production … However, new technology requires greater capital investment, so irrigators are often reluctant to adopt new systems unless they can be convinced of the likely benefits. Where water costs are low, sugar cane growers have little incentive to switch technology to improve efficiency unless there are other externalities that might influence their ability to maximise net crop return. But rising energy, labour and water costs, the need to increase water productivity, less water available for abstraction due to expansion of cropped areas, increasing competition for limited resources, climate change risks and demands for greater environmental protection are now the driving forces influencing technology choice in irrigated sugar cane production.

Tea

Research institutes serving an industry such as tea face difficult challenges. Namely, the producers who are funding the research, often through a statutory levy, as well as donors, usually want answers to short-term problems that are

Continued

Box 11.1 (*cont.*)

often time and location specific. At the same time, researchers are meant to serve an industry located in diverse ecological areas, with both large-scale and small-scale producers, often with conflicting and constantly changing priorities and resources. Fundamental information is needed to enable results of experiments to be extrapolated from one location to another where conditions may be very different. It is also necessary for researchers to anticipate future problems and not to concentrate only on the immediate challenges facing the industry. Somehow a balance has to be struck within the financial and skill constraints available.

Technical issues

The priority technical issues can be summarised for each crop as follows:

- *Banana*: to specify the design and operating criteria for irrigation systems (e.g. regulated deficit irrigation).
- *Cocoa*: to develop practical recommendations on the benefits, or otherwise, of irrigation.
- *Coconut*: to ensure that research is designed to be of value away from the immediate location of the experiments.
- *Coffee*: to quantify the effects of water stress on productivity.
- *Oil palm*: to define the amount of water needed to maintain yields.
- *Rubber*: to understand better the role that water plays in productivity.
- *Sisal*: to undertake some fundamental research specific to sisal.
- *Sugar cane*: to demonstrate convincingly to farmers the value of investing in modern water-saving technologies.
- *Tea*: to meet the challenge of serving large-scale and small-scale producers with different capital assets in different ecological areas.

In short, the design and operating criteria for irrigation systems, usually the preserve of engineers, need to be specified with precision in order to maximise the crop-yield/water-use efficiencies that advanced methods of irrigation can offer growers with minimum adverse effects on the water environment. A key component of this process is the requirement for representative and reliable meteorological data. Unfortunately this is sometimes not the case, with poorly sited, poorly equipped, and often poorly maintained weather stations, together with poorly trained and motivated recorders (often the most junior member of staff). It is surprising how often it never rains at weekends, the wick on the wet bulb thermometer is dry, the anemometer is impeded, there are no bright sunshine recording cards, and the evaporation pan is empty (Figure 11.1).

Figure 11.1 A selection of weather stations of variable quality from around the world. They provide the data on which so much depends (MKVC). (See also colour plate.)

Research

For research on long-term plantation crops to be effective, researchers must also seek:

- to understand the structure of the industry they are serving; e.g. estate and/or smallholder dominated?
- to recognise the stage reached in the evolution of the industry; e.g. how profitable is the crop?
- to understand the expectations of the research sponsors/stakeholders
- to identify both long-term and short-term research priorities
- to set a reasonable balance between fundamental research and answering immediate practical problems
- to ensure that the research outputs can be interpreted in commercially useful ways
- to be able to extrapolate the results from one location to another ecological area with confidence
- to seek opportunities for international collaboration
- to communicate the results of the research to the industry, and to the wider scientific community, in the most appropriate ways.

Research can only be effective, and the chances of its uptake improved, if the policy environment is supportive, if farmers are in a position to implement the outputs of research (and have confidence in their reliability and value), and if there is an effective delivery mechanism through an appropriate extension service, or direct to the farmer by researchers. It also needs to be a two-way process, with information flows in the opposite direction from the grower to the researcher. Researchers need to understand policy issues, marketing constraints and opportunities, the livelihood systems of farming households (e.g. access to capital in its various forms[1]), and hence the relevance of the research in terms of the needs of farmers.

While there is always pressure to respond to the short-term requirements of an industry, researchers must not lose sight of the need to take a longer-term view, and to anticipate the questions that the industry may be asking in 10 years' time. Progressive smallholders as well as the estate sector will be demanding the most up-to-date techniques.

Research is not just about new discoveries. It is also about reviewing, reinterpreting and representing old information in ways that are of value in today's contexts. For this reason, and so that potentially valuable information is not lost, easy access to the literature through modern information retrieval systems is essential.

The challenge

Plantation agriculture (smallholders as well as estates) makes a major contribution to the livelihoods of the millions of people who are involved in the production and processing of these crops, and to their national economies. They are fascinating crops to work with, often grown in beautiful places and are cultivated and managed by committed people.

Although, as this book shows, research has contributed a great deal to the development and to our understanding of plantation crops over a relatively short period of time, much remains to be done. As the world's population approaches nine billion people, and the competition for land and water become even greater, and the demand for the products of these crops increases, the challenge for the next generation of students is *to learn from history and build on what has already been discovered*. One of the main aims of this book is to contribute to that process.

Endnote

1 Human, social, physical, natural and financial.

Further reading

Al-Amoodi, L.K., Kasper, P. and Lascano, R.J. (eds.) (2007). *Irrigation of Agricultural Crops*. Madison, WI: American Society of Agronomy.

Allen, R.G., Pereira, L.S., Raes, D. and Smith, M. (1998). *Crop evapotranspiration: guidelines for computing crop water requirements*. Food and Agricultural Organisation of the United Nations, Irrigation and Drainage Paper 56. Rome: FAO.

Azam-Ali, S.N. and Squire, G.R. (2002). *Principles of Tropical Agronomy*. Wallingford, UK: CAB International.

Molden, D. (ed.) (2007). *Water for Food Water for Life: a Comprehensive Assessment of Water Management in Agriculture*. London: International Water Management Institute/ Earthscan.

Squire, G.R. (1990). *The Physiology of Tropical Crop Production*. Wallingford, UK: CAB International.

Publications in Experimental Agriculture on which this book is based

Carr, M.K.V. (2001). The water relations and irrigation requirements of coffee. *Experimental Agriculture* **37**:1–36.

Carr, M.K.V. (2009). The water relations and irrigation requirements of banana (*Musa* spp.). *Experimental Agriculture* **45**:333–371.

Carr, M.K.V. (2010a). The role of water in the growth of the tea (*Camellia sinensis* L.) crop: a synthesis of research in eastern Africa. 1. Plant water relations. *Experimental Agriculture* **46**:329–349.

Carr, M.K.V. (2010b). The role of water in the growth of the tea (*Camellia sinensis* L.) crop: a synthesis of research in eastern Africa. 2. Water productivity. *Experimental Agriculture* **46**:351–379.

Carr, M.K.V. (2011). The water relations and irrigation requirements of coconut (*Cocos nucifera* L.): a review. *Experimental Agriculture* **47**:27–51.

Carr, M.K.V. (2011). The water relations and irrigation requirements of oil palm (*Elaeis guineensis*): a review. *Experimental Agriculture* **47**:629–652.

Carr, M.K.V. The water relations of rubber (*Hevea brasiliensis*): a review. *Experimental Agriculture*; published online 12 October 2011. DOI: 10.1017/S0014479711000901.

Carr, M.K.V. and Knox, J.W. (2011). The water relations and irrigation requirements of sugar cane (*Saccharum officinarum* L.): a review. *Experimental Agriculture* **47**:1–25.

Carr, M.K.V. and Lockwood, G. (2011). The water relations and irrigation requirements of cocoa (*Theobroma cacao* L.): a review. *Experimental Agriculture* **47**:653–676.

References

Abeywardena, V. (1968). Forecasting coconut crops using rainfall data – a preliminary study. *Ceylon Coconut Quarterly* **19**:161–176.

Acheampong, K. (2010). A physiological study on field establishment of cacao clones through the improvement of agro-ecological conditions. PhD thesis, University of Reading, UK, 297 pp.

Acheampong, K., Daymond, A.J. and Hadley, P. (2009). The physiological basis of the shade requirement of young clonal cocoa. Paper presented at the International Cocoa Research Conference, Bali, November 2009.

Akunda, E.M.W. and Kumar, D. (1981). A simple technique for timing irrigation in coffee using cobalt chloride paper discs. *Irrigation Science* **3**:57–62.

Ali, F.M. (1969). Effects of rainfall on yield of cocoa in Ghana. *Experimental Agriculture* **5**:209–213.

Allen, R.G., Pereira, L.S., Raes, D. and Smith, M. (1998). *Crop Evapotranspiration: Guidelines for Computing Crop Water Requirements*. FAO Irrigation and Drainage Paper 56. Rome: Food and Agriculture Organization of the United Nations.

Almeida, A.-A.F. de and Valle, R.R. (2007). Ecophysiology of the cacao tree. *Brazilian Journal of Plant Physiology* **19**:425–448.

Almeida, A.-A.F. de, Brito, R.C.T., Aguilar, M.A.G. and Valle, R.R. (2001). Some water relations aspects of *Theobroma cacao* clones. In *Proceedings 13th International Cocoa Research Conference, Kota Kinabalu, Malaysia, October 2000*, pp. 349–363. Lagos, Nigeria: Cocoa Producers' Alliance.

Alvim, P. de T. (1958). Stomatal opening as a practical indicator of moisture deficiency in cacao. In *Séptima Conferencia Interamericana de Cacao, Palmira, Colombia, July 1958*, pp. 283–293.

Alvim, P. de T. (1960). Moisture stress as a requirement for flowering of coffee. *Science* **132**:354.

Alvim, P. de T. (1973). Factors affecting flowering of coffee. *Journal of Plantation Crops* **1**:37–43.

Alvim, P. de T. (1977). Cacao. In *Ecophysiology of Tropical Crops*, pp. 279–313 (ed. T.T. Kozlowski). London: Academic Press.

Alvim, P. de T. (1988). Relaçôes entre fatores climaticos e produção do cacaueiro. In *Proceedings of the 10th International Cocoa Research Conference, Santo Domingo, Dominican Republic, May 1987*, pp. 159–167. Lagos, Nigeria: Cocoa Producers' Alliance.

Alvim, R. and Alvim, P. de T. (1977). Hydroperiodicity in cocoa tree. In *Proceedings of the 5th International Cocoa Research Conference, Ibadan, Nigeria, September 1975*, pp. 204–209. Ibadan, Nigeria: Cocoa Research Institute of Nigeria.

Angelocci, L.R. and Magalhaes, A.C. (1977). Estimating leaf water potential of coffee with the pressure bomb. *Turrialba* **27**:305–306.

Anim-Kwapong, G.J. and Frimpong, E.B. (2006). Vulnerability of agriculture to climate change – impact of climate change on cocoa production. In *Report on Vulnerability and Adaptation Assessment under the Netherlands Climate Change Studies Assistance Programme Phase 2*. Tafo, Ghana: Cocoa Research Institute of Ghana.

Arachchi, L.P.V. (1998). Preliminary requirements to design a suitable drip irrigation system for coconut (*Cocos nucifera* L.) in gravelly soils. *Agricultural Water Management* **38**:169–180.

Araya, M. (2005). Stratification and spatial distribution of the banana (*Musa* AAA, Cavendish subgroup, cvs. 'Valery' and 'Grand Nain') root system. In *Banana Root System: Towards a Better Understanding for its Productive Management*, pp. 83–103 (ed. D.W. Turner and F.E. Rosales). Montpellier, France: International Network for the Improvement of Banana and Plantain.

Araya, M., Vargas, A. and Cheves, A. (1998). Changes in distribution of roots of banana (*Musa* AAA cv. Valery) with plant height, distance from the pseudostem and soil depth. *Journal of Horticultural Science and Biotechnology* **73**:437–440.

Arscott, T.G., Bhangoo, M.S. and Karon, M.L. (1965). Irrigation investigations of the Giant Cavendish banana. 1. Consumption of water applied to banana plantings in the Upper Aguan Valley, Honduras, as influenced by temperature and humidity. *Tropical Agriculture (Trinidad)* **42**:139–144.

Astegiano, E.D., Maestri, M. and Estevao, M. de M. (1988). Water stress and dormancy release in flower buds of *Coffea arabica* L.: water movement into the buds. *Journal of Horticultural Science* **63**:529–533.

Avilan, L.A., Rivas, N. and Sucre, R. (1984). Estudio del sistema radical del cocotero (*Cocos nucifera* L.). *Oléagineux* **39**:13–23.

Azevedo, P.V., Sausa, I.F. and Silva, B.B. (2006). Water-use efficiency of dwarf green coconut (*Cocos nucifera* L.) orchards in north-east Brazil. *Agricultural Water Management* **84**:259–264.

Azizuddin, M., Krishnamurthy Rao, W., Anantha Naik, S., Manjunath, A.N. and Hariyappa, N. (1994). Drip irrigation: effect on *C. arabica* var. Cauvery (Catimor). *Indian Coffee* **58**:3–8.

Bae, H., Kim, S-H., Kim, M.S. *et al.* (2008). The drought response of *Theobroma cacao* (cacao) and the regulation of genes involved in polyamine biosynthesis by drought and other stresses. *Plant Physiology and Biochemistry* **46**:174–188.

Bae, H., Sicher, R.C., Kim, M.S. *et al.* (2009). The beneficial endophyte *Trichoderma hamatum* isolate DIS219b promotes growth and delays the onset of the drought response in *Theobroma cacao*. *Journal of Experimental Botany* **60**(11):3279–3296.

Baggio, A.J., Caramore, P.H., Androceli Filho, A. and Montoya, L. (1997). Productivity of southern Brazilian coffee plantations shaded by different stockings of *Grevillea robusta*. *Agricultural Systems* **37**:111–120.

Baillie, I.C. and Burton, R.G.O. (1993). *Ngwazi Estate, Mufindi, Tanzania. Report on land and water resources with special reference to the development of irrigated tea. Part 1: Soils and land suitability*. Cranfield, UK: Soil Survey and Land Research Centre, Cranfield University.

Balasimha, D. (1999). Stress physiology of cocoa. *Journal of Plantation Crops* **27**:1–8.

Balasimha D., Anil Kumar V., Viraktamath B.C. and Ananda, K.S. (1999). Leaf water potential and stomatal resistance in cocoa hybrids and parents. *Plantations, Recherche, Développement* **6**:116–118.

Balasimha, D. and Daniel, E.V. (1988). A screening method for drought tolerance in cocoa. *Current Science* **57**(7):395.

Balasimha, D., Subramonan, N. and Chenchu Subbaiah, C. (1985). Leaf characteristics in cocoa (*Theobroma cacao* L.) accessions. *Café Cacao Thé* **29**:95–98.

Balasimha, D., Rajagopal, V., Daniel, E.V., Nair, R.V. and Bhagavan, S. (1988). Comparative drought tolerance of cacao accessions. *Tropical Agriculture (Trinidad)* **65**:271–274.

Balasimha, D., Daniel, E.V. and Bhat, P.G. (1991). Influence of environmental factors on photosynthesis in cocoa trees. *Agricultural and Forest Meteorology* **55**:15–21.

Balasubramanian, R., Ramanathian, T. and Vijayaraghavan, H. (1985). Certain aspects of moisture conservation in coconut gardens. *Indian Coconut Journal* **16**(2):13–15.

Balasuriya, J. (2000). The partitioning of net total dry matter to roots of clonal tea (*Camellia sinensis*) at different altitudes in the wet zone of Sri Lanka. *Tropical Agriculture (Trinidad)* **77**:163–168.

Baligar, V.C., Bunce, J.A., Machado, R.C.R. and Elson, M.K. (2008). Photosynthetic photon flux density, carbon dioxide concentration, and vapour pressure deficit effects on photosynthesis in cacao seedlings. *Photosynthetica* **46**:216–221.

Banerjee, B. (1992). Botanical classification of tea. In *Tea: Cultivation to Consumption*, pp. 25–51 (ed. K.C. Willson and M.N. Clifford). London: Chapman and Hall.

Barker, W.G. and Steward, F.C. (1962). Growth and development of the banana plant. I. The growing regions of the vegetative shoot. *Annals of Botany* **26**:389–411.

Barlow, C. (1997). Growth, structural change and plantation tree crops; the case of rubber. *World Development* **25**(10):1589–1607.

Barman, T.S., Baruah, U. and Saikia, J.K. (2008). Irradiance influences tea leaf (*Camellia sinensis* L.) photosynthesis and transpiration. *Photosynthetica* **48**:618–621.

Barradas, V.L. and Fanjul, L. (1986). Microclimatic characterization of shaded and open-grown coffee (*Coffea arabica* L.) plantations in Mexico. *Agricultural and Forest Meteorology* **38**:101–112.

Barros, R.S., Maestri, M. and Rena, A.B. (1995). Coffee crop ecology. *Tropical Ecology* **36**:1–19.

Barros, R.S., da Silva e Mota, J.W., Da Matta, F.M. and Maestri, M. (1997). Decline of vegetative growth in *Coffea arabica* L. in relation to leaf temperature, water potential and stomatal conductance. *Field Crops Research* **54**:65–72.

Basavaraju, T.B. and Hanumanthappa, M. (2009). Drip irrigation requirement of coconut in *Maidan* tract of Karnataka. *Mysore Journal of Agricultural Sciences* **43**:75–730.

Bassette, C. and Bussière, F. (2008). Partitioning of splash and storage during raindrop impacts on banana leaves. *Agricultural and Forest Meteorology* **148**:991–1004.

Bassoi, L.H., Moura e Silva, J.A., Gomes da Silva, E.E., Ramos, C.M.C. and Sediyama, G.C. (2004a). Guidelines for scheduling of banana crop in Sao Francisco Valley, Brazil. I. Root distribution and activity. *Revista Brasileira de Fruiticultura* **26**:459–463.

Bassoi, L.H., Teixeira, A.H. de C., Filho, J.M.P.L. *et al*. (2004b). Guidelines for scheduling of banana crop in Sao Francisco Valley, Brazil. II. Water consumption, crop coefficient, and physiological behaviour. *Revista Brasileira de Fruiticultura* **26**:464–467.

Batchelor, C.R., Soopramanien, G.C., Bell, J.P., Nayamuth, R. and Hodnett, M.G. (1990). Importance of irrigation regime, dripline placement and row spacing in the drip irrigation of sugarcane. *Agricultural Water Management* **17**:75–94.

Bauer, H., Wierer, R., Hatheway, W.H. and Larcher, W. (1985). Photosynthesis of *Coffea arabica* after chilling. *Physiologia Planta* **64**:449–454.

Bauer, H., Comploj, A. and Bodner, M. (1990). Susceptibility to chilling of some central-African cultivars of *Coffea arabica*. *Field Crops Research* **24**:119–129.

Beer, J. (1987). Advantages, disadvantages and desirable characteristics of shade trees for coffee, cacao and tea. *Agroforestry Systems* **5**:3–13.

Bell, J.P., Wellings, S.R., Hodnett, M.G. and Ah Koon, P.D. (1990). Soil water status: a concept for characterising soil water conditions beneath a drip irrigated row crop. *Agricultural Water Management* **17**:171–187.

Bénard, G. and Daniel, C. (1971). Économie de l'eau en jeunes palmeraies sélectionnées du Dahomey castration et sol nu. *Oléagineux* **26**:225–232.

Bhaskaran, U.P. and Leela, K. (1978). Response of coconut to irrigation in relation to production status of palms and soil type. In *Proceedings of First Annual Symposium on Plantation Crops (PLACROSYM 1)*, pp. 200–206 (ed. E.V. Nelliat). Kasaragod, India: Indian Society for Plantation Crops.

Bierhuizen, J.F., Nunes, M.A. and Ploegman, L. (1969). Studies on productivity of coffee. II. Effect of soil moisture on photosynthesis and transpiration of *Coffea arabica* L. *Acta Botanica Neerlandica* **18**:367–374.

Blackie, J.R. (1979). The water balance of the Kericho catchments. *East African Agriculture and Forestry Journal* **43**:55–84.

Blomme, G. and Ortiz, R. (2000). Preliminary assessment of root systems morphology in *Musa*. In *Proceedings of the First International Conference on Banana and Plantain in Africa* (ed. K. Craenen, R. Ortiz, E.B. Karamura and D.R. Vuylsteke). *Acta Horticulturae* **540**:259–266.

Blomme, G., Teugels, K., Blanckeart, I. *et al.* (2005). Methodologies for root system assessment in banana and plantains (*Musa* spp.). In *Banana Root System: Towards a Better Understanding for its Productive Management*, pp. 43–57 (ed. D.W. Turner and F.E. Rosales). Montpellier, France: International Network for the Improvement of Banana and Plantain.

Blore, T.W.D. (1966). Further studies of water use by irrigated and unirrigated Arabica coffee in Kenya. *Journal of Agricultural Science (Cambridge)* **67**:145–154.

Bonneau, X. and Subagio, K. (1999). Culture du cocotier en zone exposée au risque de sécheresse. *Plantations, Recherche, Développement* **6**:432–439.

Bonneau, X., Ochs, R., Kitu, W.T. and Yuswohadi. (1993). Chlorine: an essential element in the mineral nutrition of hybrid coconuts in Lampung (Indonesia). *Oléagineux* **48**(4):179–189.

Bonneau, X., Boulin, D., Bourgoing, G. and Sugarianto, J. (1997). Le chlorure de sodium, fertilisant idéal du cocotier en Indonésie. *Plantations, Recherche, Développement* **4**(5):336–346.

Borland, A.M., Griffiths, H., Hartwell, J. and Smith, J.A.C. (2009). Exploiting the potential of plants with crassulacean acid metabolism for bioenergy production on marginal lands. *Journal of Experimental Botany* **60**(10):2879–2806.

Bovee, A.C.J. (1975). Lysimeteronderzoek naar de verdamping van bananen in Libanon. *Landbouwkundig Tijdschrift (Netherlands)* **87**:174–180.

Boyer, J. (1969). Etude expérimentale des effets du régime d'humidité du soil sur la croissance végétative, la floraison et la fructification des cafiers Robusta. *Café, Cacao Thé* **13**:187–200.

Braconnier, S. and Bonneau, X. (1998). Effects of chlorine deficiency in the field on leaf gas exchanges in the PB121 coconut hybrid. *Agronomie* **18**:563–572.

Braconnier, S. and d'Auzac, J. (1990). Chloride and stomatal conductance in coconut. *Oléagineux* **45**(6):259–264.

Bragança, S.M. (2005). Crescimento e accúmulo de nutrients pelocafeeiro (*Coffea canephora* Pierre). PhD thesis, Universiada Federal de Viçosa, Brazil.

Braudeau, J. (1969). *Le Cacaoyer*. Paris: G.-P. Maisonneuve et Larose.

Bridgland, L.A. (1953). Study of the relationship between cacao yield and rainfall. *The Papua and New Guinea Agricultural Gazette* **8**(2):7–14.

Browning, G. (1975a). Shoot growth in *Coffea arabica* L. 1. Responses to rainfall when soil moisture status and giberellin supply are not limiting. *Journal of Horticultural Science* **50**:1–11.

Browning, G. (1975b). Environmental control of flower bud development in *Coffea arabica* L. In *Environmental Effects on Crop Physiology*, pp. 321–331 (ed. J.J. Landsberg and C.V. Cutting). London: Academic Press.

Browning, G. and Fisher, N.M. (1975). Shoot growth in *Coffea arabica* L. II. Growth flushing stimulated by irrigation. *Journal of Horticultural Science* **50**:207–218.

Brun, W.A. (1961). Photosynthesis and transpiration from upper and lower surfaces of intact banana leaves. *Plant Physiology* **36**:399–405.

Bull, R.A. (1963). Studies on the effect of yield and irrigation on root and stem development in *Coffea arabica* L. 1. Changes in the root systems induced by mulching and irrigation. *Turrialba* **13**:96–115.

Bull, T. and Glasziou, K.T. (1976). Sugar cane. In *Crop Physiology*, pp. 51–72 (ed. L.T. Evans). Cambridge, UK: Cambridge University Press.

Burgess, P.J. (1992a). Responses of tea clones to drought in southern Tanzania. PhD thesis, Cranfield University, UK.

Burgess, P.J. (1992b). Yield responses of clone 6/8 to fertiliser and irrigation. *Ngwazi Tea Research Unit, Dar es Salaam, Tanzania, Quarterly Report* **9**:3–10.

Burgess, P.J. (1992c). Dry matter production and partitioning by clone 6/8. *Ngwazi Tea Research Unit, Dar es Salaam, Tanzania, Quarterly Report* **9**:11–16.

Burgess, P.J. (1993a). Irrigation scheduling for mature and young tea. *Ngwazi Tea Research Unit, Dar es Salaam, Tanzania, Quarterly Report* **11**:3–7.

Burgess, P.J. (1993b). Economic analysis of irrigation for mature tea. *Ngwazi Tea Research Unit, Dar es Salaam, Tanzania, Quarterly Report* **14**:10–16.

Burgess, P.J. (1994a). Annual increases in the height of the plucking table. *Ngwazi Tea Research Unit, Dar es Salaam, Tanzania, Quarterly Report* **15**:3–10.

Burgess, P.J. (1994b). Methods of determining the water requirements of mature tea. *Ngwazi Tea Research Unit, Dar es Salaam, Tanzania, Quarterly Report* **17**:11–21.

Burgess, P.J. (1994c). N9. Responses of clonal tea to fertiliser and irrigation. *Ngwazi Tea Research Unit, Dar es Salaam, Tanzania, Annual Report* **1993**/94:6–7.

Burgess, P.J. and Carr, M.K.V. (1996a). Responses of young tea (*Camellia sinensis*) clones to drought and temperature. I. Yield and yield distribution. *Experimental Agriculture* **32**:357–372.

Burgess, P.J. and Carr, M.K.V. (1996b). Responses of young tea (*Camellia sinensis*) clones to drought and temperature. II. Dry matter production and partitioning. *Experimental Agriculture* **32**:377–394.

Burgess, P.J. and Carr, M.K.V. (1997). Responses of young tea (*Camellia sinensis*) clones to drought and temperature. III. Shoot extension and development. *Experimental Agriculture* **33**:367–383.

Burgess, P.J. and Carr, M.K.V. (1998). The use of leaf appearance rates estimated from measurements of air temperature to determine harvest intervals for tea. *Experimental Agriculture* **34**:207–218.

Burgess, P.J. and Sanga, N.K. (1994a). Dry weight and root distribution of mature tea. *Ngwazi Tea Research Unit, Dar es Salaam, Tanzania, Quarterly Report* **16**:14–18.

Burgess, P.J. and Sanga, N.K. (1994b). Dry weight and root distribution of six-year old tea. *Ngwazi Tea Research Unit, Dar es Salaam, Tanzania, Quarterly Report* **18**:12–16.

Burgess, P.J., Carr, M.K.V., Mizambwa, F.C.S. *et al.* (2006). Evaluation of simple hand-held mechanical systems for harvesting tea (*Camellia sinensis*). *Experimental Agriculture* **42**:165–187.

Burle, L. (1961). *Le Cacaoyer*. Paris: G.-P. Maisonneuve et Larose.

Butler, D.R. (1977). Coffee leaf temperatures in a tropical environment. *Acta Botanica Neerlandica* **26**:129–140.

Cai, C.-T., Cai, Z.-Q., Yao, T.-Q. and Qi, X. (2007). Vegetative growth and photosynthesis in coffee plants under different watering and fertilization managements in Yunnan, SW China. *Photosynthetica* **45**(3):455–461.

Caliman, J.P. (1992). Palmier à huile et déficit hydrique production, techniques culturales adaptées. *Oléagineux* **47**:205–216.

Caliman, J.P. and Southworth, A. (1998). Effect of drought and haze on the performance of oil palm. In *Proceedings of 1998 International Oil Palm Conference 'Commodity of the Past, Today and the Future'*, pp. 250–274 (ed. A. Jatmika), Medan, Indonesia: Indonesian Oil Palm Research Institute.

Cannell, M.G.R. (1971). Production and distribution of dry matter in trees of *Coffea arabica* L. in Kenya as affected by seasonal climatic differences and the presence of fruit. *Annals of Applied Biology* **67**:99–120.

Cannell, M.G.R. (1972). Primary production, fruit production and assimilate partition in Arabica coffee: a review. In *Annual Report for 1971/2*, pp. 6–24. Ruiru, Kenya: Coffee Research Foundation.

Cannell, M.G.R. (1973). Effects of irrigation, mulch and N-fertilisers on yield components of Arabica coffee in Kenya. *Experimental Agriculture* **9**:225–232.

Cannell, M.G.R. (1974). Factors affecting Arabica coffee bean size in Kenya. *Journal of the Horticultural Science* **49**:65–76.

Cannell, M.G.R. (1985). Physiology of the coffee crop. In *Coffee: Botany, Biochemistry and Production of Beans and Beverage*, pp. 108–134 (ed. N.M. Clifford and K.C. Wilson). London: Chapman and Hall.

Cannell, M.G.R., Harvey, F.J., Smith, R.I. and Deans, J.D. (1990). Genetic improvement of tea. *Final Report of Project TO1057d*. Edinburgh, UK: Institute of Terrestrial Ecology.

Caramori, P.H., Androcioli Filho, A. and Leal, A.C. (1996). Coffee shade with *Mimosa scabrella* Benth. for frost protection in southern Brazil. *Agroforestry Systems* **33**:205–214.

Cardinal, A.B.B., Goncalves, P.D. and Martins, A.L.M. (2007). Stock-scion interactions on growth and rubber yield of *Hevea brasiliensis*. *Scientia Agricola (Piracicara, Brazil)* **64**(3):235–240.

Carr, M.K.V. (1968). Report on research into the water requirements of tea in East Africa. Paper presented to the Specialist Committee in Applied Meteorology, Nairobi, Kenya.

Carr, M.K.V. (1969). The water requirements of the tea crop. PhD thesis, University of Nottingham, UK.

Carr, M.K.V. (1970). The role of water in the growth of the tea crop. In *Physiology of Tree Crops*, pp. 287–305 (ed. L.C. Luckwill and C.V. Cutting). London: Academic Press.

Carr, M.K.V. (1971a). An assessment of some of the results of tea/soil/water studies in southern Tanzania. In *Water and the Tea Plant*, pp. 21–47 (ed. M.K.V. Carr and S. Carr). Kericho, Kenya: Tea Research Institute of East Africa.

Carr, M.K.V. (1971b). The internal water status of the tea plant (*Camellia sinensis*): some results illustrating the use of the pressure chamber technique. *Agricultural Meteorology* **9**:447–460.

Carr, M.K.V. (1972). The climatic requirements of the tea plant. *Experimental Agriculture* **8**:1–14.

Carr, M.K.V. (1974). Irrigating seedling tea in Southern Tanzania: effects on total yields, distribution of yield and water use. *Journal of Agricultural Science (Cambridge)* **83**:363–378.

Carr, M.K.V. (1976). Methods of bringing tea into bearing in relation to water status during dry weather. *Experimental Agriculture* **12**:341–351.

Carr, M.K.V. (1977a). Changes in the water status of tea clones during dry weather in Kenya. *Journal of Agricultural Science (Cambridge)* **89**:297–307.

Carr, M.K.V. (1977b). Responses of seedling tea bushes and their clones to water stress. *Experimental Agriculture* **13**:317–24.

Carr, M.K.V. (1985). Some effects of shelter on the yield and water use of tea. In *Effects of Shelter on the Physiology of Plants and Animals* (ed. J. Grace). Lisse, Netherlands: Swets and Zeitlinger B.V. *Progress in Biometeorology* **2**:127–144.

Carr, M.K.V. (1992). Plantation crop research in Sri Lanka. Unpublished report, Silsoe College, Cranfield University, UK, pp. 12–20.

Carr, M.K.V. (1999). Evaluating the impact of research for development: tea in Tanzania. *Experimental Agriculture* **35**:247–264.

Carr, M.K.V. (2000a). Irrigation research: developing a holistic approach. In *Proceedings of the Third International Symposium on Irrigation of Horticultural Crops* (ed. M.I. Ferreira and H.G. Jones). *Acta Horticulturae* **537**:733–739.

Carr, M.K.V. (2000b). *1. Shoot growth plus plucking equals profit, and 2. Definitions of terms used in crop harvesting.* TRIT Occasional Publication No. 1. Dar es Salaam, Tanzania: Tea Research Institute of Tanzania (available at: www.trit.or.tz/).

Carr, M.K.V. (2001). The water relations and irrigation requirements of coffee. *Experimental Agriculture* **37**:1–36.

Carr, M.K.V. and Punchihewa, P.G. (2002). *External Review Report.* Montpellier, France: BUROTROP, pp. 13–46.

Carr, M.K.V. and Stephens, W. (1992). Climate, weather and the yield of tea. In *Tea: Cultivation to Consumption*, pp. 87–135 (ed. K.C. Willson and M. N. Clifford). London: Chapman and Hall.

Carr, M.K.V., Dale, M.O. and Stephens, W. (1987). Yield distribution at two sites in eastern Africa. *Experimental Agriculture* **23**:75–85.

Carr, M.K.V., Ndamugoba, D.M., Burgess, P.J. and Myinga, G.R. (1992). An overview of tea research in Tanzania with special reference to the Southern Highlands. In *Agricultural Research, Training and Technology Transfer in the Southern Highlands of Tanzania: Past Achievements and Future Prospects*, pp. 237–252 (ed. J.A. Ekpere, D.J. Rees, M.B. Mbwile and N.G. Lyimo). Mbeya, Tanzania: Uyole Agricultural Centre.

Cassidy, D.S.M. and Kumar, D. (1984). Root distribution of *Coffea arabica* L. in Zimbabwe. 1. The effect of plant density, mulch and cova planting in Chipinge. *Zimbabwe Journal of Agricultural Research* **22**:119–132.

CDB (2010). *Coconut Development Board of India.* www.coconutboard.nic.in/package.htm (accessed 26 June 2010).

CFC (2005). *Product and Market Development of Sisal and Henequen.* Project completion report, Kenya – Tanzania, January 1997 – December 2005, 64 pp. Vienna: Common Fund for Commodities, UNIDO. www.fao.org/es/esc/common/ecg/330/en/SummaryReport.pdf (accessed October 2011).

Chabot, R., Bouarfa, S., Zimmer, D., Chaumont, C. and Moreau, S. (2005). Evaluation of the sap flow determined with a heat balance method to measure the transpiration of a sugarcane canopy. *Agricultural Water Management* **75**:10–24.

Chaillard, H., Daniel, C., Houeto, V. and Ochs, R. (1983). L'irrigation du palmier à huile et du cocotier. Expérience sur 900 ha en République populaire du Benin. *Oléagineux* **38**:519–529.

Chandrashekar, T.R., Nazeer, M.A., Marattukulam, J.G., *et al.* (1998). An analysis of growth and drought tolerance in rubber during the immature phase in a dry subhumid climate. *Experimental Agriculture* **34**:287–300.

Chang, Jen-hu, Campbell, R.B. and Robinson, F.E. (1963). On the relationship between water and sugar cane yield in Hawaii. *Agronomy Journal* **55**:450–453.

Cheesman, E.E. (1944). Notes on the nomenclature, classification and possible relationships of cacao populations. *Tropical Agriculture (Trinidad)* **21**:144–159.

Chen, Ching-Yih (1971) Study on water relations in banana. *Journal of the Agricultural Association of China* **76**:24–38.

Chen, J-W., Zhang, Q., Li, X-S. and Cao, K-F. (2010). Gas exchange and hydraulics in seedlings of *Hevea brasiliensis* during water stress and recovery. *Tree Physiology* **30**(7):876–885.

Cintra, F.L.D., Leal, M. de L. da S.L. and Passos, E.E. de M. (1992). Evaluation of root system of dwarf coconut cultivars. *Oléagineux* **47**:225–234.

Cintra, F.L.D., Passos, E.E. de M. and Leal, M. de L. da S.L. (1993). Evaluation de la distribution du système racinaire de cultivars de cocotier Grands. *Oléagineux* **48**:453–461.

Clowes, M. St. J. and Allison, J.C.S. (1982). A review of the coffee plant *Coffea arabica* L., its environment and management in relation to coffee growing in Zimbabwe. *Zimbabwe Journal of Agricultural Research* **20**:1–19.

Clowes, M. St. J. and Logan, W.J.C. (ed.) (1985). *Advances in Coffee Management and Technology in Zimbabwe.* Harare, Zimbabwe: Coffee Growers Association.

Clowes, M. St. J. and Wilson, J.H. (1974). Physiological factors influencing irrigation management of coffee in Rhodesia. *Rhodesian Agricultural Journal* **71**:54–55.

Colas, H., Mouchet, S., Rey, H. and Kitu, W.T. (1999). Une approche du comportement hybrique du cacaoyer (*Theobroma cacao* L.) par des mesures de flux de sève brute: comparison entre une culture pure et une culture associée sous cocotier (*Cocos nucifera* L.). In *Proceedings of the 12th International Cocoa Research Conference, Salvador, Bahia, Brazil, November 1996,* pp. 637–644. Lagos, Nigeria: COPAL.

Cooper, J.D. (1979). Water use of a tea estate from soil moisture measurements. *East African Agricultural and Forestry Journal* **43**:102–121.

Copeland, E. H. (1906). On the water relations of the coconut palm (*Cocos nucifera*). *Journal of Science, Manila* **1**:6–57.

Corley, R.H.V. (1973). Midday closure of stomata in the oil palm in Malaysia. *MARDI Research Bulletin* **1**(2):1–4.

Corley, R.H.V. (1976a). The genus *Elaeis.* In *Oil Palm Research,* pp. 3–5 (ed. R.H.V. Corley, J.J. Hardon and B.J. Wood). Amsterdam: Elsevier.

Corley, R.H.V. (1976b). Photosynthesis and productivity. In *Oil Palm Research*, pp. 55–76 (ed. R.H.V. Corley, J.J. Hardon and B.J. Wood). Amsterdam, the Netherlands: Elsevier.

Corley, R.H.V. (1983). Potential productivity of tropical perennial crops. *Experimental Agriculture* **19**:217–237.

Corley, R.H.V. (1985). Yield potentials of plantation crops. In *Potassium in the Agricultural Systems of the Humid Tropics*, pp. 61–80. Proceedings of the 19th Colloquium of the Potash Institute. Berne, Switzerland: International Potash Institute.

Corley, R.H.V. (1996). Irrigation of oil palms – a review. *Journal of Plantation Crops* **24** (supplement):45–52.

Corley, R.H.V. and Gray, B.S. (1976a). Growth and morphology. In *Oil Palm Research*, pp. 7–21 (ed. R.H.V. Corley, J.J. Hardon and B.J. Wood). Amsterdam, the Netherlands: Elsevier.

Corley, R.H.V. and Gray, B.S. (1976b). Yield and yield components. In *Oil Palm Research*, pp. 77–86 (ed. R.H.V. Corley, J.J. Hardon and B.J. Wood). Amsterdam, Netherlands: Elsevier.

Corley, R.H.V. and Hong T.K. (1982). Irrigation of oil palms in Malaysia. In *The Oil Palm in Agriculture in the Eighties*, Vol. 2, pp. 343–346 (ed. E. Pushparajah and P. S. Chew). Kuala Lumpur: Incorporated Society of Planters.

Corley, R.H.V. and Tinker, P.B. (2003). *The Oil Palm*, 4th edn. Oxford, UK: Blackwell Publishing.

Cowan, I.R. and Innes, R.F. (1956). Meteorology, evaporation and the water requirements of sugar cane. *Proceedings of the International Society of Sugar-cane Technologists* **9**:215–232.

CPCRI (2010). Research/Achievements. Kerala, India: Central Plantation Crops Research Institute. http://cpcri.gov.in/achievements.htm (accessed October 2011).

Crisosto, C.H., Grantz, D.A. and Meinzer, F.C. (1992). Effects of water deficit on flower opening in coffee (*Coffea arabica* L.). *Tree Physiology* **10**:127–139.

Cuenca, G., Aranguren, J. and Herrera, R. (1983). Root growth and litter decomposition in a coffee plantation under shade trees. *Plant and Soil* **71**:477–486.

Dagg, M. (1970). A study of the water use of tea in East Africa using an hydraulic lysimeter. *Agricultural Meteorology* **7**:303–320.

Dale, M.O. (1971). Progress with irrigation experiments in Malawi. In *Water and the Tea Plant*, pp. 59–78 (ed. M.K.V. Carr and S. Carr). Kericho, Kenya: Tea Research Institute of East Africa.

DaMatta, F.M. (2004a). Exploring drought tolerance in coffee: a physiological approach with some insights for plant breeding. *Brazilian Journal of Plant Physiology* **16**(1):1–6.

DaMatta, F.M. (2004b). Ecophysiological constraints on the production of shaded and unshaded coffee: a review. *Field Crops Research* **86**:99–114.

DaMatta, F.M. and Ramalho, J.D. (2006). Impacts of drought and temperature stress on coffee physiology and production: a review. *Brazilian Journal of Plant Physiology* **18**(1):55–81.

DaMatta, F.M., Maestri, M., Barros, R.S. and Regazzi, A.J. (1993). Water relations of coffee leaves (*Coffea arabica* and *C. canephora*) in response to drought. *Journal of Horticultural Science* **68**:741–746.

DaMatta, F.M., Maestri, M. and Barros, R.S. (1997). Photosynthetic performance of two coffee species under drought. *Photosynthetica* **34**:257–264.

DaMatta, F.M., Ronchi, C.P., Maestri, M. and Barros, R.S. (2010). Coffee: environment and crop physiology. In *Ecophysiology of Tropical Tree Crops*, pp. 181–216 (ed. F. DaMatta). New York: Nova Science Publishers.

Dancer, J. (1963). The response of seedling Arabica coffee to moisture deficits. *Euphyitica* **12**:294–298.

Daniells, J. (2004). *Banana: methods of irrigation*. The State of Queensland, Department of Primary Industries and Fisheries. www2.dpi.qld.gov.au/horticulture/5269.html (accessed 29/10/08).

Daniells, J.W. (1986). Determining patterns of soil water use by bananas. In *Proceedings of Symposium on Physiology of Productivity of Subtropical and Tropical Tree Fruits* (ed. B.W. Cull and P.E. Page). *Acta Horticulturae* **175**:357–361.

Daniells, J., Jenny, C., Karamura, D. and Tomekpe, K. (2001). *Musalogue: a catalogue of Musa germplasm. Diversity in the genus Musa* (compilers E. Arnaud and S. Sharrock). Montpellier, France: International Network for the Improvement of Banana and Plantain.

Dart, I.L., Baille, C.P., and Thorburn, P.J. (2000). Assessing nitrogen application rates for subsurface trickle irrigated sugarcane at Bundaberg. *Proceedings of Australian Society of Sugar Cane Technologists Conference* (ed. D.M. Hogarth) **22**:230–235.

Daymond, A.J. and Hadley, P. (2004). The effects of temperature and light integral on early vegetative growth and chlorophyll fluorescence of four contrasting genotypes of cacao (*Theobroma cacao*). *Annals of Applied Biology* **145**:257–262.

Daymond, A.J. and Hadley, P. (2008). Differential effects of temperature on fruit development and bean quality of contrasting genotypes of cacao (*Theobroma cacao*). *Annals of Applied Biology* **153**:175–185.

Daymond, A.J., Hadley, P., Machado, R.C.R. and Ng, E. (2002a). Canopy characteristics of contrasting clones of cocoa (*Theobroma cacao*). *Experimental Agriculture* **38**:359–367.

Daymond, A.J., Hadley, P., Machado, R.C.R. and Ng, E. (2002b). Genetic variability in partitioning to the yield component of cacao (*Theobroma cacao* L.). *HortScience* **37**:799–801.

Daymond, A.J., Tricker, P.J. and Hadley, P. (2009). Genotypic variation in photosynthetic and leaf traits in cocoa. Paper presented at the International Cocoa Research Conference, Bali, November 2009.

De Costa, W.A.J.M., Mohotti, A.J. and Wijeratne, M.A. (2007). Ecophysiology of tea. *Brazilian Journal of Plant Physiology* **19**:299–332.

Delvaux, B. (1995). Soils. In *Bananas and Plantains*, pp. 230–257 (ed. S. Gowen). London: Chapman and Hall.

Devakumar, A.S., Sathik. M.B.M., Jacob. J. and Annamalainathan, K. (1998). Effects of atmospheric and soil drought on growth and development of *Hevea brasiliensis. Journal of Rubber Research* **1**(3):190–198.

Devakumar, A.S., Prakash, P., Sathik. M.B.M. and Jacob. J. (1999). Drought alters the canopy architecture and micro-climate of *Hevea brasiliensis* trees. *Trees* **13**:161–167.

Diczbalis, Y., Lemin, C., Richards, N. and Wicks, C. (2010). *Producing Cocoa in Northern Australia*. Australian Government, Rural Industries Research and Development Corporation Report 09/092, 279 pp.

Do, F. and Rocheteau, A. (2002). Influence of natural temperature gradients on measurements of xylem sap flow with thermal dissipation probes. 2. Advantages and calibration of a non-continuous heating system. *Tree Physiology* **22**:649–654.

Dodsworth, G.H., Nixon, D.J. and Sweet, C.P.M. (1990). An assessment of drip irrigation of sugarcane on poorly structured soils in Swaziland. *Agricultural Water Management* **17**:325–335.

Donaldson, R.A. and Bezuidenhout, C.N. (2000). Determining the maximum drying off periods for sugarcane grown in different regions of the South African industry. *Proceedings of the South African Sugar Technologists' Association* **74**:162–166.

Doorenbos, J. and Kassam, A.H. (1979). *Yield Response to Water*. Food and Agricultural Organization of the United Nations, Irrigation and Drainage Paper 33. Rome: FAO.

Doorenbos, J. and Pruitt, W.D. (1977). *Crop Water Requirements* (revised). Food and Agricultural Organisation of the United Nations, Irrigation and Drainage Paper 24. Rome: FAO.

dos Santos, C.A.C., da Silva, B.B., Rao, T.V.R. and Neale, C.M.U. (2009). Energy balance measurements over a banana orchard in the semiarid region of north east Brazil. *Pesquisa Agropecuária Brasileira* **44**(11):1365–1373.

Draye, X., Lecompte, F. and Pagès, L. (2005). Distribution of banana roots in time and space: new tools for an old science. In *Banana Root System: Towards a Better Understanding for its Productive Management*, pp. 58–74 (ed. D.W. Turner and F.E. Rosales). Montpellier, France: International Network for the Improvement of Banana and Plantain.

Drinnan, J.E. (1995). Managing bearing trees. In *Coffee Growing in Australia: a Machine Harvesting Perspective*, pp. 61–78 (ed. R. Lines-Kelly). Barton, Australia: Rural Industries Research and Development Corporation.

Drinnan, J.E. and Menzel, C.M. (1994). Synchronisation of anthesis and enhancement of vegetative growth in coffee (*Coffea arabica* L.) following water stress during floral initiation. *Journal of Horticultural Science* **69**: 841–849.

Drinnan, J.E. and Menzel, C.M. (1995). Temperature affects vegetative growth and flowering of coffee (*Coffea arabica* L.). *Journal of Horticultural Science* **70**:25–34.

Dufour, O., Frère, J.L., Caliman, J.P. and Horns, P. (1988). Présentation d'une méthode simplifée de prévision de la production d'une plantation de palmiers à huile à partir de la climatologie. *Oléagineux* **43**:271–278.

Dufrêne, E. (1989). Photosynthèse, consummation en eau et modélisation de la production chez le palmier à huile (*Elaeis guineensis* Jacq.). Thesis, Universite de Paris-Sud, Orsay.

Dufrêne, E. and Saugier, B. (1993). Gas exchange of oil palm in relation to light, vapour pressure deficit, temperature and leaf age. *Functional Ecology* **7**:97–104.

Dufrêne, E., Dubos, B., Rey, J., Quencez, P. and Sauugier, B. (1992). Changes in evapotranspiation from an oil palm stand (*Elaeis guineensis* Jacq.) exposed to seasonal water deficit. *Acta Oecologia* **13**: 299–314.

Dunlop, W.R. (1925). Rainfall correlations in Trinidad. *Nature (London)* **115**:192–193.

Eckstein, K. and Robinson, J.C. (1995a). Physiological responses of banana (*Musa* AAA; Cavendish subgroup) in the subtropics. 1. Influence of internal plant factors on gas exchange of banana leaves. *Journal of Horticultural Science* **70**:147–156.

Eckstein, K. and Robinson, J.C. (1995b). Physiological responses of banana (*Musa* AAA; Cavendish subgroup) in the subtropics. II. Influence of climatic conditions on seasonal and diurnal variations in gas exchange of banana leaves. *Journal of Horticultural Science* **70**:157–167.

Eckstein, K. and Robinson, J.C. (1996). Physiological responses of banana (*Musa* AAA; Cavendish subgroup) in the subtropics. VI. Seasonal responses of leaf gas exchange to short term water stress. *Journal of Horticultural Science* **71**:679–692.

Eckstein, K., Robinson, J.C. and Fraser, C. (1996). Physiological responses of banana (*Musa* AAA; Cavendish subgroup) in the subtropics. V. Influence of leaf tearing on assimilation potential and yield. *Journal of Horticultural Science* **71**:503–514.

Eckstein, K., Fraser, C., Botha, A. and Husselmann, J. (1998). Evaluation of various irrigation systems for highest economical yield and optimum water use for bananas. In *Proceedings of an International Symposium on Banana in the Subtropics* (ed. V. Galan Sauco). Acta Horticulturae **490**:147–158.

Edwards, D.F. (1973). Pollination studies on Upper Amazon cocoa clones in Ghana in relation to the production of hybrid seed. *Journal of Horticultural Science* **48**:247–259.

Edwin, J. and Masters, W.A. (2005). Genetic improvement and cocoa yields in Ghana. *Experimental Agriculture* **41**:491–503.

Ekanayake, I.J., Ortiz, R. and Vuylsteke, D.R. (1994). Influence of leaf age, soil moisture, VPD and time of day on leaf conductance of various *Musa* genotypes in a humid forest–moist savannah transition site. *Annals of Botany* **74**:173–178.

Ekanayake, I.J., Ortiz, R. and Vuylsteke, D.R. (1998). Leaf stomatal conductance and stomatal morphology of *Musa* germplasm. *Euphytica* **99**:221–229.

Ellis, R. and Nyirenda, H.E. (1995). A successful plant improvement programme on tea (*Camellia sinensis*). *Experimental Agriculture* **31**:307–323.

Ellis, R.D. and Lankford, B.A. (1990). The tolerance of sugarcane to water stress during the main development phases. *Agricultural Water Management* **17**:117–128.

Ellis, R.D., Wilson, J.H. and Spies P.M. (1985). Development of an irrigation policy to optimise sugar production during seasons of water shortage. *Proceedings of the South African Sugar Technologists' Association* **59**:142–147.

Fanjul, L., Arreola-Rodriguez, R. and Mendez-Castrejou, M.P. (1985). Stomatal responses to environmental variables in shade and sun grown coffee plants in Mexico. *Experimental Agriculture* **21**:249–258.

FAO (2009a). Food and agricultural commodities production for 2007. http://faostat.fao.org/site/339/default.aspx (accessed December 2010).

FAO (2009b). Jute, kenaf, sisal, abaca, coir and allied fibres: statistics. www.fao.org/es/ESC/common/ecg/323/en/STAT_BULL_2009.pdf (accessed October 2011).

FAO (2010a). Food and agricultural commodities production for 2008. http://faostat.fao.org/site/339/default.aspx (accessed December 2010).

FAO (2010b). Production/Crops. Data for 2008. http://faostat.fao.org/site/567/Desktop Default.aspx?PageID = 567#ancor (accessed December 2010).

FAO (2011a). Production/Crops. Data for 2009. http://faostat.fao.org/site/567/(accessed January 2011).

FAO (2011b). Food and agricultural commodities production for 2009. http://faostat.fao.org/site/339/default.aspx (accessed December 2010).

Fereres, E. and Soriano, M.A. (2006). Deficit irrigation for reducing agricultural water use. *Journal of Experimental Botany* **58**:147–159.

Filho, A.A., Siqueira, R., Caramori, P.H., Pavan, M.A., Sera, T. and Soderholm, P.K. (1986). Frost injury and performance of coffee at 23°S in Brazil. *Experimental Agriculture* **22**:71–74.

Filho, J.P. de L., Nova, N.A.V. and Pinto, H.S. (1993). Base temperature and heat units for leaf flushing emission and growth of *Hevea brasiliensis* Muell. Arg. *International Journal of Biometeorology* **37**:65–67.

Filho, L.C. de A. L., de Mello, C.R. and de Carvalho, L.C. (2010). Spatial-temporal analysis of water requirements of coffee crop in Minas Gerais State, Brazil. *Revista Brasileira Engenharia Agricola e Ambiental* **14**(2):165–172.

Finkel, H.J. (1983). Irrigation of sugar crops. In *CRC Handbook of Irrigation Technology*, Vol. II (ed. H.J. Finkel). Boca Raton, FL: CRC Press.

Fisher, N.M. and Browning, G. (1979). Some effects of irrigation and plant density on the water relations of high density coffee (*Coffea arabica* L.) in Kenya. *Journal of Horticultural Science* **54**:13–22.

Fleuret, A and Fleuret, P. (1985). The banana in the Usambara. In *Food Energy in Tropical Ecosystems*, pp. 145–166 (ed. D.J. Cattle and K.H. Schwern). Food and Nutrition in History and Anthropology Volume 4. New York: Gordon and Breach.

Flowers, C. (1996). Evaluating irrigation uniformity on a tea plantation in East Africa. *Irrigation News* **25**:37–40.

Foong, S.F. (1993). Potential evapotranspiration, potential yield and leaching losses of oil palm. In *Proceedings of 1991 PORIM International Palm Oil Conference, Module-Agriculture*, pp. 105–119. Kuala Lumpur, Malaysia: Palm Oil Research Institute.

Fordham, R. (1970). Factors affecting tea yields in Malawi. PhD thesis, Bristol University, UK.

Fordham, R. (1971). Stomatal physiology and the water relations of the tea bush. In *Water and the Tea Plant*, pp. 21–47 (ed. M.K.V. Carr and S. Carr). Kericho, Kenya: Tea Research Institute of East Africa.

Fordham, R. (1972a). The water relations of cacao. In *Proceedings of the IV International Cocoa Research Conference, St. Augustine, Trinidad, January 1972*, pp. 320–325. Lagos, Nigeria: COPAL.

Fordham, R. (1972b). Observations on the growth of roots and shoots of tea (*Camellia sinensis*, L.) in southern Malawi. *Journal of Horticultural Science* **47**:221–229.

Fordham, R. (1977). Tea. In *Ecophysiology of Tropical Crops*, pp. 333–349 (ed. P. de T. Alvim and T.T. Kozlowski). New York: Academic Press.

Fordham, R. and Palmer-Jones, R.W. (1977). Simulation of intraseasonal yield fluctuations of tea in Malawi. *Experimental Agriculture* **13**:33–42.

Franco, L.M. (1939). Relation between chromosome number and stomata in *Coffea*. *Botanical Gazette* **100**:817–827.

Friend, D. and Corley, R.H.V. (1994). Measuring coconut palm dry matter production. *Experimental Agriculture* **30**:223–235.

Garriz, P.I. (1979). Distribución radicular de tres cultivares de *Coffea arabica* L. en un suelo limo-arcilloso. *Agronomía Tropical* **29**:91–103.

Gathaara, M.P.H. and Kiara, J.M. (1984). Factors that influence yield in close-spaced coffee. I. Light, dry matter production and plant water status. *Kenya Coffee* **49**:159–167.

Gathaara, M.P.H. and Kiara, J.M. (1985). Factors that influence yield in close-spaced coffee. II. Yield components. *Kenya Coffee* **50**:387–392.

Gathaara, M.P.H. and Kiara, J.M. (1988). Effects of irrigation rates and frequency on the growth and yield of *Arabica* coffee. *Kenya Coffee* **53**:309–312.

Gathaara, M.P.H., Kiara, J.M. and Gitau, K.M. (1993). The influence of drip irrigation and tree density on the yield and quality of Arabica coffee. *Kenya Coffee* **58**:1599–1603.

George, B.R.F (1988). A simple field method of scheduling irrigation. *Proceedings of the South African Sugar Technologists' Association*, **1988**:149–151.

George, S., Suresh, P.R., Wahid, P.A., Nair, R.B. and Punnoose, K.I. (2009). Active root distribution pattern of *Hevea brasiliensis* determined by radioassay of latex serum. *Agroforestry Systems* **76**:275–281.

Gerritsma, W. and Wessel, M. (1997). Oil palm: domestication achieved? *Netherlands Journal of Agricultural Science* **45**:463–475.

Ghavami, M. (1973). Determining water needs of the banana plant. *Transactions of the American Society of Agricultural Engineers* **16**:598–600.

Ghavami, M. (1974). Irrigation of Valery bananas in Honduras. *Tropical Agriculture (Trinidad)* **51**:443–446.

Glover, J. (1939). The root system of *Agave sisalana* in certain East African soils. *Empire Journal of Experimental Agriculture* **7**:11–20.

Glover, J. (1967). The simultaneous production of sugar cane roots and tops in relation to soil and climate. *Proceedings of the South African Sugar Technologists' Association* **41**:143–159.

Goenaga, R. and Irizarry, H. (1995). Yield performance of banana irrigated with fractions of Class A pan evaporation in a semiarid environment. *Agronomy Journal* **87**:172–176.

Goenaga, R. and Irizarry, H. (1998). Yield of banana grown with supplementary drip irrigation on an ultisol. *Experimental Agriculture* **34**:439–448.

Goenaga, R. and Irizarry, H. (2000). Yield and quality of banana irrigated with fractions of Class A pan evaporation on an oxisol. *Agronomy Journal* **92**:1008–1012.

Goenaga, R., Irizarry, H. and Gonzalez, E. (1993). Water requirement of plantains (*Musa acuminata* × *Musa balbisiana* AAB) grown under semi-arid conditions. *Tropical Agriculture (Trinidad)* **70**:3–7.

Goenaga, R., Irizarry, H., Coleman, B. and Ortiz, E. (1995). Drip irrigation recommendations for plantain and banana grown on the semiarid southern coast of Puerto Rico. *Journal of the Agricultural University of Puerto Rico* **79**:13–27.

Gomes, A.R.S. and Kozlowski, T.T. (1988). Stomatal characteristics, leaf waxes, and transpiration rates of *Theobroma cacao* and *Hevea brasiliensis* seedlings. *Annals of Botany* **61**:425–432.

Gomes, A.R.S. and Kozlowski, T.T. (1989). Responses of seedlings of two varieties of *Theobroma cacao* to wind. *Tropical Agriculture (Trinidad)* **66**:137–141.

Gomes, F.P. and Prado, C.H.B.A. (2007). Ecophysiology of coconut palm under water stress. *Brazil Journal of Plant Physiology* **19**:377–391.

Gomes, F.P., Mielke, M.S. and Almeida, A-AF. (2002a). Leaf gas exchange of green dwarf coconut (*Cocos nucifera* L. var. *nana*) in two contrasting environments of the Brazilian north-east region. *Journal of Horticultural Science and Biotechnology* **77**:766–772.

Gomes, F.P., Mielke, M.S., Almeida, A.-A.F. and Muniz, W.S. (2002b). Leaf gas exchange in two dwarf coconut genotypes in the southeast of Bahia State, Brazil. *Coconut Research and Development* **18**(2):37–55.

Gomes, F.P., Oliva, M.A., Mielke, M.S. *et al.* (2008). Photosynthetic limitations in leaves of young Brazilian Green Dwarf coconut (*Cocos nucifera* L. 'nana') palm under well-watered conditions or recovering from drought stress. *Environmental and Experimental Botany* **62**:195–204.

Gomes, F.P., Oliva, M.A., Mielke, M.S. *et al.* (2009). Is abscisic acid involved in the drought responses of Brazilian Green Dwarf coconut? *Experimental Agriculture* **45**:189–198.

Gomez, J.B. and Hamzah, S. (1980). Variation in leaf morphology and anatomy between clones of *Hevea*. *Journal of the Rubber Research Institute of Malaysia* **28**:157–182.

Gómez, L.F., López, J.C., Riaño, N.M., López, Y. and Montoya, E.C. (2005). Diurnal changes in leaf gas exchange and validation of a mathemtaical model for coffee (*Caffea arabica* L.) canopy photosynthesis. *Photosynthetica* **43**(4):575–582.

Gonkhamdee, S., Pierret, A., Maeght, J.-L. and Do, F.C. (2009). Growth dynamics of fine *Hevea brasilensis* roots along a 4.5 m soil profile. *Khon Kaen Agriculture Journal* **3**:265–276.

Gowen, S. (1988). Bananas. *Biologist* **35**:187–191.

Greathouse, D.C., Laetsch, W.M. and Phinney, B.O. (1971). The shoot growth rhythm of a tropical tree, *Theobroma cacao. Australian Journal of Botany* **58**:281–286.

Green, G., Sunding, D., Zilberman, D., and Parker, D. (1996). Explaining irrigation technology choices: a microparameter approach. *American Journal of Agricultural Economics* **78**:1064–1072.

Greenwood, M. and Posnette, A.F. (1950). The growth flushes of cacao. *Journal of Horticultural Science* **25**:164–174.

Gregory, P.J. (1990). Soil physics and irrigation: tapping the potential for drip. *Agricultural Water Management* **17**:159–169.

Guardiola-Claramonte, M., Troch, P.A., Ziegler, A.D. *et al.* (2008). Local hydrologic effects of introducing non-native vegetation in a tropical environment. *Ecohydrology* **1**(1):13–22.

Guardiola-Claramonte, M., Troch, P.A., Ziegler, A.D. *et al.* (2010). Hydrologic effects of the expansion of rubber (*Hevea brasiliensis*) in a tropical environment. *Ecohydrology* **3**(3):306–314.

Gunasekara, H.K.L.K., De Costa, W.A.J.M. and Nugawela, F.A. (2007a). Genotypic variation in canopy photosynthesis, leaf gas exchange characteristics and their response to tapping in rubber (*Hevea brasiliensis*). *Experimental Agriculture* **43**:223–239.

Gunasekara, H.K.L.K., Nugawela, F.A., De Costa, W.A.J.M. and Attanayake D.P.S.T.G. (2007b). Possibility of early commencement of tapping in rubber (*Hevea brasiliensis*) using different genotypes and tapping systems. *Experimental Agriculture* **43**:201–221.

Gururaja Rao, G., Sanjeeva Rao, P., Devakumar, A.S., Vijayakumar, K.R. and Sethuraj, M.R. (1990). Influence of soil, plant and meteorological factors on water relations and yield of *Hevea brasiliensis. International Journal of Biometeorology* **34**:175–180.

Gutierrez, M.V. and Meinzer, F.C. (1994a). Energy balance and latent heat flux partitioning in coffee hedgerows at different stages of canopy development. *Agricultural and Forest Meteorology* **68**:173–186.

Gutierrez, M.V. and Meinzer, F.C. (1994b). Estimating water use and irrigation requirements of coffee in Hawaii. *Journal of the American Society of Horticultural Science* **119**:652–657.

Gutierrez, M.V., Meinzer, F.C. and Grandtz, D.A. (1994). Regulation of transpiration in coffee hedgerows: co-variation of environmental variables and apparent responses of stomata to wind and humidity. *Plant, Cell and Environment* **17**:1305–1313.

Guzman, O. and Gomaz, L. (1987). Permanence of free water on coffee leaves. *Experimental Agriculture* **23**:213–220.

Hadley, P. and Yapp, J.H.H. (1993). Measurement of physiological parameters with respect to yield. In *Proceedings International Workshop on Conservation and Utilization of Cocoa Genetic Resources in the 21st Century*, pp. 121–138. Trinidad: Cocoa Research Unit, University of the West Indies.

Hadley, P., End, M., Taylor, S.J. and Pettipher, G.L. (1994). Environmental regulation of vegetative and reproductive growth in cocoa grown in controlled glasshouse conditions. In *Proceedings of the International Cocoa Conference: Challenges in the 90s*, held Kuala Lumpur, Malaysia, September 1991, pp. 319–331 (ed. E.B. Tay). Kuala Lumpur: Malaysian Cocoa Board.

Haines, M.G., Inman-Bamber, N.G., Attard, S.J. and Linedale, A.I. (2010). Enhancing irrigation management planning with EnviroScan and WaterSense. www.irrigation.org.

au/assets/pages/762A58E3–1708–51EB-A69E09D5747B3C06/82%20-%20Haines%20Paper. pdf (accessed July 2010).

Hanks, R.J., Keller, J., Ramussen, V.P. and Wilson, G.D. (1976). Line source sprinkler for continuously variable irrigation-crop production studies. *Soil Science Society America Journal* **44**:426–429.

Hardon, J.J. (1995). Oil palm *Elaeis guineensis* (Palmae). In *Evolution of Crop Plants*, 2nd edn., pp. 395–399 (ed. J. Smart and N.W. Simmonds). Harlow, UK: Longman.

Hardwick, K., Machado, R.C.R., Smith, J. and Veltkamp, C.J. (1988a). Apical bud activity in cocoa. In *Proceedings of the 9th International Cocoa Research Conference*, held Lomé, Togo, February 1984, pp. 153–158. Lagos, Nigeria: COPAL.

Hardwick, K., Robinson, A.W. and Collin, H.A. (1988b). Plant water status and the control of leaf production in cocoa. In *Proceedings of the 9th International Cocoa Research Conference*, held Lomé, Togo, February 1984, pp. 111–116. Lagos, Nigeria: COPAL.

Hardy, F. (1958). The effects of air temperature on growth and production in cacao. *Cacao* **3**(17):1–15.

Harries, H.C. (1978). The evolution, dissemination and classification of *Cocos nucifera* L. *Botanical Review* **44**:265–320.

Harries, H.C. (1995). Coconut *Cocos nucifera* L. In *Evolution of Crop Plants*, 2nd edn., pp. 389–394 (ed. J. Smart and N.W. Simmonds). Harlow, UK: Longman.

Harris, D. (1997). The partitioning of rainfall by a banana canopy in St. Lucia, Windward Islands. *Tropical Agriculture (Trinidad)* **74**:198–202.

Hartemink, A.E. (2005). Plantation agriculture in the tropics. *Outlook on Agriculture* **34**(1):11–21.

Hartemink, A.E. and Wienk, J.F. (1995). Sisal production and soil fertility decline in Tanzania. *Outlook on Agriculture* **24**(2):91–96.

Hartley, C.W.S. (1988). *The Oil Palm*, 3rd edn. London: Longman.

Hegde, D.M. and Srinivas, K. (1989a). Irrigation and nitrogen fertility influences on plant water relations, biomass, and nutrient accumulation and distribution in banana cv. Robusta. *Journal of Horticultural Science* **64**:91–98.

Hegde, D.M. and Srinivas, K. (1989b). Effect of soil matric potential and nitrogen on growth, yield, nutrient uptake and water use of banana. *Agricultural Water Management* **16**:109–117.

Hegde, D.M. and Srinivas, K. (1991). Growth, yield, nutrient uptake and water use of banana crops under drip and basin irrigation with N and K fertilization. *Tropical Agriculture (Trinidad)* **68**:331–334.

Henson, I.E. (1991a). Age-related changes in stomatal and photosynthetic characteristics of leaves of oil palm (*Elaeis guineensis*). *Elaeis* **3**:336–348.

Henson, I.E. (1991b). Limitations to gas exchange, growth and yield of young oil palm by soil water supply and atmospheric humidity. *Transactions of the Malaysian Society of Plant Physiology* **2**:39–45.

Henson, I.E. (1991c). Use of leaf temperature measurements for detection of stress conditions in oil palm. *Transactions of the Malaysian Society of Plant Physiology* **2**:51–57.

Henson, I.E. (1995). Carbon assimilation, water-use and energy balance of an oil palm plantation assessed using micrometeorlogical techniques. In *Proceedings 1993 PORIM International Palm Oil Conference – Agriculture*, pp. 137–158 (ed. Y. Basiron). Kuala Lumpur: Palm Oil Research Institute of Malaysia.

Henson, I.E. (1998). Notes on oil palm productivity. III. The use of sap flux probes to monitor palm responses to environmental conditions. *Journal of Oil Palm Research* **10**:39–44.

Henson, I.E. (1999). Notes on oil palm productivity. IV. Carbon dioxide gradients and evapotranspiration, above and below the canopy. *Journal of Oil Palm Research* **11**:33–40.

Henson, I.E. (2006). Modelling the impact of some oil palm crop management options. *MPOB Technology* **29**:52–59.

Henson, I.E. (2009a). Oil palm: ecophysiology of growth and production. In *Ecophysiology of Tropical Tree Crops*, pp. 253–286 (ed. F. DaMatta). New York: Nova Science Publishers Inc.

Henson, I.E. (2009b). Comparative ecophysiology of oil palm and tropical rain forest. In *Sustainable Production of Palm Oil – a Malaysian Experience*, pp. 1–51 (ed. G. Singh, K.H. Lim and K.W. Chan). Kuala Lumpur: Malaysian Palm Oil Association.

Henson, I.E. and Chang, K.C. (1990). Evidence for water as a factor limiting performance of field palms in West Malaysia. In *Proceedings of 1989 PORIM International Palm Oil Development Conference – Agriculture*, pp. 487–498. Kuala Lumpur: Palm Oil Research Institute of Malaysia.

Henson, I.E. and Chang, K.C. (2000). Oil palm productivity and its component processes. In *Advances in Oil Palm Research*, Vol. 1, pp. 97–145 (ed. Y. Basion, B.S. Jalani and K.W. Chan). Kuala Lumpur: Malaysian Palm Oil Board.

Henson, I.E. and Harun, M.H. (2005). The influence of climatic conditions on gas and energy exchanges above a young oil palm stand in north Kedah, Malaysia. *Journal of Oil Palm Research* **17**:73–91.

Henson, I.E. and Harun, M.H. (2007). Short-term responses of oil palm to an interrupted dry season in north Kedah, Malaysia. *Journal of Oil Palm Research* **19**:364–372.

Henson, I.E., Jamil, Z.M. and Dolmat, M.T. (1992). Regulation of gas exchange and abscisic acid concentrations in young oil palm (*Elaeis guineensis*). *Transactions of the Malaysian Society of Plant Physiology* **3**:29–34.

Henson, I.E., Noor, M.R.M., Harun, M.H., Yahya, Z. and Mustakim, S.N.A. (2005). Stress development and its detection in young oil palms in North Kedah Malaysia. *Journal of Oil Palm Research* **17**:11–26.

Henson, I.E., Yahya, Z., Noor, M.R.M., Harun, M.H. and Mohammed, A.T. (2007). Predicting soil water status, evapotranspiration, growth and yield of young oil palm in a seasonally dry region of Malaysia. *Journal of Oil Palm Research* **19**:398–415.

Henson, I.E., Harun, M.H. and Chang, K.C. (2008). Some observations on the effects of high water tables and flooding on oil palm, and a preliminary model of oil palm water balance and use in the presence of a high water table. *Oil Palm Bulletin* **56**:14–22.

Herd, E.M. and Squire, G.R. (1976). Observations on the winter dormancy in tea (*Camellia sinensis* L.) in Malawi. *Journal of Horticultural Science* **51**:267–279.

Hernandez, A.P., Cock, J.H. and El-Sharkawy, M.A. (1989). The responses of leaf gas exchange and stomatal conductance to air humidity in shade-grown coffee, tea and cacao plants as compared with sunflower. *Revista Brasileira de Fisiologia Vegetal* **1**:155–161.

Heslop-Harrison, J.S. and Schwarzacher, T. (2007). Domestication, genomics and the future for banana. *Annals of Botany* **100**:1073–1084.

Hess, T.M., Stephens, W., Weatherhead, E.K., Knox, J.W. and Kay, M.G. (1998). Management of irrigation for tea and coffee. Unpublished report, Cranfield University, Silsoe, UK.

Hobhouse, H. (1985). *Seeds of Change: Five Plants that Transformed Mankind*. London: Sidgwick and Jackson Ltd.

Hodnett, M.G., Bell, J.P., Al Koon, P.D. Soopramanten, G.C. and Batchelor, C.M. (1990). The control of drip irrigation of sugarcane using 'index' tensiometers, some comparisons with control by the water budget method. *Agricultural Water Management* **17**:180–207.

Hoffman, H.P. and Turner, D.W. (1993). Soil water deficits reduce the elongation rate of emerging banana leaves but the night/day elongation ratio remains the same. *Scientia Horticulturae* **54**:1–12.

Holden, J.R. (1998). *Irrigation of Sugarcane*. Brisbane, Australia: Bureau of Sugar Experiment Stations.

Holder, G.D. and Gumbs, F.A. (1982). Effects of water supply during floral initiation and differentiation on female flower production by Robusta bananas. *Experimental Agriculture* **18**:183–193.

Holder, G.D. and Gumbs, F.A. (1983a). Effects of irrigation on the growth and yield of banana. *Tropical Agriculture (Trinidad)* **60**:25–30.

Holder, G.D. and Gumbs, F.A. (1983b). Agronomic assessment of the relative suitability of the banana cultivars 'Robusta' and 'Giant Cavendish' (Williams hybrid) to irrigation. *Tropical Agriculture (Trinidad)* **60**:17–24.

Holder, G.D. and Gumbs, F.A. (1983c). Effects of nitrogen and irrigation on the growth and yield of bananas. *Tropical Agriculture (Trinidad)* **60**:179–183.

Hong, T.K. and Corley, R.H.V. (1976). Leaf temperature and photosynthesis of a tropical C_3 plant, *Elaeis guineensis*. *MARDI Research Bulletin* **4**(1):16–20.

Huan, L.K., Yee, H.C. and Wood, B.J. (1986). Irrigation of cocoa on coastal soils in Peninsular Malaysia. In *Cocoa and Coconuts: Progress and Outlook*, pp. 117–132. Kuala Lumpur: Incorporated Society of Planters.

Humbert, R.P. (1968). *The Growing of Sugar Cane*. Amsterdam, the Netherlands: Elsevier.

Hutcheon, W.V. (1977a). Growth and photosynthesis of cocoa in relation to environmental and internal factors. In *Proceedings of the 5th International Cocoa Research Conference*, held Ibadan, Nigeria, September 1975, pp. 222–232. Lagos, Nigeria: COPAL.

Hutcheon, W.V. (1977b). Water relations and other factors regulating the seasonal periodicity and productivity of cocoa in Ghana. In *Proceedings of the 5th International Cocoa Research Conference*, held Ibadan, Nigeria, September 1975, pp. 233–244. Lagos, Nigeria: COPAL.

Hutcheon, W.V. (1981a). Physiological studies of cocoa (*Theobroma cacao*) in Ghana. PhD thesis, University of Aberdeen, UK.

Hutcheon, W.V. (1981b). *The Cocoa Swollen Shoot Research Project at the Cocoa Research Institute, Tafo Ghana. 1969–78*. Technical Report Volume III, Section 5, Physiology Studies, pp. 11–35. A joint research project funded by the Overseas Development Administration, London, UK and the Ministry of Cocoa Affairs, Accra, Ghana.

Hutcheon, W.V., Smith, R.W. and Asomaning, E.J.A. (1973). Effect of irrigation on the yield and physiological behaviour of mature Amelonado cocoa in Ghana. *Tropical Agriculture (Trinidad)* **50**(4):261–272.

Huxley, P. (1999). *Tropical Agroforestry*. Oxford, UK: Blackwell Science.

Huxley, P.A. and Turk, A. (1975). Preliminary investigations with Arabica coffee in root observation laboratory in Kenya. *East African Agricultural and Forestry Journal* **40**:300–312.

Huxley, P.A., Patel, R.Z., Kabaara, A.M. and Mitchell, H.W. (1974). Tracer studies with 32P on the distribution of functional roots of Arabica coffee in Kenya. *Annals of Applied Biology* **77**:159–180.

ICCO (2010). *Quarterly Bulletin of Cocoa Statistics XXXVI No 2* (Cocoa Year 2009/10). London: International Cocoa Organisation.

ICSM (2008). *Canegro Sugarcane Plant Module*. International Consortium for Sugarcane Modelling. http://sasri.sasa.org.za/misc/DSSAT%20Canegro%20SCIENTIFIC%20 documentation_20081215.pdf (accessed October 2011).

Inman-Bamber, N.G. (1994). Temperature and seasonal effects on canopy development and light interception of sugarcane. *Field Crops Research* **36**:41–51.

Inman-Bamber, N.G. (1995). Automatic plant extension measurement in sugarcane in relation to temperature and soil moisture. *Field Crops Research* **42**:135–142.

Inman-Bamber, N.G. (2004). Sugarcane water stress criteria for irrigation and drying off. *Field Crops Research* **89**:107–122.

Inman-Bamber, N.G. and De Jager, J.M. (1986). The reaction of two varieties of sugarcane to water stress. *Field Crops Research* **14**:15–28.

Inman-Bamber, N.G. and McGlinchey, M.G. (2003). Crop coefficients and water-use estimates for sugarcane based on long-term Bowen ratio energy balance measurements. *Field Crops Research* **83**:125–138.

Inman-Bamber, N.G. and Smith, D.M. (2005). Water relations in sugarcane and response to water deficits. *Field Crops Research* **92**:185–202.

Inman-Bamber, N.G. and Spillman, N.F. (2002). Plant extension, soil water extraction and water stress in sugarcane. *Proceedings of the Australian Society of Sugar Cane Technologists* **24**:242–256.

Inman-Bamber, N.G., Schuurs, M. and Muchow, R.C. (1999). Advances in the science and economics of supplementary irrigation of sugarcane. *Proceedings of the South African Sugar Technologists' Association* **73**:9–15.

Inman-Bamber, N.G., Muchow, R.C. and Robertson, M.J. (2002). Dry matter partitioning of sugarcane in Australia and South Africa. *Field Crops Research* **76**:71–84.

Inman-Bamber, N.G., Attard, S.J., Haines, M.G. and Linedale, A.I. (2008). Deficit irrigation in sugarcane using the WaterSense scheduling tool. In *Share the Water, Share the Benefits*. Proceedings of the Irrigation Australia Congress, held Melbourne, May 2008. CDROM. www.irrigation.org.au/assets/pages/762A58E3–1708–51EB-A69E09D5747B3C06/ 79%20Inman%20Paper2.pdf (accessed July 2010).

Inman-Bamber, N.G., Bonnett, G.D., Spillman, M.F., Hewitt, M.L. and Xu, J. (2009). Source-sink differences in genotypes and water regimes influencing sucrose accumulation in sugarcane stalks. *Crop and Pasture Science* **60**:316–327.

IRSG (2010). Statistics. International Rubber Study Group. www.rubberstudy.com/statistics. aspx (accessed October 2011).

Isarangkool Na Ayutthaya, S. (2010). Change of whole-tree transpiration of mature *Hevea brasiliensis* under soil and atmospheric droughts: analysis in intermittent and seasonal droughts under the framework of the hydraulic limitation hypothesis. PhD twin thesis between Khon Kaen University (Thailand) and Blaise Pascal University, Clermont-Ferrand, France, 142 pp.

Isarangkool Na Ayutthaya, S., Do, F.C., Pannengpetch, K. *et al.* (2009). Transient thermal dissipation method of xylem sap flow measurement: multi-species calibration and field evaluation. *Tree Physiology* **30**:139–148.

Isarangkool Na Ayutthaya, S., Do, F.C., Pannengpetch, K. *et al.* (2011). Water loss regulation in mature *Hevea brasiliensis*: effects of intermittent drought in rainy season and hydraulic limitations. *Tree Physiology* **31**:751–762.

Israeli, Y. and Nameri, N. (1986). Water use and banana plant yields. *Water and Irrigation Review* **6**:14–17.

Israeli, Y. and Nameri, N. (1987). Seasonal changes in banana water use. *Water and Irrigation Review* **7**:10–14.

Jadin, P. and Jacquemart, J-P. (1978). Effet de l'irrigation sur la précocité des jeunes cacaoyers. *Café Cacao Thé* **22**:31–35.

Jadin, P. and Snoeck, J. (1981). Evolution du stock d'eau sous une cacaoyere – relation avec le climat. In *Proceedings of the 8th International Cocoa Research Conference*, held Cartagena, Colombia, October 1981, pp. 127–135. Lagos, Nigeria: COPAL.

Jadin, P., Chauchard, A. and Bois, J.F. (1976). Alimentation hydrique des jeunes cacaoyers influence de l'irrigation. *Café Cacao Thé* **20**:173–199.

James, G. (ed.) (2004). *Sugarcane*, 2nd edn. World Agriculture Series. Oxford, UK: Blackwell Publishing.

Jayakumar, M., Saseendran, S.A. and Hemapraba, M. (1988). Crop coefficient for coconut (*Cocos nucifera* L.): a lysimetric study. *Agricultural and Forest Meteorology* **43**:235–240.

Jayasekara, C., Ranasinghe, C.S. and Mathes, D.T. (1993). Screening for high yield and drought tolerance in coconut. In *Advances in Coconut Research and Development*, pp. 209–218 (ed. M.K. Nair, H.H. Khan, P. Gopalasundaram and E.V.V. Bhaskara Rao). New Delhi: IBH Publishing.

Jessy, M.D., Mathew, M., Jacob, S. and Punnoose, K.I. (1994). Comparative evaluation of basin and drip irrigation systems of irrigation in rubber. *Indian Journal Natural Rubber Research* **7**(1):51–56.

Joly, R.J. (1988). Physiological adaptations for maintaining photosynthesis under water stress in cacao. In *Proceedings of the 10th International Cocoa Conference*, held Santo Domingo, Dominican Republic, May 1987, pp. 199–203. Lagos, Nigeria: COPAL.

Joly, R.J. and Hahn, D.T. (1989). Net CO_2 assimilation of cacao seedlings during periods of plant water deficit. *Photosynthesis Research* **21**:151–159.

Jones, C.A., Santo, L.T., Kingston, G. and Gascho, G.J. (1990). Sugarcane. In *Irrigation of Agricultural Crops* (ed. B.A. Stewart and D.R. Nielsen). Agronomy No. 30. Madison, WI: American Society of Agronomy.

Jones, H.G., Lakson, A.N. and Syvertsen, J.P. (1985). Physiological control of water status in temperate and subtropical fruit trees. *Horticultural Reviews* **7**:301–344.

Josis, P., Ndayishimiye, V. and Renard, C. (1983). Etude des relations hydriques chez *Coffea arabica* L. II. Evaluation de la resistance à la sécheresse de divers cultivars à Gisha (Burundi). *Café Cacao Thé* **27**:275–282.

Jourdan, C. and Rey, H. (1997a). Architecture and development of the oil palm (*Elaeis guineensis* Jacq.) root system. *Plant and Soil* **189**:33–48.

Jourdan, C. and Rey, H. (1997b). Modelling and simulation of the architecture and development of the oil palm (*Elaeis guineensis* Jacq.) root system. *Plant and Soil* **190**:235–246.

Julien, M.H.R., Irvine, J.E. and Benda, G.T.A. (1989). Sugarcane anatomy, morphology and physiology. In *Diseases of Sugarcane* (ed. C. Ricaud, B.T. Egan, A.G. Gillaspie Jr. and C.G. Hughes), pp. 1–20. Amsterdam, Netherlands: Elsevier.

Juma, M. and Fordham, R. (1998). The effect of environmental stress on coconut (*Cocos nucifera* L.) growth in Zanzibar. In *Proceedings of the International Cashew and Coconut Conference: Trees for Life – the Key to Development*, pp. 342–347 (ed. C.P. Topper *et al.*). Reading, UK: BioHybrids International Ltd.

Kallarackal, J., Jeyakumar, P. and George, S.J. (2004). Water use of irrigated oil palm at three different arid locations in peninsular India. *Journal of Oil Palm Research* **16**:45–53.

Kallarackal, J., Milburn, J.A. and Baker, D.A. (1990). Water relations of the banana. III. Effects of controlled water stress on water potential, transpiration, photosynthesis and leaf growth. *Australian Journal of Plant Physiology* **17**:79–90.

Kalorizou, H.A., Gowen, S.R. and Wheeler, T.R. (2007). Genotypic differences in the growth of bananas (*Musa* spp.) infected with migratory endoparasitic nematodes. 1. Roots. *Experimental Agriculture* **43**:331–342.

Kanechi, M., Uchida, N., Yasuda, T. and Yamaguchi, T. (1995). Water stress effects on leaf transpiration and photosynthesis of *Coffea arabica* L. under different irradiance conditions. In *Proceedings 16th International Scientific Colloquium on Coffee*, pp. 520–527. Paris: Association Scientifique Internationale du Café.

Karamura, E.B. and Karamura, D.A. (1995). Banana morphology – part II: the aerial shoot. In *Bananas and Plantains*, pp. 190–205 (ed. S. Gowen). London: Chapman and Hall.

Karunaratne, P.M.A.S., Wijeratne, M.A. and Sangakkara, U.R. (1999). Osmotic adjustments and associated water relations of clonal tea (*Camellia sinensis* L.). *Sabaragamuwa University Journal* **2**:77–85.

Kashaija, I.N., McIntyre, B.D., Ssali, H. and Kizito, F. (2004). Spatial distribution of roots, nematodes and root necrosis in highland banana in Uganda. *Nematology* **6**:7–12.

Kasturi Bai, K.V., Voleti, S.R. and Rajagopal, V. (1988). Water relations of coconut palms as influenced by environmental variables. *Agricultural and Forest Meteorology* **43**:193–199.

Kasturi Bai, K.V., Naresh Kumar, S., Rajagopal, V. and Vijayakumar, K. (2008). Principal component analysis of chlorophyll fluorescent transients for tolerance to drought stress in coconut seedlings. *Indian Journal of Horticulture* **65**:471–476.

Kay, M.G. (1990). Recent developments for improving water management in surface and overhead irrigation. *Agricultural Water Management* **17**:7–22.

Kayange, C.W., Scarborough, I.P. and Nyirenda, H.E. (1981). Rootstock influence on yield and quality of tea (*Camellia sinensis* L.). *Journal of Horticultural Science* **56**:117–120.

Ke, Lih-shang. (1979). Studies on the physiological characteristics of banana in Taiwan. 1. Studies on the stomatal behaviour of banana leaves. *Journal of the Agricultural Association of China* **108**:1–10.

Keller, J., Sivanappan R.K. and Varadan K.M. (1992). Design logic for deficit drip irrigation of coconut trees. *Irrigation and Drainage Systems* **6**:1–7.

Kerfoot, O. (1961). Tea root systems. *Pamphlet Tea Research Institute of East Africa, Kericho, Kenya* **19**:61–72.

Kerfoot, O. (1962). Tea root systems. *World Crops* **14**:140–143.

Kiara, J.M. and Stolzy, L.H. (1986). The effects of tree density and irrigation on coffee growth and production in Kenya. *Applied Agricultural Research* **1**:26–31.

Kigalu, J.M. (1997). Effects of planting density on the productivity and water use of young tea (*Camellia sinensis* L.) clones in southern Tanzania. PhD thesis, Cranfield University, UK.

Kigalu, J.M. (2002). Experiment N12: effects of plant density and drought on clonal tea productivity. *Annual Report for 2000/2001*, pp. 16–24. Dar es Salaam: Tea Research Institute of Tanzania.

Kigalu, J.M. (2007a). Effects of planting density on the productivity and water use of tea (*Camellia sinensis* L.) clones. I. Measurement of water use in young tea using sap flow meters with a stem heat balance method. *Agricultural Water Management* **90**:224–232.

Kigalu, J.M. (2007b). Effects of planting density and drought on the productivity of tea clones (*Camellia sinensis* L.): yield responses. *Physics and Chemistry of the Earth* **32**:1098–1106.

Kigalu, J.M. and Nixon, D.J. (1997). Responses of young clonal tea to planting density. 1. Annual yields. In *Quarterly Report 28*, pp. 3–12. Dar es Salaam: Ngwazi Tea Research Unit.

Kigalu, J.M., Kimambo, E.I., Msite, I. and Gembe, M. (2008). Drip irrigation of tea (*Camellia sinensis* L.). I. Yield and crop water productivity responses to irrigation. *Agricultural Water Management* **95**:1253–1260.

Kimambo, E. (2005a). Experiment N9: responses of clonal tea to nitrogen and irrigation. *Annual Reports for 1996/1997, 1997/1998, 1999/2000, 2000/2001, 2001/2002, 2003/2004.* Dar es Salaam: Tea Research Institute of Tanzania.

Kimambo, E. (2005b). Experiment N14: responses clones to drought and irrigation. *Annual Report for 2003/2004*, pp. 13–16. Dar es Salaam: Tea Research Institute of Tanzania.

Kimaro, D.N., Msanya, B.M. and Takamura, Y. (1994). Review of sisal production and research in Tanzania. *African Study Monographs* **15**(4):227–242.

Kingdom-Ward, F. (1950). Does wild tea exist? *Nature* **165**:297–299.

Knox, J.W., Rodríguez Díaz, J.A., Nixon, D. and Mkhwanazi, M. (2010). A preliminary assessment of climate change impacts on sugarcane in Swaziland. *Agricultural Systems* **103**(2): 63–72.

Koehler, P.H., Moore, P.H., Jones, C.A., Dela Cruz, A. and Maretzki, A. (1982). Response of drip-irrigated sugarcane to drought stress. *Agronomy Journal* **74**:906–911.

Köhler, M., Schwendenmann, L. and Hölscher, D. (2010). Throughfall reduction in a cacao agroforest: tree water use and soil water budgeting. *Agricultural and Forest Meteorology* **150**:1079–1089.

Krishna, T.H., Bhaskar, C.V.S., Rao, P.S. *et al.* (1991). Effect of irrigation on physiological performance of immature plants of *Hevea brasliensis* in North Konkan. *Indian Journal of Natural Rubber Research* **4**:36–45.

Krishnan, B.M. and Shanmugavelu, K.G. (1980). Effect of different soil moisture depletion levels on the root distribution of banana cv. 'Robusta'. *South Indian Horticulture* **28**:24–25.

Kumar, D. and Tieszen, L.L. (1980a). Photosynthesis in *Coffea arabica*. I. Effects of light and temperature. *Experimental Agriculture* **16**:13–19.

Kumar, D. and Tieszen, L.L. (1980b). Photosynthesis in *Coffea arabica* L. II. Effects of water stress. *Experimental Agriculture* **16**:21–27.

Kummerow, J., Kummerow, A. and Alvim, P. de T. (1981). Root biomass in a mature cacao (*Theobroma cacao* L.) plantation. *Révista Theobroma* **12**(1):77–85.

Kushwah, B.L., Nelliat, E.V., Markose, V.T. and Sunny A.F. (1973). Rooting pattern of coconut (*Cocos nucifera* L.). *Indian Journal of Agronomy* **18**:71–74.

Laclau, P.B. and Laclau, J.-P. (2009). Growth of the whole root system for a plant crop of sugarcane under rainfed and irrigated conditions in Brazil. *Field Crops Research* **114**:351–360.

Lahav, A. and Kalmar, D. (1981). Shortening the irrigation interval as a means of saving water in a banana plantation. *Australian Journal of Agricultural Research* **32**:465–477.

Lahav, A. and Kalmar, D. (1988). Response of banana to drip irrigation, water amounts and fertilisation regimes. *Communications in Soil Science and Plant Analysis* **19**:25–46.

Lamade, E. and Setiyo, E (1996). Variation in maximum photosynthesis of oil palm in Indonesia: comparison of three morphologically contrasting clones. *Plantations, Recherche, Développement* **3**:429–435.

Lamade, E., Setiyo, E. and Purba, A. (1998). Gas exchange and carbon allocation of oil palm seedlings submitted to water logging in interaction with N fertiliser application. In *Proceedings of the 1998 International Oil Palm Conference. Commodity of the Past, Today and the Future*, pp. 573–584 (ed. A. Jatmika, D. Bangun, D. Asmono, *et al.*). Medan, Indonesia: Indonesian Oil Palm Research Institute.

Lassoudiere, A. (1978). Quelques aspects de la croissance et du développement du bananier 'Poyo' en Côte d'Ivoire, 4 en 5 partie. *Fruits* **33**:314–338.

Laycock, D.H. (1964). An empirical correlation between weather and yearly tea yields in Malawi. *Tropical Agriculture (Trinidad)* **41**:277–291.

Laycock, D.H. and Wood, R.A. (1963). Some observations of soil moisture use under tea in Malawi. 1. The effect of pruning mature tea. *Tropical Agriculture (Trinidad)* **40**:35–42.

Lebrun, P., N'cho, Y.P., Seguin, M., Grivet, L. and Baudouin, L. (1998). Genetic diversity in coconut (*Cocos nucifera* L.) revealed by restriction fragment length polymorphism (RFLP) markers. *Euphytica* **101**:103–108.

Lecompte, F. and Pagès, L. (2007). Apical diameter and branching density affect lateral root elongation rates in banana. *Environmental and Experimental Botany* **59**:243–251.

Lecompte, F., Vaucelle, A., Pagès, L. and Ozier-Lafontaine, H. (2002). Number, position, diameter and initial direction of growth of primary roots in *Musa*. *Annals of Botany* **90**:43–51.

Lee, G.R. (1975). Irrigated Upper Amazon cacao in the Lower Shire Valley of Malawi. II A water rates trial. *Tropical Agriculture (Trinidad)* **52**:179–182.

Legros, S., Mialet-Serra, I., Caliman, J.P. *et al.* (2009a). Phenology and growth adjustments of oil palm (*Elaeis guineensis*) to photoperiod and climate variability. *Annals of Botany* **104**(6):1171–1182.

Legros, S., Mialet-Serra, I., Clement-Vidal, A. *et al.* (2009b). Role of transitory carbon reserves during adjustment to climate variability and source-sink imbalances in oil palm (*Elaeis guineensis*). *Tree Physiology* **29**(10):1199–1211.

Lejju, B.J., Robertshaw, P. and Taylor, D. (2006). Africa's earliest bananas? *Journal of Archaeological Science* **33**:102–113.

Lim, K.H., Goh, K.J., Kee, K.K. and Henson, I.E. (2008). Climatic requirements of the oil palm. Paper presented at Seminar on Agronomic Principles and Practices of Oil Palm Cultivation. The Agricultural Group Trust, Sibu, Sarawak, Malaysia, October 2008.

Lima, E.P. and da Silva, E.L. (2008). Temperatura base, coeficientes de cultura e graus-dia para cafeeiro arábica em fase de implantação. *Revista Brasileira Engenharia Agricola e Ambiental* **12**(3):266–273.

Lisson, S.N., Inman-Bamber, N.G., Robertson, M.J. and Keating B.A. (2005). The historical and future contribution of crop physiology and modelling research to sugarcane production systems. *Field Crops Research* **92**:321–335.

Liu, D.L., Kingston, G. and Bull, T.A. (1998). A new technique for determining the thermal parameters of phenological development in sugarcane, including suboptimum and supraoptimum temperature regimes. *Agricultural and Forest Meteorology* **90**:119–139.

Liu, H.J., Cohen, S., Tanny, J., Lemcoff, J.H. and Huang, G. (2008a). Transpiration estimation of banana (*Musa* sp.) plants with the thermal dissipation method. *Plant and Soil* **308**:227–238.

Liu, H.J., Cohen, S., Tanny, J., Lemcoff, J.H. and Huang, G. (2008b). Estimation of banana (*Musa* sp.) plant transpiration using a standard 20 cm pan in a greenhouse. *Irrigation and Drainage Systems* **22**:311–323.

Lock, G.W. (1962). *Sisal*, 1st edn. London: Longman.

Lock, G.W. (1969). *Sisal*, 2nd edn. London: Longman.

Lockwood, G. and Pang, J.T.Y. (1996).Yields of cocoa clones in response to planting density in Malaysia. *Experimental Agriculture* **32**:41–47.

Logan, W.J.C. and Biscoe, J. (1987). Irrigation. In *Coffee Handbook*, pp. 70–82. Harare, Zimbabwe: Coffee Growers' Association.

Lu, P., Woo, K.C. and Liu, Z.T. (2002). Estimation of whole plant transpiration of bananas using sap flow measurements. *Journal of Experimental Botany* **53**:1771–1779.

Machin, D. (2009). Sisal: small farmers and plantation workers. In *Proceedings of the Symposium on Natural Fibres*, held 20 October 2008, pp. 39–42. Rome: Common Fund for Commodities, FAO. ftp://ftp.fao.org/docrep/fao/011/i0709e/i0709e.pdf (accessed October 2011).

Madramootoo, C.A. and Jutras, P.J. (1984). Supplemental irrigation of bananas in St. Lucia. *Agricultural Water Management* **9**:149–156.

Madurapperuma, W.S., de Costa, W.A.J.M., Sangakkara, U.R. and Jayescara, C. (2009a). Estimation of water use of mature coconut (*Cocos nucifera* L.) cultivars (CRIC 60 and CRIC 65) grown in the low country intermediate zone using the compensation heat pulse method (CHPM). *Journal of the National Science Foundation of Sri Lanka* **37**:175–186.

Madurapperuma, W.S., Bleby, T.M. and Burgess, S.S.O. (2009b). Evaluation of sap flow methods to determine water use by cultivated palms. *Environmental and Experimental Botany* **66**:372–390.

Maestri, M. and Barros, R.S. (1977). Coffee. In *Ecophysiology of Tropical Crops*, pp. 249–278 (ed. T.T. Kozlowski). New York: Academic Press.

Maestri, M., DaMatta, F.M., Regazzi, A.J. and Barros, R.S. (1995). Accumulation of proline and quaternary ammonium compounds in mature leaves of water stressed coffee plants (*Coffea arabica* and *C. canephora*). *Journal of Horticultural Science* **70**:229–233.

Magalhaes, A.C. and Angelocci, L.R. (1976). Sudden alterations in water balance associated with flower bud opening in coffee plants. *Journal of Horticultural Science* **51**:419–423.

Magambo, M.J.S. and Cannell, M.G.R. (1981). Dry matter production and partition in relation to yield of tea. *Experimental Agriculture* **17**:33–38.

Magnaye, A.B. (1969). Studies on the root system of healthy and cadang-cadang affected coconut trees. *Philippine Journal of Plant Industry* **34**(3/4):143–154.

Mahindapala, R. and Pinto, J.L.J.G. (1991). *Coconut Cultivation*. Lunuwila, Sri Lanka: Coconut Research Institute of Sri Lanka.

Mahouachi, J. (2007). Growth and mineral nutrient content of developing fruit on banana plants (*Musa acuminata* AAA, 'Grand Nain') subject to later stress and recovery. *Journal of Horticultural Science and Biotechnology* **82**:839–844.

Maidment, W.T.O. (1928). Correlation between rainfall and cacao yields in the Gold Coast with special reference to effect of April rains on the following cacao crop. *Department of Agriculture Gold Coast Year-Book, 1927, Bulletin* **13**:83–91.

Maillard, G., Daniel, C. and Ochs, R. (1974). Analyse des effets de la sécheresse sur le palmier à huile. *Oléagineux* **29**:397–404.

Mair, V. H. and Hoh, E. (2009). *The True History of Tea*. London: Thames and Hudson.

Manivong, V. (2007). The economic potential for smallholder rubber production in Northern Laos. MPhil thesis, University of Queensland, Brisbane, Australia.

Manthriratna, M.A.P.P. and Sambasivam, S. (1974). Stomatal density in varieties and forms of the coconut. *Ceylon Coconut Journal* **25**:105–108.

Marar, M.M.K. and Kunhiraman, C.A. (1957). Husk burial for the improvement of coconut gardens. *Indian Coconut Journal* **10**(3):3–11.

Marin, F.R., Angelocci, L.R., Righi, E.Z. and Sentelhas, P.C. (2005). Evapotranspiration and irrigation requirements of a coffee plantation in Brazil. *Experimental Agriculture* **41**:187–197.

Marinho, F.J.L. Gheyi, H.R., Fernandes, P.D., de Holanda, J.S. and Neto, M.F. (2006). Cultivo de coco 'Anão Verde' irrigado com águas salinas. *Pesquisa Agropecuária Brasileira, Brasília* **41**(8):1277–1284.

Martin, E.C., Stephens, W., Wiedenfeld, R. *et al.* (2007). Sugar, oil and fibre. In *Irrigation of Agricultural Crops*, pp. 279–236. (eds. R.J. Lascano and R.E. Sojka).Madison, WI: American Society of Agronomy.

Martins, M.B.G. and Zieri, R. (2003). Leaf anatomy of rubber tree clones. *Scientia Agricola* **60**(4):709–713.

Mathew, C. (1981). Water relations of coconut palm affected by root (wilt) disease. *Journal of Plantation Crops* **9**:51–55.

Matthews, R.B. and Stephens, W. (1997). Evaluating irrigation strategies for tea using the 'CUPPA-Tea' model. *Quarterly Report 27*, pp. 3–11. Dar es Salaam, Tanzania: Ngwazi Tea Research Unit.

Matthews, R.B. and Stephens, W. (1998a). CUPPA-Tea: a simulation model describing seasonal yield variation and potential production of tea. 1. Shoot development and extension. *Experimental Agriculture* **34**:345–367.

Matthews, R.B. and Stephens, W. (1998b). CUPPA-Tea: a simulation model describing seasonal yield variation and potential production of tea. 2. Biomass production and water use. *Experimental Agriculture* **34**:369–389.

Matthews, R.B. and Stephens, W. (1998c). The role of photoperiod in determining seasonal yield variation in tea. *Experimental Agriculture* **34**:323–340.

McCreary, C.W.R., McDonald, J.A., Muldoon, V.I. and Hardy, F. (1943). The root system of cacao: results of some preliminary investigations in Trinidad. *Tropical Agriculture (Trinidad)* **20**:207–220.

McCulloch, J.S.G. (1965). Tables for the rapid computation of the Penman estimate of evaporation. *East African Agricultural and Forestry Journal* **30**:286–295.

McKelvie, A.D. (1956). Cherelle wilt of cacao. 1. Pod development and its relation to wilt. *Journal of Experimental Botany* **7**(20):252–263.

Meguro, N.E. and Magalhaes, A.C. (1983). Water stress affecting nitrate reduction and leaf diffusive resistance in *Coffea arabica* L. cultivars. *Journal of Horticultural Science* **58**:147–152.

Meinzer, F.C. and Grantz, D.A. (1989). Stomatal control of transpiration from a developing sugarcane canopy. *Plant, Cell and Environment* **12**:635–642.

Meinzer, F.C., Grantz, D.A., Goldstein, G. and Saliendra, N.Z. (1990). Water relations and maintenance of gas exchange in coffee cultivars grown in a drying soil. *Plant Physiology* **94**:1781–1787.

Meinzer, F.C., Saliendra, N.Z. and Crisosto, C.H. (1992). Carbon isotope discrimination and gas exchange in *Coffea arabica* during adjustment to different soil moisture regimes. *Australian Journal of Plant Physiology* **19**:171–184.

Mendes, M.E.G., Villagra, M.M., de Souza, M.D., Bacchi, O.O.S. and Reichardt, K. (1992). Water relations in a rubber-tree plantation of Piracicaba, SP. *Scientia Agricola* **49**(1):103–109.

Merry, R.E. (2003). Dripping with success: the challenges of an irrigation redevelopment project. *Irrigation and Drainage* **52**:71–83.

Meyer, J.P. and Schoch, P.G. (1976). Besoin en eau du bananier aux Antilles. Mesure de l'evapotranspiration maximale. *Fruits* **31**:3–19.

Meyer, W.S. (1997). The irrigation experience from Australia. In *Intensive Sugarcane Production*, pp. 437–454 (eds. B.A. Keating and J.R. Wilson). Wallingford, UK: CAB International.

Mhlanga, B.F.N., Ndlovub, L.S. and Senzanje, A. (2006). Impacts of irrigation return flows on the quality of the receiving waters: a case of sugarcane irrigated fields at the Royal Swaziland Sugar Corporation (RSSC) in the Mbuluzi River Basin (Swaziland). *Physics and Chemistry of the Earth* **31**:804–813.

Mialet-Serra, I., Clement-Vidal, A., Roupsard, O., Jourdan, C. and Dingkuhn, M. (2008). Whole plant adjustments in coconut (*Cocos nucifera*) in response to sink-source relationships. *Tree Physiology* **28**:1199–1209.

Milburn, J.A. and Zimmerman, M.H. (1977). Preliminary studies on sapflow in *Cocos nucifera* L. 1. Water relations and xylem transport. *New Phytologist* **79**:535–541.

Milburn, J.A., Kallarackal, J. and Baker, D.A. (1990). Water relations of the banana. I. Predicting the water relations of the field grown banana using the exuding latex. *Australian Journal of Plant Physiology* **17**:57–68.

Miranda, F.R. de, Gomes, A.R.M., Oliveira, C.H.C. de, Montenegro, A.A.T. and Bezerra, F.M.L. (2007). Evapotranspiration and crop coefficients for green-dwarf coconut in the coastal area of Ceara State, Brazil. *Revista Ciencia Agronomica* **38**(2):129–135.

Miyaji, K.-I., da Silva, W.S. and Alvim, P. de T. (1997). Productivity of leaves of a tropical tree, *Theobroma cacao*, grown under shade, in relation to leaf age and light conditions within the canopy. *New Phytologist* **137**:463–472.

Mizambwa, F.C.S. (1997). N9: responses of clonal tea to fertiliser and irrigation. In *Annual Report for 1997*, pp. 3–4. Dar es Salaam, Tanzania: Ngwazi Tea Research Unit.

Mizambwa, F.C.S. (2002a). Responses of composite tea plants to drought and irrigation in the Southern Highlands of Tanzania. PhD thesis, Cranfield University, UK.

Mizambwa, F.C.S. (2002b). Experiment N13: responses of composite tea plants to drought and irrigation. In *Annual Report for 2000/01*, pp. 25–28. Dar es Salaam: Tea Research Institute of Tanzania.

Mizambwa, F.C.S. (2003). Experiment N13: responses of composite tea plants to drought and irrigation. In *Annual Report for 2001/02*, pp. 17–19. Dar es Salaam: Tea Research Institute of Tanzania.

Mizambwa, F.C.S. (2004a). Experiment N10: responses of clones to drought and irrigation. In *Annual Report for 2002/03*, pp. 9–11. Dar es Salaam: Tea Research Institute of Tanzania.

Mizambwa, F.C.S. (2004b). Experiment N14: Responses of clones to drought and irrigation. In *Annual Report for 2002/03*, pp. 11–13. Dar es Salaam: Tea Research Institute of Tanzania.

Mohotti, A.J. and Lawlor, D.W. (2002). Diurnal variation in photosynthesis and photo-inhibition in tea: effects of irradiance and nitrogen supply during growth in the field. *Journal of Experimental Botany* **53**:313–322.

Molden, D. (Ed.) (2007). *Water for Food, Water for Life: A Comprehensive Assessment of Water Management in Agriculture*. London: Earthscan, and Colombo: International Water Management Institute.

Möller, M. and Weatherhead, E.K. (2007). Evaluating drip irrigation in commercial tea production in Tanzania. *Irrigation and Drainage Systems* **21**:17–34.

Montagnon, C. and Leroy, T. (1993). Response to drought of young *Coffea canephora* coffee trees from different genetic groups in the Côte-d'Ivoire. *Café, Cacao Thé* **37**:179–190.

Monteith, J.L. (1972). Solar radiation and productivity in tropical ecosystems. *Journal of Applied Ecology* **9**:747–766.

Monteith, J.L., Gregory, P.J., Marshall, B. *et al.* (1981). Physical measurements in crop physiology. I. Growth and gas exchange. *Experimental Agriculture* **17**:113–126.

Montenegro, A.A.T., Gomes, A.R.M., Rodrigues de Miranda, F. and Crisostomo, L.A. (2008). Evapotranspiration and crop coefficient of banana in the coastal region of the state of Ceara, Brazil. *Revistas de Ciencias Agronomia* **39**:203–208.

Monteny, B.A. and Barigah, S. (1986). Effect of leaf age and water deficit on the leaf photosynthetic rate of *Hevea brasiliensis*. In *Proceedings of the International Rubber Conference*, held Kuala Lumpur, Malaysia, October 1985. Kuala Lumpur: Rubber Research Institute of Malaysia.

Morgan, D.D.V. and Carr, M.K.V. (1988). Analysis of experiments involving line source irrigation. *Experimental Agriculture* **24**:169–176.

Moser, G., Leuschner, C., Hertel, D. *et al.* (2010). Response of cocoa trees (*Theobroma cacao*) to a 13-month desiccation period in Sulawesi, Indonesia. *Agroforestry Systems* **79**:171–187.

Moss, J.R.J. (1992). Measuring light interception and the efficiency of light utilization by the coconut palm (*Cocos nucifera* L.). *Experimental Agriculture* **28**:273–285.

Motamayor, J.C., Lachenaud, P., de Silva e Mota, J.W. *et al.* (2008). Geographic and genetic population differentiation of the Amazonian chocolate tree (*Theobroma cacao* L.). *PLoS ONE* **3**(10): e3311.doi:10.1371/journal.pone.0003311.

Mouli, C.M.R., Onsando, J.M. and Corley, R.H.V. (2007). Intensity of harvesting in tea. *Experimental Agriculture* **43**:41–50.

Murray, D.B. (1961). Soil moisture and cropping cycles in cacao. In *A Report on Cacao Research 1959–1960*, pp. 18–22. St. Augustine, Trinidad: Imperial College of Tropical Agriculture.

Murray, D.B. (1977). Coconut palm. In *Ecophysiology of Tropical Crops*, pp. 383–407 (ed. P. de T. Alvim and T.T. Kozlowski). London: Academic Press.

Nagwekar, D.D., Desai, V.S., Sawant, V.S. *et al.* (2006). Effect of drip irrigation on yield of coconut (*Cocos nucifera* L.) in sandy soil of Konkan region of Maharashtra (India). *Journal of Plantation Crops* **34**(3):344–346.

Nainanayake, A., Ranasinghe, C.S. and Tennakoon, N.A. (2008). Efects of drip irrigation on canopy and soil temperature, leaf gas exchange, flowering and nut setting of mature coconut (*Cocos nucifera* L.). *Journal of the National Science Foundation, Sri Lanka* **38**:43–50.

Nair, R.R. (1989). Summer irrigation requirements of the coconut palm. *Indian Coconut Journal* **19**(12):3–7.

Naresh Kumar, S. and Kasturi Bai, K.V. (2009). Photosynthetic characters in different shapes of coconut canopy under irrigated and rainfed conditions. *Indian Journal of Plant Physiology* **14**(3):215–223.

Naresh Kumar, S., Rajagopal, V. and Karun A. (2000). Leaflet anatomical adaptations in coconut cultivars for drought tolerance. In *Recent Advances in Plantations Crops Research*, pp. 225–229 (ed. N. Muraleedharan and R. Rajkumari). New Delhi: Allied Publishers Ltd.

Naresh Kumar, S., Rajagopal, V., Siju Thomas, T. *et al.* (2007). Variations in nut yield of coconut and dry spell in different agro-climatic zones of India. *Indian Journal of Horticulture* **64**(3):309–313.

Naresh Kumar, S., Kasturi Bai, K.V., Rajagopal, V. and Aggarwal, P.K. (2008). Simulating coconut growth, development and yield with the InfoCrop-coconut model. *Tree Physiology* **28**:1049–1058.

Nataraja, K.N. and Jacob, J. (1999). Clonal differences in photosynthesis in *Hevea brasiliensis* Muell. Arg. *Photosynthetica* **36**:89–98.

Nelliat, E.V. and Padmaja, P.K. (1978). Irrigation requirement of coconut and response to levels of fertiliser under irrigated condition during the early bearing stage. In *Proceedings of First Annual Symposium on Plantation Crops (PLACROSYM 1)*, pp. 186–199 (ed. E.V. Nelliat). Kasaragod, India: Indian Society for Plantation Crops.

Nelson, P.N., Banabas, M., Scotter, D.R. and Webb, M.J. (2006). Using soil water depletion to measure spatial distribution of root activity in oil palm (*Elaeis guineensis* Jacq.) plantations. *Plant and Soil* **286**:109–121.

Ng Kee Kwong, K.F., Paul, J.P. and Deville, J. (1999). Drip fertigation – a means for reducing fertiliser nitrogen to sugarcane. *Experimental Agriculture* **35**:31–37.

Ng'etich, W.K. and Stephens, W. (2001a). Responses of tea to environment in Kenya. 1. Genotype × environment interactions for total dry matter production and yield. *Experimental Agriculture* **37**:333–342.

Ng'etich, W.K. and Stephens, W. (2001b). Responses of tea to environment in Kenya. 2. Dry matter production and partitioning. *Experimental Agriculture* **37**:343–360.

Ng'etich, W.K. and Wachira, F.N. (2003). Variations in leaf anatomy and gas exchange in tea clones with different ploidy. *Journal of Horticultural Science and Biotechnology* **78**:173–176.

Ng'etich, W.K., Stephens, W. and Othieno, C.O. (2001). Responses of tea to environment in Kenya. 3. Yield and yield distribution. *Experimental Agriculture* **37**:361–372.

Niemenak, N., Cilas, C., Rohsius, C. *et al.* (2009). Phenological growth stages of cacao plants (*Theobroma* sp.): codification and description according to the BBCH scale. *Annals of Applied Biology* **156**:13–24.

Nijs, I., Olyslaegers, G., Kockelbergh, F. *et al.* (2000). Detecting sensitivity to extreme climatic conditions in tea: an ecophysiological vs. an energy balance approach. In *Topics in Ecology: Structure and Function in Plants and Ecosystems*, pp. 253–265 (ed. R. Ceulemans, J. Bagaert, G. Deckmyn and I. Nijs). Wilrijk, Belgium: University of Antwerp.

Nixon, D.J. (1995). Irrigation of tea in Southern Tanzania: a seminar review. In *Quarterly Report 19*, pp. 3–23. Dar es Salaam, Tanzania: Ngwazi Tea Research Unit.

Nixon, D.J. (1996a). The effects of a modified irrigation regime on the yield and yield distribution of fully irrigated clone 6/8. In *Quarterly Report 24*, pp. 3–8. Dar es Salaam, Tanzania: Ngwazi Tea Research Unit.

Nixon, D.J. (1996b). The effects of age and pruning on the development of drought resistance in maturing clones. In *Quarterly Report 26*, pp. 3–14. Dar es Salaam, Tanzania: Ngwazi Tea Research Unit.

Nixon, D.J. and Carr, M.K.V. (1995). The effects of irrigation frequency on yield and yield distribution. In *Quarterly Report 22*, pp. 3–15. Dar es Salaam, Tanzania: Ngwazi Tea Research Unit.

Nixon, D.J. and Sanga, B.N.K. (1995). Dry weight and root distribution of unirrigated mature tea. In *Quarterly Report 21*, pp. 18–23. Dar es Salaam, Tanzania: Ngwazi Tea Research Unit.

Nixon, D.J. and Workman, M. (1987). Drip irrigation of sugarcane on a poorly draining saline/sodic soil. In *Proceedings of the South African Sugar Technologists' Association*, 140–145.

Nixon, D.J., Burgess, P.J., Sanga, B.N.K. and Carr, M.K.V. (2001). A comparison of the responses of mature and young tea to drought. *Experimental Agriculture* **37**:391–402.

Njoroge, J.M. and Mwakha, E. (1985). Results of field experiments, Ruiru. I: Long term effects of various cultural practices on *Coffea arabica* L. yield and quality in Kenya. *Kenya Coffee* **50**:441–445.

Nobel, P.S. (1988). *Environmental Biology of Agaves and Cacti*. Cambridge, UK: Cambridge University Press.

Nobel, P.S. (1991). Achievable productivities of certain CAM plants: basis for high values compared with C_3 and C_4 plants. *New Phytologist* **119**:183–205.

Norman, M.J.T., Pearson, C.J. and Searle, P.G.E. (1984). Bananas (Musa spp). In *The Ecology of Tropical Food Crops*, pp. 271–285. Cambridge, UK: Cambridge University Press.

Nouy, B. Baudouin, L., Dejégul, N. and Omoré, A. (1999). Le palmier à huile en conditions hydriques limitantes. *Plantations, Recherche, Développement* **6**:31–45.

Novaes, P., Souza, J.P. and Prado, C.H.B.A. (2011). Grafting for improved net photosynthesis of *Coffea arabica* in field in southeast of Brazil. *Experimental Agriculture* **47**:53–68.

Nugawela, A., Long, S.P. and Aluthhewage, R.K. (1995). Possible use of certain physiological characteristics of young *Hevea* plants in predicting yield at maturity. *Indian Journal of Natural Rubber Research* **8**:100–108.

Nunes, M.A. (1976). Water relations of coffee. Significance of plant water deficits to growth and yield: a review. *Journal of Coffee Research* **6**:4–21.

Nunes, M. A. and Correia, M.M. (1983). Regulaçço estomatica da agua disponível no solo em *C. arabica* L. (cvs. Caturra, Catuai e Harrar). *Garcia de Orta Estudos Agronomicos* **10**:83–90.

Nunes, M.A., Bierhuizen, J.F. and Pluegman, C. (1968). Studies on the productivity of coffee. I. Effect of light, temperature and CO_2 concentration on photosynthesis of *Coffea arabica*. *Acta Botanica Neerlandica* **17**:93–102.

Nutman, F.J. (1933a). The root system of *Coffea arabica*. I. Root systems in typical soils of British East Africa. *Empire Journal of Experimental Agriculture* **1**:271–284.

Nutman, F.J. (1933b). The root system of *Coffea arabica* L. II. The effect of some soil conditions in modifying the 'normal' root system. *Empire Journal of Experimental Agriculture* **1**:285–296.

Nutman, F.J. (1934). The root system of *Coffea arabica* L. III. The spatial distribution of the absorbing area of the root. *Empire Journal of Experimental Agriculture* **2**:293–302.

Nutman, F.J. (1937a). Studies on the physiology of *Coffea arabica* L. I. Photosynthesis of coffee leases under natural conditions. *Annals of Botany (N.S.)* **1**:353–367.

Nutman, F.J. (1937b). Studies on the physiology of *Coffea arabica* L. II. Stomatal movements in relation to photosynthesis under natural conditions. *Annals of Botany (N.S.)* **1**:681–693.

Nutman, F.J. (1941). Studies on the physiology of *Coffea arabica* L. III. Transpiration rates of whole trees in relation to natural environmental conditions. *Annals of Botany (N.S.)* **5**:59–81.

Obiri, B.D., Bright, G.A., McDonald, M.A., Anglaaere, L.C.N. and Cobbina, J. (2007). Financial analysis of shaded cocoa in Ghana. *Agroforestry Systems* **71**:139–149.

Ochs, R. and Daniel, C. (1976). Research on techniques adapted to dry regions. *Oil Palm Research*, pp. 315–330 (ed. R.H.V. Corley, J.J. Hardon and B.J. Wood). Amsterdam, Netherlands: Elsevier.

Odhiambo, H.O., Nyabundi, J.O. and Chweya, J. (1993). Effects of soil moisture and vapour pressure deficits on shoot growth and the yield of tea in the Kenya highlands. *Experimental Agriculture* **29**:341–350.

Olivier, F. and Singels, A. (2004). Survey of irrigation scheduling practices in the South African sugar industry. *Proceedings of the South African Sugar Technologists' Association* **78**:239–243.

Ollagnier M., Ochs R., Pomier M. and de Taffin G. (1983). Action du chlore sur le cocotier hybride PB 121 en Côte d'Ivoire et en Indonésie. *Oléagineux* **38**(5):309–321.

Olyslaegers, G., Nijs, I., Roebben, J. *et al.* (2002). Morphological and physiological indicators of tolerance to atmospheric stress in two sensitive and two tolerant clones in South Africa. *Experimental Agriculture* **38**:397–410.

Omary, M. and Izuno, F.T. (1995). Evaluation of sugarcane evapotranspiration from water table data in the Everglades agricultural area. *Agricultural Water Management* **27**:309–319.

Orchard, J.E. (1985). The effect of the dry season on the water status of *T. cacao* in Ecuador. In *Proceedings of the 9th International Cocoa Research Conference*, held Lomé, Togo, February 1984, pp. 103–109. Lagos, Nigeria: COPAL.

Orchard, J.E. and Saltos, M. (1988). The growth and water status of cacao during its first year of establishment under different methods of soil water management. In *Proceedings of the 10th International Cocoa Conference*, held Santo Domingo, Dominican Republic, May 1987, pp. 193–198. Lagos, Nigeria: COPAL.

Othieno, C.O. (1975). Surface run-off and erosion in fields of young tea. *Tropical Agriculture (Trinidad)* **52**:299–308.

Othieno, C.O. (1977). *Annual Report 1976*. Kericho, Kenya: Tea Research Institute of East Africa. 33 pp.

Othieno, C.O. (1978a). Supplementary irrigation of young clonal tea in Kenya. I. Survival, growth and yield. *Experimental Agriculture* **14**:229–238.

Othieno, C.O. (1978b). Supplementary irrigation of young clonal tea in Kenya. II. Internal water status. *Experimental Agriculture* **14**:309–316.

Othieno, C.O. (1980). Effects of mulches on soil water content and water status of tea plants in Kenya. *Experimental Agriculture* **16**:295–302.

Othieno, C.O. (1983). Effect of weather on recovery from pruning and yield of tea in Kenya. *Journal of Plantation Crops (Supplement)* 44–52.

Othieno, C.O. and Ahn, P.M. (1980). Effects of soil mulches on soil temperatures and growth of tea plants in Kenya. *Experimental Agriculture* **16**:287–294.

Othieno, C.O. and Laycock, D.H. (1977). Factors affecting soil erosion within tea fields. *Tropical Agriculture (Trinidad)* **54**:329–329.

Othieno, C.O., Stephens, W. and Carr, M.K.V. (1992). Yield variability at the Tea Research Foundation of Kenya. *Agricultural and Forest Meteorology* **61**:237–252.

Owuor, P.O., Kamau, D.M. and Jondiko, E.O. (2009). Responses of clonal tea to location of production and plucking intervals. *Food Chemistry* **115**:290–296.

Owusu, K. and Waylen, P. (2009). Trends in spatial-temporal variability in annual rainfall in Ghana (1951–2000). *Weather* **64**(5):115–120.

Paardekooper, E.C. (1989). Exploitation of the rubber tree. In *Rubber*, pp. 349–414 (ed. C.C. Webster and W.J. Baulkwill). Harlow, UK: Longman.

Paardekooper, E.C. and Sookmark, S. (1969). Diurnal variation in latex yield and dry rubber content in relation to the saturation deficit of air. *Journal of the Rubber Research Institute of Malaysia* **21**:341–347.

Pagès, L., Le Roux, Y. and Thaler, P. (1995). Root system architecture and modelling. *Plantations, Recherche, Développement* **2**:30–34.

Palat, T., Smith, B.G. and Corley, R.H.V. (2000). Irrigation of oil palm in Southern Thailand. In *Proceedings of the International Planters Conference; Plantation Crops in the New Millenium: The Way Ahead*, pp. 303–315 (ed. E. Pushparajah). Kuala Lumpur, Malaysia: Incorporated Society of Planters.

Palat, T., Chayawat, N., Clendon, J.H. and Corley, R.H.V. (2009). A review of 15 years of oil palm irrigation research in Southern Thailand. *International Journal of Oil Palm Research* **6**:146–154.

Panda, R.K., Stephens, W. and Matthews, R.B. (2003). Modelling the influence of irrigation on the potential yield of tea (*Camellia sinensis*) in north-east India. *Experimental Agriculture* **39**:181–198.

Pang, J.T.Y. (2006). Yield efficiency in progeny trials with cocoa. *Experimental Agriculture* **42**:289–299.

Park, M. (1934). Some notes on the effects of drought on the yield of coconut palms. *Tropical Agriculturalist* **33**:141–150.

Passos, E.E.M. and Da Silva, J.V. (1990). Functionnement des stomates de cocotier (*Cocos nucifera*) au champ. *Canadian Journal of Botany* **68**:458–460.

Passos, E.E.M. and Da Silva, J.V. (1991). Détermination de l'état hydrique du cocotier par le méthode dendrométrique. *Oléagineux* **46**:233–238.

Passos, E.E.M., Prado, C.H.B.A. and Aragão, W.M. (2009). The influence of vapour pressure deficit on leaf water relations of *Cocos nucifera* in northeast Brazil. *Experimental Agriculture* **45**:93–106.

Paul, S., Wachira, F.N., Powell, W. and Waugh, R. (1997). Diversity and genetic differentiation among populations of Indian and Kenyan tea (*Camellia sinensis* (L.) O. Kuntze) revealed by AFLP markers. *Theoretical and Applied Genetics* **94**:255–263.

Peiris, T.S.G. and Thattil, R.O. (1998). The study of climate effects on the nut yield of coconut using parsimonious models. *Experimental Agriculture* **34**:189–206.

Peiris, T.S.G., Thattil, R.O. and Mahindapala, R. (1995). An analysis of the effect of climate and weather on coconut (*Cocos nucifera*). *Experimental Agriculture* **31**:451–460.

Pereira, H.C. (1957). Field measurements of water use for irrigation control in Kenya coffee. *Journal of Agricultural Science (Cambridge)* **49**:459–467.

Pereira, H.C. (1963). Studies on the effect of mulch and irrigation on root and stem development in *Coffea arabica* L. 2. A five year water budget for a coffee irrigation experiment. *Turrialba* **13**:227–230.

Pereira, H.C. (1967). The irrigation of coffee. In *Irrigation of Agricultural Lands* (ed. R.M. Hagan, H.R. Haise and T.W. Edminster). American Society of Agronomy **11**:738–768.

Pereira, H.C. and Jones, R.A. (1954). Field responses of Kenya coffee to fertilisers, manures and mulches. *Empire Journal of Experimental Agriculture* **22**:23–36.

Persley, G.J. (1992). *Replanting the Tree of Life: Towards an International Agenda for Coconut Palm Research.* Oxford, UK: CAB International.

Pilditch, A.G. and Wilson, G.J. (1978). Irrigation systems used for coffee. *Rhodesia Agricultural Journal* **75**:123–124.

Pinheiro, H.A., DaMatta, F.M., Chaves, A.R.M., Lureiro, M.E. and Ducatti, C. (2005). Drought tolerance is associated with rooting depth and stomatal control of water use in clones of *Coffea canephora*. *Annals of Botany* **96**:101–108.

Ploetz, R.C., Kepler, A.K., Daniells, J. and Nelson, S.C. (2007). Banana and plantain – an overview with emphasis on Pacific island cultivars. In *Species Profiles for Pacific Island Agroforestry* (ed. C.R. Elvitch). Holualo, Hawai'i: Permanent Agriculture Resources (PAR). www.traditionaltree.org (accessed October 2011).

Pollok, J.G., Geldard, G.W. and Street, C.P.M. (1990). Experience with approximately 600 hectare drip irrigation at Simunye sugar estate, Swaziland. *Agricultural Water Management* **17**:151–158.

Pomier, M. and Bonneau, X. (1987). Développement du système racinaire du cocotier en function de milieu en Côte d'Ivoire. *Oléagineux* **42**(11):409–421.

Portmann, F., Siebert, S., Bauer, C. and Döll, P. (2008). *Global dataset of monthly growing areas of 26 irrigated crops: version 1.0.* Frankfurt Hydrology Paper 6. Frankfurt, Germany: Institute of Physical Geography, University of Frankfurt (Main).

Prado, C.H.B.A., Passos, E.E.M. and de Moraes, J.A.P.V. (2001). Photosynthesis and water relations of six tall genotypes of *Cocos nucifera* in wet and dry seasons. *South African Journal of Botany* **67**:169–176.

Premachandra, G.S. and Joly, R.J. (1994). Leaf water relations, net CO_2 assimilation, stomatal conductance and osmotic concentration as affected by water deficit in cacao seedlings. In *Proceedings of the International Cocoa Conference: Challenges in the 90s*, held Kuala Lumpur, Malaysia, September 1991, pp. 351–359 (ed. E.B. Tay). Kuala Lumpur: Malaysian Cocoa Board.

Price, N.S. (1995). Banana morphology. 1: Roots and rhizomes. In *Bananas and Plantains*, pp. 179–189 (ed. S. Gowen). London: Chapman and Hall.

Prioux, G., Jaquemard, J.C., Franqueville, H. de. and Caliman, J.P. (1992). Oil palm irrigation. Initial results obtained by PHCI (Côte d'Ivoire). *Oléagineux* **47**:497–509.

Priyadarshan, P.M. (2003). Breeding *Hevea brasiliensis* for environmental constraints. *Advances in Agronomy* **79**:351–400.

Priyadarshan, P.M., Hoa, T.T.T., Huasun, H. and Gonçalves, P. de S. (2005). Yielding potential of rubber (*Hevea brasiliensis*) in sub-optimal environments. In *Genetic and Production Innovations in Field Crop Technology: New Developments in Theory and Practice*, pp. 221–247 (ed. M.S. Kang). London: Haworth Press.

Purseglove, J.W. (1968). *Tropical Crops: Dicotyledons.* London: Longman.

Purseglove, J.W. (1972). *Tropical Crops: Monocotyledons.* London, Longman.

Qureshi, M.E. Wegener, M.K., Harrison, S.R. and Bristow, K.L. (2001). Economic evaluation of alternative irrigation systems for sugarcane in the Burdekin delta in north Queensland, Australia. In *Water Resources Management*, pp. 47–57 (ed. C.A. Brebbia, P. Anagnostopoulos, K. Katsifarakis and A.H-D. Cheng). Boston, MA: WIT Press.

Rada, F., Jaimez, R. E., Garcia-Núñez, C., Azócar, A. and Ramirez, M.E. (2005). Water relations and gas exchange in *Theobroma cacao* var. Guasare under periods of water deficit. *Revista de la Facultad de Agronomia*, **22**(2):112–120.

Radersma, S. and Ridder, N. de (1996). Computed evapotranspiration of annual and perennial crops at different temporal and spatial scales using published parameter values. *Agricultural Water Management* **31**:17–34.

Raja Harun, R.M. and Hardwick, K. (1988). The effect of different temperature and vapour pressure deficits on photosynthesis and transpiration of cocoa leaves. In *Proceedings of the 10th International Cocoa Conference*, held Santo Domingo, Dominican Republic, May 1987, pp. 211–214. Lagos, Nigeria: COPAL.

Rajagopal, V. and Kasturi Bai, K.V. (2002). Drought tolerance mechanism in coconut. *Burotrop Bulletin* **17**:21–22.

Rajagopal, V., Patil, K.D. and Sumathykuttyamma, B. (1986). Abnormal stomatal opening in coconut palms affected with root (wilt) disease. *Journal of Experimental Botany* **37**:1398–1405.

Rajagopal, V., Sumathykuttyamma and Patil, K.D. (1987). Water relations of coconut palms affected with root (wilt) disease. *New Phytologist* **105**:289–293.

Rajagopal, V., Shivishankar, S. and Kasturi Bai, K.V. and Voleti, S.R. (1988). Leaf water potential as an index of drought tolerance in coconut (*Cocos nucifera* L.). *Plant Physiology and Biochemistry (India)* **15**:80–86.

Rajagopal, V., Ramadasan, A., Kasturi Bai, K.V. and Balasimha, D. (1989). Influence of irrigation on leaf water relations and dry matter production in coconut palms. *Irrigation Science* **10**:73–81.

Rajagopal, V., Kasturi Bai, K.V. and Voleti, S.R. (1990). Screening of coconut genotypes for drought tolerance. *Oléagineux* **45**:215–223.

Rajagopal, V., Shivishankar, S. and Kasturi Bai, K.V. (1993). Characterisation of drought tolerance in coconut. In *Advances in Coconut Research and Development*, pp. 191–199 (ed. M.K. Nair, H.H. Khan, P. Gopalasundaram and E.V.V. Bhaskara Rao). New Delhi: IBH Publishing Co.

Rajagopal, V., Kasturi Bai, K.V., Kumar, S.N. and Niral, V. (2007). Genetic analysis of drought responsive physiological characters in coconut. *Indian Journal of Horticulture* **64**(2):181–189.

Ram, G., Reddy, A.G.S. and Ramaiah, P.K. (1992). Effect of drip irrigation on flowering, fruit set retention and yield of *Coffea canephora*: a preliminary study. *Indian Coffee* **56**: 9–13.

Ranasinghe M.S. and Milburn, J.A. (1995). Xylem conduction and cavitation in *Hevea brasiliensis*. *Journal of Experimental Botany* **46**:1693–1700.

Rao, A.S. (1989). Water requirements of young coconut palms in a humid tropical climate. *Irrigation Science* **10**:245–249.

Rao, B.S. (1971). Controlled wintering of *Hevea brasiliensis* for avoiding secondary leaf fall. In *Proceedings of the Conference, Crop Protection in Malaysia*, pp. 204–211 (ed. R.L. Wastie and B.J. Wood). Kuala Lumpur, Malaysia: Incorporated Society of Planters.

Rao, V., Palat, T., Chayawat, N., and Corley, R.H.V. (2009). The Univanich oil palm breeding programme and progeny trial results from Thailand. *International Journal of Oil Palm Research* **6**:50–60.

Razi, M.I., Halim, A.H., Kamariah, D. and Noh, M.J. (1992). Growth, plant water relations and photosynthesis rates of young *Theobroma cacao* as influenced by water stress. *Pertanika* **15**(2):93–98.

Rees, A.R. (1961). Midday closure of stomata in the oil palm *Elaeis guineensis* Jacq. *Journal of Experimental Botany* **12**:129–146.

Renard, C. and Karamaga, P. (1984). Etude des relations hydriques chez *Coffea arabica* L. III. Evolution de la conductance stomatique et des composantes du potentiel hydrique chez deux cultivars soumis à la sécheresse en conditions controlées. *Café Cacao Thé* **28**:155–163.

Renard, C. and Ndayishimie, V. (1982). Etude des relations hydriques chez *Coffea arabica* L. I. Comparaison de la presse à membrane et de la chambre à pression pour la mesure du potentiel hydrique foliaire. *Café Cacao Thé* **26**:27–29.

Renard, C., Flemal, J. and Barampama, D. (1979). Evaluation of the resistance of the tea bush to drought in Burundi. *Café Cacao Thé* **23**:175–182.

Repellin, A., Zuily-Fodil, Y. and Daniel, C. (1993). Merits of physiological tests for characterizing the performance of different coconut varieties subjected to drought. In *European Research Working for Coconut*, Seminar Proceedings September 1993, pp. 71–89. Montpellier, France: CIRAD-CP. Also published as Repellin, A., Daniel, C. and Zuily-Fodil, Y. (1994). *Oleagineux* **49**:155–168.

Repellin, A., Laffray, D., Daniel, C., Braconnier, S. and Zuily-Fodil, Y. (1997). Water relations and gas exchange in young coconut palm (*Cocos nucifera* L.) as influenced by water deficit. *Canadian Journal of Botany* **75**:18–27.

Rey, H., Quencez, P., Dufrêne, E. and Dubos, B. (1998). Profils hydriques et alimentation en eau du plumier à huile en Côte d'Ivoire. *Plantations, Recherche, Développement* **5**:47–57.

Roberts, J.M., Nayamuth, R.A., Batchelor, C.H. and Sooprramaten, G.C. (1990). Plant–water relations of sugarcane (*Saccharum officinarum* L.) under a range of irrigated treatments. *Agricultural Water Management* **17**:95–115.

Robertson, M.J. and Donaldson, R.A. (1998). Changes in the components of cane and sucrose yield in response to drying-off of sugarcane before harvest. *Field Crops Research* **55**:201–208.

Robertson, M.J., Wood, A.W. and Muchow, R.C. (1996). Growth of sugar cane under high input conditions in tropical Australia. I. Radiation use, biomass accumulation and partitioning. *Field Crops Research* **48**:11–25.

Robertson, M.J., Inman-Bamber, N.G. and Muchow, R.C. (1997). Opportunities for improving the use of limited water by the sugarcane crop. In *Intensive Sugarcane Production*, pp. 287–304 (ed. B.A. Keating and J.R. Wilson). Wallingford, UK: CAB International.

Robertson, M.J., Inman-Bamber, N.G., Muchow, R.C. and Wood, A.W. (1999). Physiology and productivity of sugarcane with early and mid-season water deficit. *Field Crops Research* **64**:211–227.

Robinson J.B.D. and Mitchell, H.W. (1964). Studies on the effect of mulch and irrigation on root and stem development in *Coffea arabica* L. 3. The effects of mulch and irrigation on yield. *Turrialba* **14**:24–28.

Robinson, J.C. (1982). The problem of November-dump fruit with Williams banana in the subtropics. *Subtropica* **3**:11–16.

Robinson, J.C. (1987). Root growth characteristics in banana. *Institute for Tropical and Subtropical Crops, South Africa, Information Bulletin* **183**:7–9.

Robinson, J.C. (1995). Systems of cultivation and management. In *Bananas and Plantains*, pp.15–65 (ed. S. Gowen). London: Chapman and Hall.

Robinson, J.C. (1996). *Bananas and Plantains*. Wallingford, UK: CAB International.

Robinson, J.C. and Alberts, A.J. (1986). Growth and yield responses of banana (cultivar 'Williams') to drip irrigaton under drought and normal rainfall conditions in the subtropics. *Scientia Horticulturae* **30**:187–202.

Robinson, J.C. and Alberts, A.J. (1987). The influence of undercanopy sprinkler and drip irrigation systems on growth and yield of bananas (cultivar 'Williams') in the subtropics. *Scientia Horticulturae* **32**:49–66.

Robinson, J.C. and Alberts, A.J. (1989). Seasonal variations in crop water-use coefficient of banana (cultivar 'Williams') in the subtropics. *Scientia Horticulturae* **40**:215–225.

Robinson, J.C. and Bower, J.P. (1987). Transpiration characteristics of banana leaves (cultivar 'Williams') in response to progressive depletion of available soil moisture. *Scientia Horticulturae* **30**:289–300.

Robinson, J.C. and Bower, J.P. (1988). Transpiration from banana leaves in the subtropics in response to diurnal and seasonal factors and high evaporative demand. *Scientia Horticulturae* **37**:129–143.

Robinson, J.C. and Nel, D.J. (1984). Crop forecasting with Williams banana. *Subtropica* **5**:15–18.

Robinson, J.C. and Nel, D. J. (1988). Plant density studies with banana (cv. Williams) in a subtropical climate. I. Vegetative morphology, phenology and plantation microclimate. *Journal of Horticultural Science* **63**:303–313.

Robinson, J.C. and Nel, D. J. (1989). Plant density studies with banana (cv. Williams) in a subtropical climate. II. Components of yield and seasonal distribution of yield. *Journal of Horticultural Science* **64**:211–222.

Robinson, J.C., Nel, D.J. and Bower, J. P. (1989). Plant density studies with banana (cv. Williams) in a subtropical climate. III. The influence of spatial arrangement. *Journal of Horticultural Science* **64**:513–519.

Robinson, J.C., Anderson, T. and Eckstein, K. (1992). The influence of functional leaf removal at flower emergence on components of yield and photosynthetic compensation in banana. *Journal of Horticultural Science* **67**:403–410.

Rodrigo, V.H.L. (2007). Ecophysiological factors underpinning productivity of *Hevea brasiliensis*. *Brazilian Journal of Plant Physiology* **19**(4):245–255.

Rodrigo, V.H.L., Stirling, C.M., Teklehaimanot, Z., Samarasekera, R.K. and Pathirana, P.D. (2005). Interplanting banana at high densities with immature rubber crop for improved water use. *Agronomy for Sustainable Development* **25**:45–54.

Ronquim, J.C., Prado, H.B.A., Novaes, P., Fahl, J.I. and Ronquim, C.C. (2006). Carbon gain in *Coffea arabica* during clear and cloudy days in the wet season. *Experimental Agriculture* **42**(2):147–164.

Roupsard, O., Bonnefond, J.-M., Irvine, M. *et al.* (2006). Partitioning energy and evapotranspiration above and below a tropical palm canopy. *Agricultural and Forest Meteorology* **139**:252–268.

Ruf, F. (1995). From forest rents to tree-capital: basic 'laws' of cocoa supply. In *Cocoa Cycles: The Economies of Cocoa Supply*, pp. 1–53 (ed. F. Ruf and P.S. Siswoputranto). Cambridge, UK: Woodhead Publishing Ltd.

Sale, P.J.M. (1970a). Growth, flowering and fruiting of cacao under controlled soil moisture conditions. *Journal of Horticultural Science* **45**:99–118.

Sale, P.J.M. (1970b). Growth and flowering of cacao under controlled atmospheric relative humidities. *Journal of Horticultural Science* **45**:129–132.

Samson, R., Vandenberghe, J., Vanassche, F. *et al.* (2000). Detecting sensitivity to extreme climatic conditions in tea: sap flow as a potential indicator of drought sensitivity between clones. In *Topics in Ecology: Structure and Function in Plants and Ecosystems*, pp. 267–278 (ed. R. Ceulemans, J. Bagaert, G. Deckmyn and I. Nijs). Wilrijk, Belgium:, University of Antwerp.

Samsuddin, Z. (1980). Differences in stomatal density, dimension and stomatal conductances to water vapour diffusion in seven *Hevea* species. *Biologia Plantarum* **22**:154–156.

Samsuddin, Z. and Impens, I. (1978). Comparative net photosynthesis of four *Hevea brasiliensis* clonal seedlings. *Experimental Agriculture* **14**:337–340.

Samsuddin, Z. and Impens, I. (1979a). Relationship between leaf age and some carbon dioxide exchange characteristics of four *Hevea brasiliensis* Muel. Agr. clones. *Photosynthetica* **13**:208–210.

Samsuddin, Z. and Impens, I. (1979b). The development of photosynthetic rate with leaf age in *Hevea brasiliensis* Muel. Agr. clonal seedlings. *Photosynthetica* **13**:267–270.

Sanders, C.L. (1997). The water requirements of coffee and a review of coffee irrigation water management at Ngpani Estate, Malawi. A report to the Commonwealth Development Corporation, London. Cranfield University, Silsoe, UK, 40 pp.

Sanga, B.N.K. and Kigalu, J.M. (2005). Experiment M5: soil and water conservation in young tea. In *Annual Report*, pp. 57–59. Dar es Salaam, Tanzania: Tea Research Institute of Tanzania.

Sanga, B.N.K. and Kigalu, J.M. (2006). Experiment M5: soil and water conservation in young tea. In *Annual Report*, pp. 40–46. Dar es Salaam, Tanzania: Tea Research Institute of Tanzania.

Sangsing, K., Kasemsap, P., Thanisawanyangkura, S. *et al.* (2004a). Xylem embolism and stomatal regulation in two rubber clones (*Hevea brasiliensis* Muell. Arg.). *Trees, Sructure and Function* **18**:109–114.

Sangsing, K., Kasemsap, P., Thanisawanyangkura, S. *et al.* (2004b). Is growth performance in rubber (*Hevea brasiliensis*) clones related to xylem hydraulic efficiency? *Canadian Journal of Botany* **82**(7):886–891.

Santana, J.L., Suarez, C.L. and Fereres, E. (1993). Evapotranspiration and crop coefficients in banana. In *Irrigation of Horticultural Crops* (ed. J. Lopez-Galvez). *Acta Horticulturae* **335**:341–348.

SASA (1977). *Irrigation of Sugarcane.* Bulletin 17 (revised), 28 pp. Mount Edgecombe, South Africa: Experiment Station of the South African Sugar Association.

Schuch, U.K., Fuchigami, L.H. and Nagao, M.A. (1992). Flowering, ethylene production, and ion leakage of coffee in response to water stress and gibberellic acid. *Journal of the American Society of Horticultural Science* **117**:158–163.

Sealy, J. (1958). *A Revision of the Genus Camellia.* London: Royal Horticultural Society.

Senanayake, V.D. and Samaranayake, P. (1970). Intraspecific variation in stomatal density in *Hevea brasilensis*. *Quarterly Journal of the Rubber Research Institute of Ceylon* **46**:61–65.

Senevirathna, A.M.W.K., Stirling, C.M. and Rodrigo, V.H.L. (2003). Growth, photosynthetic performance and shade adaptation of rubber (*Hevea brasiliensis*) grown in natural shade. *Tree Physiology* **23**:705–712.

Shalhevet, J., Mentell, A., Bielorai, H. and Shimshi, D. (eds.) (1979). *Irrigation of Field and Orchard Crops under Semi-arid Conditions.* Publication No. 1. Bet Dagan, Israel: International Irrigation Information Centre.

Shamte, S. (2001). Overview of the sisal and henequen industry: a producer's perspective. www.fao.org/docrep/004/y1873e/y1873e05.htm (accessed 25 October 2010).

Shanmugan, K.S. (1973). Moisture management for coconut. *Coconut Bulletin* **4**(7):2–10.

Shivashankar, S., Kusturi Bai, K.V. and Rajagopal, V. (1991). Leaf water potential, stomatal resistance and activities of enzymes during the develoment of moisture stress in the coconut palm. *Tropical Agriculture (Trinidad)* **68**:106–110.

Shmueli, E. (1953). Irrigation studies in the Jordan Valley. 1. Physiological activity of the banana in relation to soil moisture. *Bulletin of the Research Council of Israel* **3**:228–237.

Silpi, U., Thaler, P., Kasemsap, P. *et al.* (2006). Effect of tapping activity on the dynamics of radial growth of *Hevea brasiliensis* trees. *Tree Physiology* **26**:1579–1587.

Simmonds, N.W. (1962). *The Evolution of Bananas*. London: Longman.

Simmonds, N.W. (1966). *Bananas*, 2nd edn. London: Longman.

Simmonds, N.W. (1995). Bananas. In *Evolution of Crop Plants*, 2nd edn., pp. 370–375 (ed. J. Smart and N.W. Simmonds). Harlow, UK: Longman.

Simmonds, N.W. (1998). Tropical crops and their improvement. In *Agriculture in the Tropics*, 3rd edn., pp. 257–293 (ed. C.C. Webster and P.N. Wilson). Oxford, UK: Blackwell Science.

Simmonds, N.W. and Shepherd, K. (1955). The taxonomy and origins of the cultivated bananas. *Journal of the Linnean Society* **55**:302–312.

Singels, A. and Smith, M.T. (2006). Provision of irrigation scheduling advice to small-scale sugarcane farmers using a web-based crop model and cellular technology: a South African case study. *Irrigation and Drainage* **55**:363–372.

Singels, A., van den Berg, M., Smit, M.A., Jones, M.R. and van Antwerpen, R. (2010). Modelling water uptake, growth and sucrose accumulation of sugarcane subjected to water stress. *Field Crops Research* **117**:59–69.

Skidmore, C.L. (1929). Indications of existing correlation between the rainfall and the number of pods harvested at Aburi and Asuansi. *Department of Agriculture Gold Coast Year-Book, 1928, Bulletin* **16**:114–120.

Sleigh, P.A., Hardwick, K. and Collin, H.A. (1981). A study of growth periodicity in cocoa seedlings, with particular emphasis on the root system. *Café Cacao Thé* **25**(3):169–172.

Smit, M.A. and Singels, A. (2006). The response of sugar cane canopy development to water stress. *Field Crops Research* **98**:91–97.

Smith, B.G. (1989). The effects of soil water and atmospheric vapour pressure deficit on stomatal behaviour and photosynthesis in the oil palm. *Journal of Experimental Botany* **40**:647–651.

Smith, B.G., Burgess, P.J. and Carr, M.K.V. (1993a). Responses of tea (*Camellia sinensis*) clones to drought. II. Stomatal conductance, photosynthesis and water potential. *Aspects of Applied Biology, Physiology of Varieties*, **34**:259–268.

Smith, B.G., Stephens, W., Burgess, P.J. and Carr, M.K.V. (1993b). Effects of light, temperature, irrigation and fertiliser on photosynthetic rate in tea (*Camellia sinensis*). *Experimental Agriculture* **29**:291–306.

Smith, B.G., Burgess, P.J. and Carr, M.K.V. (1994). Effects of clone and irrigation on the stomatal conductance and photosynthetic rate of tea (*Camellia sinensis*). *Experimental Agriculture* **30**:1–16.

Smith, D.M., Inman-Bamber, N.G. and Thorburn, P.J. (2005). Growth and function of the sugarcane root system. *Field Crops Research* **92**:169–183.

Smith, R.I., Harvey, F.J. and Cannell, M.G.R. (1993c). Clonal responses of tea shoot extension to temperature in Malawi. *Experimental Agriculture* **29**:47–60.

Smith, R.W. (1964). The establishment of cocoa under different soil moisture regimes. *Empire Journal of Experimental Agriculture* **32**:249–256.

Soong, N.K. (1976). Feeder root development of *Havea brasiliensis* in relation to clones and environment. *Journal of the Rubber Research Institute of Malaysia* **24**:283–298.

Soopramanien, G.C. and Batchelor, C.R. (eds) (1987). *MSIRI-IH Drip Irrigation Research Project: Second Ratoon Crop Interim Report*, 95 pp. Reduit, Mauritius: Mauritius Sugar Industry Research Institute.

Squire, G.R. (1976). Xylem water potential and yield of tea (*Camellia sinensis*) clones in Malawi. *Experimental Agriculture* **12**:289–297.

Squire, G.R. (1977). Seasonal changes in photosynthesis of tea (*Camellia sinensis* L.). *Journal of Applied Ecology* **14**:303–316.

Squire, G.R. (1979). Weather, physiology and seasonality of tea (*Camellia sinensis*) yields in Malawi. *Experimental Agriculture* **15**:321–330.

Squire, G.R. (1984). Techniques in environmental physiology of oil palm. 2. Partitioning of rainfall above ground. *PORIM Bulletin* **12**:12–31.

Squire, G.R. (1990). *The Physiology of Tropical Crop Production*. Wallingford, UK: CAB International.

Squire, G.R. and Callander, B.A. (1981). Tea plantations. In *Water Deficits and Plant Growth*, Vol. 6, pp. 471–510 (ed. T.T. Kozlowski). New York: Academic Press.

Squire, G.R., Black, C.R. and Gregory, P.J. (1981). Physical measurements in crop physiology. II. Water relations. *Experimental Agriculture* **17**:225–242.

Squire, G.R., Obaga, S.M.O. and Othieno, C.O. (1993). Altitude, temperature and shoot production of tea in the Kenyan Highlands. *Experimental Agriculture* **29**:107–120.

Stanhill, G. (1986). Water use efficiency. *Advances in Agronomy* **39**:53–85.

Stephens, W. and Carr, M.K.V. (1989). A water stress index for tea (*Camellia sinensis*). *Experimental Agriculture* **25**:545–558.

Stephens, W. and Carr, M.K.V. (1990). Seasonal and clonal differences in shoot extension rates and numbers in tea (*Camellia sinensis*). *Experimental Agriculture* **26**:83–98.

Stephens, W. and Carr, M.K.V. (1991a). Responses of tea (*Camellia sinensis*) to irrigation and fertiliser. I. Yield. *Experimental Agriculture* **27**:177–191.

Stephens, W. and Carr, M.K.V. (1991b). Responses of tea (*Camellia sinensis*) to irrigation and fertiliser. II. Water use. *Experimental Agriculture* **27**:193–210.

Stephens, W. and Carr, M.K.V. (1993). Responses of tea (*Camellia sinensis*) to irrigation and fertiliser. III. Shoot extension and development. *Experimental Agriculture* **29**:323–339.

Stephens, W. and Carr, M.K.V. (1994). Responses of tea (*Camellia sinensis*) to irrigation and fertiliser. IV. Shoot population density, size and mass. *Experimental Agriculture* **30**:189–205.

Stephens, W., Othieno, C.O. and Carr, M.K.V. (1992). Climate and weather variability at the Tea Research Foundation of Kenya. *Agricultural and Forest Meteorology* **61**:219–235.

Stephens, W., Burgess, P.J. and Carr, M.K.V. (1994). Yield and water use of tea in Southern Tanzania. *Aspects of Applied Biology* **38**:223–230.

Stephens, W., Hamilton, A.P. and Carr, M.K.V. (1998). Plantation crops. In *Agriculture in the Tropics*, 3rd edn., pp. 200–221 (ed. C.C. Webster and P.N. Wilson). Oxford, UK: Blackwell Science.

Stover, R.H. and Simmonds, N.W. (1987). *Bananas*, 3rd edn. Harlow, UK: Longman.

Sumner, M.E. (1997). Opportunities for amelioration of soil physical and chemical constraints under intensive cropping. In *Intensive Sugarcane Production*, pp. 305–326 (ed. B.A. Keating and J.R. Wilson). Wallingford, UK: CAB International.

Taffin, G. de and Daniel, C. (1976). Premiers résultats d'un essai d'irrigation lente sur palmier à huile. *Oléagineux* **31**:413–419.

Tanton, T.W. (1981). Growth and yield of the tea bush. *Experimental Agriculture* **17**:323–331.

Tanton, T.W. (1982a). Environmental factors affecting the yield of tea (*Camellia sinensis*). I. Effects of air temperature. *Experimental Agriculture* **18**:47–52.

Tanton, T.W. (1982b). Environmental factors affecting the yield of tea (*Camellia sinensis*). II. Effects of soil temperature, day length and dry air. *Experimental Agriculture* **18**:53–63.

Tanton, T.W. (1992). Tea crop physiology. In *Tea: Cultivation to Consumption*, 173–199 (ed. K.C. Willson and M. N. Clifford). London: Chapman and Hall.

Tausend, P.C., Goldstein, G. and Meinzer, F.C. (2000). Water utilization, plant hydraulic properties and xylem vulnerability in three contrasting coffee (*Coffea arabica*) clones. *Tree Physiology* **20**:159–168.

Taylor, S.E. and Sexton, O.J. (1972). Some implications of leaf tearing in Musaceae. *Ecology* **53**:143–149.

Taylor, S.J. and Hadley, P. (1988). Relationships between root and shoot growth in cocoa (*Theobroma cacao* L.) grown under different shade regimes. In *Proceedings of the 10th International Cocoa Research Conference*, held Santa Domingo, Dominican Republic, May 1987, pp. 177–183. Lagos, Nigeria: COPAL.

Teeluck, M. (1997). Development of the centre pivot irrigation system in Mauritius. www.gov.mu/portal/sites/ncb/moa/farc/amas97/html/p02.htm (accessed October 2011).

Templeton, J.K. (1969). Where lies the yield summit for *Hevea*? *Planters' Bulletin for the Rubber Research Institute of Malaya* **104**:220–225.

Teoh, K.C., Chan, K.S. and Chew, P.S. (1986). Dry matter and nutrient composition in hybrid coconuts (MAWA) and cocoa on coastal clay soils. In *Cocoa and Coconuts: Progress and Outlook*, pp. 819–835 (ed. E. Pushparajah and P. S. Chew), Proceedings of the International Conference held Kuala Lumpur, Malaysia, October 1984. Kuala Lumpur, Malaysia: Incorporated Society of Planters.

Tesfaye, S.G., Razi, I.M. and Maziah, M. (2008). Effects of deficit irrigation and partial root zone drying on growth, dry matter partitioning and water use efficiency in young coffee (*Coffea arabica* L.) plants. *Journal of Food Agriculture & Environment* **6**:312–317.

Tesha, A.J. and Kumar, D. (1979). Effects of soil moisture, potassium and nitrogen on mineral absorption and growth of *Coffea arabica* L. *Turrialba* **29**:213–218.

Thaler, P. and Pagès, L. (1996). Periodicity in the development of the root system of young rubber trees (*Hevea brasilienis* Muell. Arg.): relationship with shoot development. *Plant, Cell and Environment* **19**:56–64.

Thaler, P. and Pagès, L. (1998). Modelling the influence of assimilate availability on root growth and architecture. *Plant and Soil* **201**:307–320.

Thamban, C., Mathew, A.C. and Arulraj, S. (2006). Field performance of drip irrigation system in coconut gardens. *Journal of Plantation Crops* **34**(2):98–102.

Thomas, D.S. and Turner, D.W. (1998). Leaf gas exchange of droughted and irrigated banana cv. Williams (*Musa* spp.) growing in hot, arid conditions. *Journal of Horticultural Science and Biotechnology* **73**:419–429.

Thomas, D.S. and Turner, D.W. (2001). Banana (*Musa* sp.) leaf gas exchange and chlorophyll fluorescence in response to soil drought, shading and lamina folding. *Scientia Horticulturae* **90**:93–108.

Thomas, D.S., Turner, D.W. and Eamus, D. (1998). Independent effects of the environment on the leaf gas exchange of three banana (*Musa* sp.) cultivars of different genomic constitution. *Scientia Horticulturae* **75**:41–57.

Thompson, G.D. (1976). Water use by sugarcane. Review paper No. 8. *The South African Sugar Journal* **60**:593–600, 627–635.

Thompson, G.D. and Boyce, J.P. (1967). Daily measurements of potential evapotranspiration from fully canopied sugarcane. *Agricultural Meteorology* **4**:267–279.

Thompson, G.D. and Boyce, J.P. (1971). Comparisons of measured evapotranspiration of sugarcane from large and small lysimeters. *Proceedings of the South African Sugar Technologists' Association* **45**:169–176.

Thompson, G.D. and Boyce, J.P. (1972). Estimating water use by sugarcane from meteorological and crop parameters. *Proceedings of the International Society of Sugar Cane Technologists* **14**:813–826.

Thompson, G.D. and De Robillard, P.J.M. (1968). Water duty experiments with sugarcane on two soils in Natal. *Experimental Agriculture* **4**:295–310.

Thompson, G.D., Pearson, G.H.O. and Cleasby, T.G (1963).The estimation of the water requirements of sugarcane in Natal. *Proceedings of the South African Sugar Technologists' Association* **37**:134–141.

Thompson, G.D., Gosnell, J.M. and de Robillard, P.J.M. (1967). Responses of sugarcane to supplementary irrigation on two soils in Natal. *Experimental Agriculture* **3**:223–238.

Thong, K.C. and Ng, W.L. (1980). Growth and nutrients composition of monocrop cocoa plants on inland Malaysian soils. In *Proceedings of the International Conference on Cocoa and Coconuts 1978*, pp. 262–286. Kuala Lumpur, Malaysia: Incorporated Society of Planters.

Tiffen, M. and Mortimore, M. (1990). *Theory and Practice in Plantation Management*. London: Overseas Development Institute, .

Tilley, L. and Chapman, L. (1999). *Benchmarking crop water index for the Queensland sugar industry*. Brisbane, Australia: Bureau of Sugar Experiment Stations.

Ting, I.P. (1987). Stomata in plants with crassulacean acid metabolism. In *Stomatal Function*, pp. 353–366 (ed. E. Zeiger, G.D. Farquhar and I.R. Cowan). Stanford, CA: Stanford University Press.

Tinker, P.B. (1976). Soil requirements of the oil palm. In *Oil Palm Research*, pp. 165–181 (ed. R.H.V. Corley, J.J. Hardon and B.J. Wood). Amsterdam, Netherlands: Elsevier.

Tomlinson, P.B. (2006). The uniqueness of palms. *Botanical Journal of the Linnean Society* **151**:5–14.

Torres, J.S. (1998). A simple visual aid for sugarcane irrigation scheduling. *Agricultural Water Management* **38**:77–83.

Toxopeus, H. (1985). Botany, types and populations. In *Cocoa*, 4th edn., pp. 17–18 (ed. G.A.R. Wood and R.A. Lass). Oxford, UK: Blackwell Science.

Toxopeus, H. and Wessel, M. (1970). Studies on pod and bean values of *Theobroma cacao* L. in Nigeria. 1. Environmental effects on West African Amelonado with particular attention to annual rainfall distribution. *Netherlands Journal of Agricultural Science* **18**:132–139.

TRFCA (2000). *Clonal catalogue*. Mulanje, Malawi: Tea Research Foundation of Central Africa.

TRIT (2000). Experiment N10. Responses of clones to drought and irrigation. *Annual Report for 1997/98 and 1998/1999*, pp. 10–13. Dar es Salaam, Tanzania: Tea Research Institute of Tanzania.

Trochoulias, T. (1973). The yield response of bananas to supplementary watering. *Australian Journal of Experimental Agriculture and Animal Husbandry* **13**:470–472.

Trochoulias, T. and Murison, R.D. (1981). Yield response of bananas to trickle irrigation. *Australian Journal of Experimental Agriculture and Animal Husbandry* **21**:448–452.

Turner, D.W. (1987). Nutrient supply and water use of bananas in a subtropical environment. *Fruits* **42**:89–93.

Turner, D.W. (1990). Modelling demand for nitrogen in the banana. *Acta Horticulturae* **275**:497–503.

Turner, D.W. (1994). Bananas and plantains. In *Handbook of Environmental Physiology of Fruit Crops*, Vol. II: *Sub-tropical and Tropical Crops*, pp. 37–64 (ed. B. Schaffer and P.C. Andersen). Boca Raton, FL: CRC Press.

Turner, D.W. (1995). The response of the plant to the environment. In *Bananas and Plantains*, pp. 206–229 (ed. S. Gowen). London: Chapman and Hall.

Turner, D.W. (1998). The impact of environmental factors on the development and productivity of bananas and plantains. In *Proceedings of the 13th ACORBAT meeting, Guayaquil, Ecuador*, pp. 635–663 (ed. L.H. Arizaga). Guayaquil, Ecuador: CONABAN.

Turner, D.W. and Lahav, E. (1983). The growth of banana plants in relation to temperature. *Australian Journal of Plant Physiology* **10**:43–53.

Turner, D.W. and Rosales, F.E. (Eds.) (2005). *Banana Root System: towards a better understanding for its productive management*. Montpellier, France : International Network for the Improvement of Banana and Plantain.

Turner, D.W. and Thomas, D.S. (1998). Measurements of plant and soil water status and their association with leaf gas exchange in banana (*Musa* spp.): a laticiferous plant. *Scientia Horticulturae* **77**:177–193.

Turner, D.W., Fortescue, J.A. and Thomas, D.S. (2007). Environmental physiology of the bananas (*Musa* spp.). *Brazilian Journal of Plant Physiology* **19**:463–484.

Turner, N.C. (1986). Crop water deficits: a decade of progress. *Advances in Agronomy* **39**:1–51.

Turner, N.C. (1990). Plant water relations and irrigation management. *Agricultural Water Management* **17**:59–73.

Turner, P.D. (1977). The effects of drought on oil palm yields in south-east Asia and the South Pacific region. In *International Developments in Oil Palm, Proceedings of the Malaysian International Agricultural Oil Palm Conference*, pp. 673–694 (ed. D.A. Earp and W. Newall). Kuala Lumpur, Malaysia: The Incorporated Society of Planters.

Tuwei, G., Kaptich, F.K.K., Langat, M.C., Chomboi, K.C. and Corley, R.H.V. (2008a). Effects of grafting on tea. 1. Growth, yield and quality. *Experimental Agriculture* **44**:521–535.

Tuwei, G., Kaptich, F.K.K., Langat, M.C., Smith, B.G. and Corley, R.H.V. (2008b). Effects of grafting on tea. 2. Drought tolerance. *Experimental Agriculture* **44**:521–535.

Van Antwerpen, R. (1999). Sugar cane root growth and relationships to above ground biomass. *Proceedings of the South African Sugar Technologists' Association* **73**:89–95.

van Asten, P.J.A., Fermont, A.M. and Taulya, G. (2011). Drought is a major yield loss factor for rainfed East African highland banana. *Agricultural Water Management* **98**:541–552. doi:10.1016/j.agwat.2010.10.005.

Van der Vossen, H.A.M. (2001). Agronomy. 1. Coffee breeding practices. In *Coffee: Recent Developments*, pp. 184–201 (eds. R.J. Clarke and O.G. Vitzhum). Oxford, UK: Blackwell Science.

Van der Vossen, H.A.M. (2005). A critical evaluation of the agronomic and economic sustainability of organic coffee production. *Experimental Agriculture* **41**:449–473.

van Himme, M. (1959). Étude de système radiculaire du cacaoyer. *Bulletin Agriculture du Congo Belge et du Rwanda-Urundi* **50**(6):1541–1600.

Van Vosselen, A., Verplancke, H. and Van Ranst, E. (2005). Assessing water consumption of banana: traditional versus modelling approach. *Agricultural Water Management* **74**:201–218.

Varghese, Y.A. and Abraham, S.T. (2005). Rubber. In *Handbook of Industrial Crops*, pp. 403–458 (ed. V.L. Chopra and K.V. Peter). New York: Haworth Press.

Venkataramana, S., Gururaja Rao, P.N. and Naidu, K.M. (1986). The effects of water stress during the formative phase on stomatal resistance and leaf water potential and its relationship with yield in ten sugarcane varieties. *Field Crops Research* **13**:345–353.

Venkataramanan, D. and Ramaiah P.K. (1987). Osmotic adjustments under moisture stress in coffee. In *Twelfth International Scientific Colloquium on Coffee*, held Montreux, Switzerland, pp. 493–500. Vevey, Switzerland: ASIC.

Venkataramanan, D., Saraswathy, V.M., D'souza, G.F., Awati, M.G. and Naidu, R. (1998). A simple physiological method to assess irrigation schedule for Robusta coffee. *Developments in Plantation Crops Research* **1998**:90–97.

Vijayakumar, K.R., Dey, S.K., Chandrasekhar, T.R. *et al*. (1998). Irrigation requirement of rubber trees (*Hevea brasiliensis*) in the subhumid tropics. *Agricultural Water Management* **35**:245–259.

Villalobos, E., Umaña, C.H. and Chinchilla, Y.C. (1992). Estado de hidratación de la palma aceitera, en respuesta a la sequía en Costa Rica. *Oléagineux* **47**:217–223.

Viswanathan, P.K. (2008). Emerging smallholder rubber farming systems in India and Thailand: a comparative economic analysis. *Asian Journal of Agriculture and Development* **5**(2):1–16.

Vogel, M. (1975). Recherche du déterminisme du rhythme de croissance du cacaoyer. *Café Cacao Thé* **19**(4):265–290.

Voleti, S.R., Kasturi Bai, K.V., Nambiar, C.K.B. and Rajagopal, V. (1993a). Influence of soil type on the development of moisture stress in coconut (*Cocos nucifera* L.). *Oléagineux* **48**:505–509.

Voleti, S.R., Kasturi Bai, K.V. and Rajagopal, V. (1993b). Water potential in the leaves of coconut (*Cocos nucifera* L.) under rainfed and irrigated conditions. In *Advances in Coconut Research and Development*, pp. 243–245 (ed. M.K. Nair, H.H. Khan, P. Gopalasundaram and E.V.V. Bhaskara Rao). New Delhi: IBH Publishing Co.

Von Uexhull, H.R. (1985). Chlorine in the nutrition of palm trees. *Oléagineux* **40**(2): 67–72.

Wachira, F., Ng'etich, W., Omolo, J. and Mamati, G. (2002). Genotype × environment interactions for tea yields. *Euphytica* **127**:289–296.

Wallis, J.A.N. (1962). Irrigating Arabica coffee in Kenya. In *Africa and Irrigation*. Proceedings of a symposium held Salisbury, Southern Rhodesia, 1961. Southampton, UK: Wright Rain Ltd.

Wallis, J.A.N. (1963). Water use by irrigated Arabica coffee in Kenya. *Journal of Agricultural Science (Cambridge)* **60**: 381–388.

Webster, C.C. and Paardekooper, E.C. (1989). The botany of the rubber tree. In *Rubber*, pp. 57–84 (ed. C.C. Webster and W.J. Baulkwill). Harlow, UK: Longman.

Wickramaratne, R.T. (1987). Breeding coconuts for adaptation to drought. *Coconut Bulletin* **4**(1):16–23.

Wiedenfeld, R.P. (2000). Water stress during different sugarcane growth periods on yield and responses to N fertilisation. *Agricultural Water Management* **43**:173–182.

Wiedenfeld, R.P. (2004). Scheduling water application on drip irrigated sugarcane. *Agricultural Water Management* **64**:169–181.

Wienk, J.F. (1970). The long fibre agaves and their improvement through breeding in East Africa. In *Crop Improvement in East Africa*, pp. 209–230 (ed. C.L.A. Leakey). Technical Communication 19. Farnham Royal, UK: Commonwealth Agricultural Bureaux.

Wight, W. (1959). Nomenclature and classification of the tea plant. *Nature* **183**:1726.

Wight, W. (1962). Tea classification revised. *Current Science* **81**:298–299.

Willatt, S.T. (1970). A comparative study of young tea under irrigation. 1. Establishment in the field. *Tropical Agriculture (Trinidad)* **47**:243–249.

Willatt, S.T. (1973). Moisture use by irrigated tea in Southern Malawi. *Ecological Studies Analysis and Synthesis* (ed. A. Hadas *et al.*) **4**:331–338.

Willey, R.W. (1975). The use of shade in coffee, cocoa and tea. *Horticultural Abstracts* **45**:791–798.

Williams, E.N.D. (1971). Investigations into certain aspects of water stress in tea. In *Water and the Tea Plant*, pp. 79–87 (ed. M.K.V. and S. Carr). Kericho, Kenya: Tea Research Institute of East Africa.

Wilson, G.J. and Pilditch, A.G. (1978). Coffee irrigation practices in Chipinge and North Mashonaland. *Rhodesia Agricultural Journal* **75**:105–113.

Wood, G.A.R. (1985a). From harvest to store. In *Cocoa*, 4th edn., pp. 444–504 (ed. G.A.R. Wood and R.A. Lass). Oxford, UK: Blackwell Science.

Wood, G.A.R. (1985b). Environment. In *Cocoa*, 4th edn., pp. 38–79 (ed. G.A.R. Wood and R.A. Lass). Oxford, UK: Blackwell Science.

Wood, G.A.R. (1985c). Establishment. In *Cocoa*, 4th edn., pp. 119–165 (ed. G.A.R. Wood and R.A. Lass). Oxford, UK: Blackwell Science.

Wood, G.H. and Wood, R.A. (1967). The estimation of cane root development and distribution using radiophosphorous. *Proceedings of the South African Sugar Technologists' Association* **41**:160–168.

World Cocoa Foundation (2010). Cocoa facts and figures. www.worldcocoafoundation. org/learn-about-cocoa/cocoa-facts-and-figures.html (accessed October 2011).

Wormer, T.M. (1965). The effects of soil moisture, nitrogen fertilisation and some meteorological factors on stomatal apertures of *Coffea arabica* L. *Annals of Botany* **29**:523–539.

Wormer, T.M. (1966). Shape of bean in *Coffea arabica* L. in Kenya. *Turrialba* **16**:221–236.

Wormer, T.M. and Gituanja, J. (1970). Floral initiation and flowering of *Coffea arabica* L. in Kenya. *Experimental Agriculture* **6**:157–170.

Wormer, T.M. and Ochs, R. (1959). Humidité du sol, ouverture des stomates et transpiration du palmier à huile et de l'arachide. *Oléagineux* **14**:571–580.

Wrigley, G. (1988). *Coffee*. Harlow, UK: Longman Scientific and Technical.

Yapp, J.H.H. and Hadley, P. (1994). Inter-relationships between canopy architecture, light interception, vigour and yield in cocoa: implications for improving production efficiency. In *Proceedings of the International Cocoa Conference: Challenges in the 90s*, held Kuala Lumpur, Malaysia, September 1991, pp. 332–350 (ed. E.B. Tay). Kuala Lumpur: Malaysian Cocoa Board.

Yates, R.A. (1984). Sugar-cane as an irrigated crop. In *Sugar Cane* (ed. F. Blackburn). Harlow, UK: Longman.

Yates, R.A. and Taylor, R.D. (1986). Water use efficiencies in relation to sugarcane yields. *Soil Use and Management* **2**:70–76.

Yegappan, T.M. and Mainstone, B.J. (1981). Comparisons between press and pressure chamber techniques for measuring leaf water potential. *Experimental Agriculture* **17**:75–84.

Young, S.C.H., Sammis, T.W. and Wu, I-Pai. (1985). Banana yield as affected by deficit irrigation and pattern of lateral layouts. *Transactions of the American Society of Agricultural Engineers* **28**:507–510.

Yusop, Z., Chong, M.H., Garusu, G.J. and Ayob, K. (2008). Estimation of evapotranspiration in oil palm catchments by short-term period water-budget method. *Malaysian Journal of Civil Engineering* **20**:160–174.

Yusuf, M. and Varadan, K.M. (1993). Water management studies on coconut in India. In *Advances in Coconut Research and Development*, pp. 337–346 (ed. M.K. Nair, H.H. Khan, P. Gopalasundaram and E.V.V. Bhaskara Rao). New Delhi: IBH Publishing Co.

Zadrazil, H. (1990). Drag-line irrigation. Practical experience with sugarcane. *Agricultural Water Management* **17**:25–35.

Zuidema, P.A., Leffelaar, P.A., Gerritsma, W, Mommer, L. and Anten, N.P.R. (2005). A physiological production model for cocoa (*Theobroma cacao*): model presentation, validation and application. *Agricultural Systems* **84**:195–225.

Index

The index has been prepared under the names of the specific crops covered in this text (listed in alphabetical order), together with entries under the generic title 'plantation crops' covering the Introduction and Synthesis chapters.